D0606433

FLY NAVY

Brian Johnson

FLY NAVY

The History of
Naval Aviation

WILLIAM MORROW AND COMPANY, INC.
New York   1981

Library of Congress Catalog Card Number 81-80194
ISBN 0-688-00621-3

First Edition
1 2 3 4 5 6 7 8 9 10

PRINTED IN GREAT BRITAIN

Contents

Acknowledgements for Illustrators
Where not given in full, the following
abbreviations have been used:
IWM for Imperial War Museum
FAAM for Fleet Air Arm Museum

Acknowledgements

In the course of my research I consulted many sources, both official and private. I am deeply indebted to the many people and organisations that gave me unstinting help, without which this book could not have been written.

Copyright material from the Public Record Office, London, is reproduced by permission of the Controller of Her Majesty's Stationery Office. The United States Navy Office of Information afforded valuable help, in particular the Assistant Head of Media Services, Anna C. Urband and Robert H. Carlisle, Head of Photojournalism. I should like also to record my gratitude to the Grumman History Center, Boeing and McDonnell Douglas for information and photographs of their aircraft.

Many photographs in this book are stills from the Imperial War Museum Film Archive and I am indebted to Ann Fleming for permission to reproduce them. The RAF Museum, Hendon, provided stills and the Keeper of Aircraft, Jack Bruce, valuable information. The Curator of the Fleet Air Arm Museum, Yeovilton, Commander Dennis White and his staff were most helpful.

I should like also to acknowledge the assistance of the MOD (Navy) and the Historical Branch of the Royal Navy in the person of David Brown, upon whose expert knowledge I drew freely. Dr Jürgen Rohwer of the Bibliothek für Zeitgeschichte, Stuttgart, provided the unique pictures of the *Königsberg* in the Rufigi Delta. British Aerospace; Gloster Aircraft; the Royal Aircraft Establishment, Farnborough; *Flight International*; *The Aeroplane*; the Coral Sea Association, also have rare photographs reproduced.

To all these and the many other people who rendered assistance, my thanks; not least to my wife, Sybil, who greatly assisted me with the preparation of the manuscript.

Introduction

This volume had its origins five years ago when I wrote and produced for BBC Television the documentary 'Pilots at Sea'. The programme told of the inception and development of a twentieth-century weapon of war, the aircraft carrier, which became in thirty-five years or so the capital ship, displacing the battleship. Such a rapid rise to pre-eminence is without precedent; no one, not even the most ardent proponent of naval aviation would, at the end of World War I in 1918, have prophesied that the greatest naval engagements ever fought would be between carriers and their aircraft, yet this was the case in the Pacific during World War II.

The television programme could deal only with the main aspects leading to those battles, though with the availability of colour combat footage, the impression was vivid. But television, by its nature, is ephemeral and I came to the conclusion that only a book could offer a fuller and permanent account.

Other commitments intervened, however, notably the BBC series, 'The Secret War', which examined the role of wartime science. After those programmes I wrote the book of the same title, the success of which encouraged me to undertake the present *Fly Navy*.

Any history which — within a single volume — sets out to record seventy years of endeavour, in whatever discipline, must of necessity be truncated. I have tried to set down the significant developments, both of the carriers and the aircraft which flew from them, and to describe the major and other decisive battles they fought. For the inevitable omissions I must ask forbearance; a book must have a finite length.

Brian Johnson
London, December 1980

I

The Aeroplane Goes to Sea

Naval aviation could be said to date from Monday, 14 November 1910, for on that day Eugene Ely flew a Curtiss biplane from a wooden ramp built on the forecastle of an American light cruiser, the USS *Birmingham*, off Hampton Roads, Virginia. Seen purely as a naval occasion the performance, though creditable, was flawed: the pilot and the aircraft were civilian, the *Birmingham* was stationary, and the take-off had very nearly ended in disaster. However, it was the first time an aircraft had flown from a ship, and was therefore a practical demonstration that aircraft could conceivably serve with the fleet.

Eugene Ely's flight from the *Birmingham* was made just seven years after the Wright brothers had made the first historic heavier-than-air flight and, during the intervening years, a good deal of pressure was applied by enthusiasts — from various motives — to interest naval powers in aviation.

Glenn Curtiss — the first American to fly after the Wright brothers and a pioneer aircraft designer and constructor, who clearly had more than passing interest in any future naval flying — had made a prize-winning flight between Albany and New York, at the conclusion of which he had been widely reported in the press as having prophesied: "The battles of the future will be fought in the air. Encumbered as [battleships] are within their turrets and masts, they cannot launch air fighters, and without these to defend them, they would be blown apart." Curtiss demonstrated the validity of his statement (which was to be proved tragically correct during World War II) under the auspices of the New York newspaper *World*, which laid out a simulated battleship target on Lake Keuka, near Hammondsport, New York. In a series of tests Curtiss dropped fifteen out of twenty-two lead 'bombs' on the target.

In the face of this much publicised demonstration, and with the added impetus provided by a rumour (unfounded) that France was about to build an aircraft carrier, in September 1910 the US Navy appointed Captain Washington Irving Chambers as Officer-In-Charge Naval Aviation. The title was a good deal grander than the office: Chambers was given neither staff nor authority, accommodation nor money and little official backing. The post was simply to head off the cranks and deal with their correspondence. Fortunately for the US Navy, Chambers took his new responsibility seriously, dutifully attending air shows, at one of which he met Glenn Curtiss and a professional pilot whom Curtiss had under contract — Eugene Ely. The enthusiasm of the two flyers and their conviction that aircraft could operate from ships inspired Chambers, who tried to convince the naval establishment of the need for an aviation branch.

It was an uphill task. The world's principal naval powers — Britain, Germany, the United States and Japan — in 1910 were engaged in an arms race precipitated by the appearance four years earlier of the first of the 'all big gun' battleships, Britain's Royal Navy *Dreadnought*, which had been designed by Admiral Fisher. This ship, with five turrets of 12 inch guns and a speed of 21 knots, had made all other battleships obsolete. The new battleships and gunnery were dominating the thinking of naval staff: all else was subservient.

It would be a very bold — or foolish — naval officer who would stake his future on the fragile aircraft of the day against the new massive battleships. Perhaps because Chambers' naval career was nearly over — he was to retire just three years after his thankless appointment — such considerations did not apply; he persevered and managed to persuade the US Navy to agree to the *Birmingham* test, using as a lever the announcement by the Germans that their Hamburg— Amerika steamship line was to carry and launch an aircraft to speed up its mail service. Chambers convinced the Navy that it was unthinkable that a foreign power should steal a march on the United States by being the first country to fly an American-invented plane from a ship — and a merchant ship at that.

The US Navy therefore gave permission for a temporary wooden take-off ramp, 83 feet long and 27 feet wide, to be erected on the forepeak of USS *Birmingham*. According to some reports, Chambers had initially contacted the Wright brothers to undertake the experiment but they declined the commission. In the event, Glenn Curtiss was approached and he agreed to provide an aircraft and the services of his contract pilot Eugene Ely. It had been in the previous May that Glenn Curtiss had made his spectacular flight from Albany to New York, a distance of some 150 miles (no mean achievement in 1910), in his Model 'D' Curtiss biplane. This aircraft, now known as the 'Albany Flyer', was hoisted aboard the light cruiser in the Navy Yard at Norfolk, Virginia. The biplane was safely secured on the ramp as the *Birmingham*, escorted by three destroyers, slipped her moorings to arrive off Hampton Roads on Monday, 14 November 1910.

The weather was bad; low cloud, rain and hail showers reduced visibility to practically zero. Ely, mindful of the Germans, was nevertheless impatient to take off. (He could have waited for the weather to improve, for the Germans, on learning of the American plans, had brought the date of their flight forward: in their haste to be first they had suffered an accident, caused — most uncharacteristically — by faulty workmanship, and were forced to postpone the venture.)

In view of the very short run available — there was only 57 feet of ramp ahead of the Curtiss — it was planned that the *Birmingham* would steam at 20 knots into wind for the take-off. By 1500 hours, in spite of the appalling weather, Ely was sitting in the aircraft, had started its single engine and signalled that he was preparing to take off. The cruiser began to weigh anchor but before it was out of the water, at 1516 hours Ely opened up his 60hp engine and gave the signal to release the aircraft.

Slowly the 'Flyer' accelerated down the ramp, aided by a 5° slope. As it left the ship, the biplane was not fully airborne, but Ely skilfully used the 37 feet drop over the ship's bow to gain speed. Even so the Curtiss barely made it; the wooden propeller, wheels and a float struck the water, damaging them and drenching the pilot, who nevertheless managed to retain control of his machine, which, owing to the damage sustained by the propeller, was vibrating alarmingly. To add to Ely's difficulties a steady rain was falling, reducing such visibility as there was. Flying blind without any instruments whatsoever, a few feet above the sea, he spotted a beach known as Willoughby Spit and prudently decided to land the damaged aircraft, safely ending his 2½ mile historic flight. Aboard the *Birmingham*, which had made the rather ambiguous signal 'Ely's just gone' as the Curtiss left the ramp, Chambers was naturally delighted with the success of the experiment: "After Ely had demonstrated his ability to leave the ship so readily, without assistance from the ship's speed, or from any starting device [catapult], such as that used by the Wright brothers, my satisfaction with the results of the experiment was increased."

Captain Chambers, having demonstrated that an aircraft could take off from a warship, now set about the far more difficult feat of landing on one. In the light of the generally favourable press to the *Birmingham* take-off, the Navy gave permission for the experiment. The cruiser USS *Pennsylvania* was selected and had a wooden landing platform, 120 x 32 feet, built over her stern, with a 15-foot 30° sloping ramp to prevent the aircraft from striking the fantail. Wooden guards were built along the sides of the platform to keep the aircraft from veering overboard. As a last ditch defence, at the end of the ramp a canvas barrier was rigged to stop the Curtiss from crashing into the ship's superstructure, should it overshoot the short deck.

The logical choice of pilot was Ely, who, in spite of his rather harrowing experience the previous November, immediately agreed to undertake the flight. The Albany Flyer was refurbished and given an increased wingspan; additional bays were let into the centre section of the upper and lower planes, the object being to decrease its wing loading, thus enabling it to make a slower landing approach. Two torpedo-like floats were mounted under the lower planes to keep the aircraft from sinking in the event of a forced landing on water. The modified aircraft was now known as the Curtiss 'D' Military.

(As the platform took shape in the Navy Yard at Mare Island, San Francisco, Ely began to doubt if it was long enough to land the 1,000lb brakeless Curtiss in light winds. Several proposals were considered and discarded, then finally a most elegant solution was found: three hooks were fitted to the undercarriage of the Curtiss to engage with one of twenty-two manilla lines which, raised about a foot high, stretched transversely across the deck. Each line was attached to a 50lb sandbag at either end: the basic concept of this brilliantly simple system was to become standard practice in aircraft carriers throughout World War II and to the present day, though the sandbags were soon replaced by more sophisticated hydraulic brakes) It is not known with any certainty who first suggested the transverse wire arrester, but Ely himself is generally credited with the idea, having previously used a similar technique to stop sprint-racing cars.

Ely tested the technique at Tanforan airfield, south of San Francisco, where he marked out a similar area to the *Pennsylvania*'s platform on the grass with a single weighted rope raised a few inches and stretched across the landing run; he quickly learned that unless the weights at each end were accurately matched the aircraft would pull violently to one side, particularly if the hook engaged the rope off centre. Ely also discovered that the hook often bounced over the rope without engaging, a fundamental discovery subsequently confirmed by generations of carrier pilots. Eventually Ely used three spring-loaded hooks and was able to engage the single rope at Tanforan consistently; if it worked with one rope then the twenty-two on the ship should offer an adequate margin. The actual arrangement on the *Pennsylvania* was that the twenty-two lines were set 3 feet apart, being raised above the deck by parallel timbers 15 feet apart, running the length of the platform, thus virtually reducing the width of the available 'runway' to that figure. It is an interesting comment on the funding of this historic experiment that Ely and Captain C. F. Pond (nicknamed 'Frog'), the *Pennsylvania*'s Commander, had to pay for the arrester gear, the manilla ropes, the sandbags and guard rails out of their own pockets.

As with the earlier *Birmingham* test, it was originally envisaged that the *Pennsylvania* would be under way for the landing, thus effectively reducing the approach speed of the Curtiss. After consideration, however, Captain Pond told Ely that San Francisco Bay was too constricted and that, if he insisted that the ship be under way, then the landing would have to be outside the Bay in the Pacific. The alternative was for the cruiser to be at anchor for the test; Ely settled for the latter course.

By the second week of January 1911 all was ready, but the famed California weather proved no better than that off the east coast: once again it rained for days on end. By the 18th the weather had cleared somewhat and Ely decided to make his attempt, planning to land aboard the *Pennsylvania* at 1100 hours local time.

There was still some mist as Ely prepared to take off from Tanforan field. An army infantry detachment was stationed there and the soldiers crowded around the Curtiss as Ely wound two bicycle inner tubes round his chest instead of a standard Navy lifejacket, which he had worn for the earlier flight and had found unduly restrictive; he also donned a leather football helmet. He then received the news that in the Bay the wind was against the flood tide; the *Pennsylvania* was riding at anchor with her stern to the wind, which meant a downwind, and therefore fast, landing. However, rather than face more delays, Ely climbed aboard the open-framed Curtiss; a mechanic swung the propeller and the engine fired. At 1045 hours the aircraft was airborne and heading for the cruiser 12 miles distant in San Francisco Bay.

The visibility was still poor, making it difficult to distinguish landmarks, but as he flew along the shoreline, it was clear enough for Ely to see the large numbers of small boats crowded with sightseers, mingling with the naval picket boats sent out from the cruiser to rescue Ely should he force-land. When about 2 miles away, he picked out the *Pennsylvania* with her bow pointing to the Golden Gate Bridge; dropping down to about 100 feet, he flew past the ship and was amazed to see that every mast, spar and deckhouse was crammed with sailors, their faces turned towards him. The tiny platform was clear, with the twenty-two lines taut. Astern of the ship, Ely banked the Curtiss, lined up with the deck and throttled his engine back. As he approached, he realised that the wind was not directly astern but blowing about 10° from the right — he would therefore drift left of the *Pennsylvania*. He corrected for the wind, arrived about 50 feet from the sloping end of the platform, and cut his engine, muttering to himself as he did so: 'This is it!'

With the Curtiss engine throttled back it was so quiet aboard the ship that an officer was heard shouting "Stand by". At that moment turbulence from the ship's structure caused the light aircraft to balloon up: witnesses later said one could hear the entire ship's company gasp. Ely eased the Curtiss down but half the twenty-two lines were behind it; then, as the wheels touched the deck, the hooks engaged first one then another of the lines, dragging the heavy sandbags, which quickly arrested the aircraft. When it finally stopped, there was just 50 feet of deck in front of the stationary machine: the time was 1059 hours, one minute early.

As Ely switched off the engine of his biplane, there was complete silence for what seemed a long time, then, spontaneously, wild cheering broke from the ship's company on the *Pennsylvania*, to be taken up by nearby ships and spectators on shore; ships' sirens added to the tumult. As Ely climbed from the Curtiss, the first person to greet him was his wife Mabel, who was aboard the cruiser as the captain's guest (she had watched the *Birmingham* take-off from one of the destroyers). "Oh Boy!", she shouted, as she rushed into his arms, "I knew you could do it". Less demonstrative, Captain Pond warmly shook Ely's hand and, as the photographers recorded the scene, said: "This is the most important landing since the dove flew back to the Ark." (As the captain guided the group to his cabin, he turned to his deck officer: "Mr. Luckey, let me know when the plane is respotted and ready for take-off." The word 'respotted' was destined to become a standard expression for preparing aircraft for take-off on United States carriers.)

Less than an hour later the officer of the deck duly reported the Curtiss was respotted. Quickly Ely strapped in and took off into what was now a headwind, making a wide circuit of the *Pennsylvania* and setting course south for Tanforan, landing there fifteen minutes later, to be carried shoulder high to their mess by the infantrymen.

The press next day made a great deal of Ely's feat: one San Francisco paper, *The Call*, had multiple headlines that were typical: "Air Monster Swoops to Warship's Deck; Ely Makes Naval History; Alights and Ascends; Conquers Wind in Greatest of Aerial Feats; Aviator Makes Daring Dive on Cruiser." Ely's own account was a good deal more sober:

> There was an appreciable wind blowing diagonally across the deck. I had to calculate the force of it and found that it was not possible to strike squarely toward the center of the platform but on a line with the windward side of the

ship. I had to take a chance that I had correctly estimated just how many feet the wind would blow me off course.

Just as I came over the stern, as I shut down the motor, I felt a sudden lift of the machine — this lift carried me a trifle further than I intended before coming in actual contact with the deck. There was never any doubt in my mind that I would effect a successful landing — I am convinced, however, that, had the ship been in motion and sailing into the wind, my landing would have been made considerably easier.

In point of fact, it is highly probable that, had the *Pennsylvania* been under way into wind, the landing would have been impossible, owing to the turbulence caused by the hot gases from the cruiser's four funnels swirling around the superstructure. Later, when the British Royal Navy began landing far heavier aircraft aboard a short deck on the stern of HMS *Furious*, it was to lead to an almost 90 per cent accident rate — some aircraft crashing overboard as pilots lost control in the severe turbulence.

An examination of several photographs of the historic Ely landing reveals an interesting insight into the early flying techniques. In those days all aircraft took off and landed on grass fields, exactly into wind. Without permanent hard runways there was no need to do otherwise, and indeed a crosswind landing would have been highly dangerous; the ultra light and slow aircraft would drift appreciably and the frail undercarriages would be quite incapable of absorbing any sideways shear. Yet Ely, owing to the ship's position on the flood tide and the wind blowing from the right (as indicated by an oversize 'Old Glory' flying from the *Pennsylvania's* aftermast and clearly visible in two existing photographs), was forced to make his landing in a crosswind of about 10°. The modern technique of compensating for such a crosswind would be to align the aircraft exactly with the centre line of the *Pennsylvania's* platform, then turn a few degrees right into the wind to counteract the drift, crabbing down towards the runway until the moment of touchdown. The pilot would then 'kick off the drift' with his rudder, landing along the centre line. Ely probably knew nothing of this technique; in any event an analysis of three consecutive photographs of the approach and arrest (probably cine frames) shows that Ely simply allowed the Curtiss to drift, and landed diagonally across the short deck. The last frame shows that his judgement was good, though not perfect — he has come to rest about 3 feet to the left of the centre line.

In the light of subsequent events it is fair to ask just what had Ely's two historic flights achieved: technically not very much. Because the ships were stationary on both occasions, all Ely had proved was that a skilled pilot could land on and take off from a very small area; in fact Ely's feat could have been performed on any convenient jetty. This is in no way to belittle his achievement, which rightly has earned him a place in naval and aeronautical history. The technique of the transverse arrester was of course fundamental to the future aircraft carriers, and for that alone Ely deserves recognition. That aside, the real value in Ely's (and Chambers') pioneering was surely that it caught the imagination of the public at a time when mass circulation papers carrying photographs could disseminate such a story. Ely's flights were, in the modern sense, news, and as such they could not be ignored even by the most conservative admirals in the naval establishments.

In any event, there can be no doubt that the successful outcome of the *Birmingham* and *Pennsylvania* tests loosened to some extent the tightly drawn Navy purse strings: in December 1910 Lt Theodore G. Ellyson was ordered to report to Glenn Curtiss Flying School on North Island, San Diego, to become the US Navy's first serving pilot, being issued with American Naval Aviation Certificate No. 1 on 1 July 1911. In March of that year $25,000 was appropriated by Congress for 'experimental work in the development of aviation for naval purposes'. In May Captain Chambers began negotiations for the purchase of two Curtiss aeroplanes, which were to become the first on US Navy charge.

Of the two pioneers who had practically forced the authorities to recognise the possibilities of naval flying, Captain Chambers was retired in December 1913 and Ely, who undertook his dangerous flights entirely without fee, went back to demonstration flying. Nine months after the *Pennsylvania* flight, while performing aerobatics at the State Fair in Macon, Georgia, Ely's aircraft suffered structural failure; it crashed and Ely was killed. The accident occurred just three days before his 25th birthday. Twenty years later a grateful country awarded him a DFC — posthumously.

That the United States Navy was the first to land and take off an aircraft from a ship there can be no doubt but, in spite of this early lead, it would be twelve years and a world war later before a US Navy plane would land on an American aircraft carrier, for the development of that most significant weapon of World War II was to be undertaken elsewhere — in Britain.

The genesis of aviation in the Royal Navy can be traced back to 1908, when the brilliant and unconventional Admiral 'Jacky' Fisher became First Sea Lord, promising on appointment "a hundred brooms to sweep away the naval cobwebs of a century". The building of the 'Dreadnoughts' was only one of the revolutionary proposals and reforms 'Great Jack' implemented, which were later to give Britain the most powerful fleet in the world. There was, however, a cloud on the eastern horizon: the Zeppelins of Germany. But these were not the only lighter-than-air craft flying: the British army had been developing observation balloons for some forty years and were manufacturing them at Farnborough (the forerunner of the present Royal Aircraft Establishment).

The superintendent of the Balloon Factory, Colonel Capper, had designed and constructed the first British airship at Farnborough, rather chauvinistically naming it *Nulli Secundus* — officially 'British Army Dirigible No. 1'. The airship was little more than an elongated balloon with a skeletal frame suspended beneath, to which an open fuselage with engine was attached. The 111 feet long airship made her maiden flight of barely 1,000 yards from Laffans Plain, Farnborough, on 10 September 1907, when her maximum speed in still air was found to be less than 20mph. Following further discreet, and rather more successful, test flights at Farnborough, Colonel Capper decided that the time had come for a more public demonstration.

Now by coincidence, the German Emperor, Kaiser Wilhelm II, was in London staying at Buckingham Palace as a guest of King Edward VII over the weekend of 5 October; the opportunity of showing the flag was too good to miss, so Colonel Capper, accompanied by the legendary pioneer American aviator Samuel F. Cody, left Farnborough that morning in the blimp and set course for London. Their progress must have been stately rather than spectacular, for it was not until midday that they completed the 27 miles to Westminster and Buckingham Palace, over which they flew at 800 feet. Disappointingly, neither the King nor his royal guest appeared, but they must have been the only people in London that Saturday who failed to see the airship. Huge cheering crowds stared up at the small dirigible, including many waving officers and staff of the War Office in Whitehall (the reaction of the occupants of the nearby Admiralty is not recorded). The two aeronauts then delighted the crowd (and themselves) by waltzing round the dome of St Paul's Cathedral before heading back to Farnborough. Rather anticlimactically, they found that a freshening westerly breeze had slowed *Nulli Secundus* down to a walking pace and it was obvious that they could not reach base before dark, so a landing was made inside the cycle track at Crystal Palace in South London. The flight, in spite of its unpremeditated ending, had proved to the public in a most dramatic way that an airship could be constructed and flown in Britain as well as in Germany.

The successful progress of the army's blimp had undoubtedly been noted by the Royal Navy. Some more farsighted naval officers, already uneasy at the threat of the German Zeppelins, began to think it was time for the Navy to take an interest in aviation. One such officer, Captain R.H.S. Bacon, then Director of Naval Ordnance, submitted to Admiral Fisher in July 1908 a proposal that a Naval Air Assistant should be appointed, and he, on assuming office, should consult the army experts at Farnborough and that Messrs Vickers, Son and Maxim (later Vickers) should be invited to tender for a rigid airship for naval use. Fisher gave the proposals his backing, and in the Naval Estimates for 1909—10 a sum of £35,000 was allocated to finance the building of a rigid airship.

In point of fact, before the formal Cabinet consent, Vickers had been asked, in August 1908, to state a price for an airship, which was presumably the £35,000 eventually allocated. That Vickers knew nothing about airships, and that their experience was almost entirely in the building of heavy armaments and warships from steel plate, seems neither to have occurred to anyone nor to have raised any doubts or misgivings.

Captain Bacon, as Director of Naval Ordnance, was to have been responsible for supervising the airship's construction, but he had retired early from the Navy, and the airship became the responsibility of the newly created post of Inspecting Captain of Aircraft, which the First Sea Lord had approved. The man selected was Captain Murray Sueter, who thus became Supervisor of Construction of the airship, though he had accepted the post — prudently as it was to turn out — on the strict understanding that he should have no responsibility for the design.

Construction of Naval Airship No. 1 began at Barrow-in-Furness in May 1909 — the origin of the unofficial name *Mayfly*. It was to prove a long and difficult task, made the more daunting by the designer, C.G. Roberton, Vickers' Marine Manager, deciding to construct the basic framework out of duralumin — then a very new material which had been patented only four years earlier by a German, Conrad Claussen, and of which neither Vickers nor anyone else in Britain had any knowledge or experience. Added to the difficulties of working an unknown alloy, which made the vital stressing of the structure a very chancy business, the Admiralty began to change its mind about the specifications, adding a heavy wireless transmitter, and demanding a full flotation capability, the ability to operate in the Arctic and structural strength requirements more appropriate to a ship (the two gondolas were to be made from solid mahogany, copper-fastened).

As the 512 feet long, 48 feet diameter airship was nearing completion it was obvious that her deadweight was far in excess of her lifting capacity. Roberton decided to lighten the structure, though warned by an assistant, H.B. Pratt — the only member of the design team who was qualified in the mathematics required — that the airship was being weakened to the point where it could collapse. He was ignored.

Mayfly was hauled out of her hangar on 22 May 1911 and rode out a gale for two days, moored to a 38 foot mast. When the wind dropped, however, it was clear that the chances of her flying seemed as remote as that of the surrounding ships (it was calculated that the airship was some 3 tons overweight), and *Mayfly* was returned to her shed to be further lightened. A suggestion that an additional bay be let into the airship to increase the gas capacity, and therefore the lift, was rejected on the grounds that the building shed was insufficiently long; cost and time were probably additional factors against this sensible proposal.

Captain Sueter must have had serious misgivings as to the airworthiness of the structure; once in May and twice in August he refused to accept the *Mayfly* on behalf of the Admiralty, on the grounds that it now no longer met the original specifications. However, the Admiralty Solicitor advised that the ship must be accepted and Murray Sueter, as Inspecting Captain of Airships, formally accepted *Mayfly* on naval charge on 22 September 1911. Unfortunately the weather was bad, with gales and rain, but by Sunday the 24th conditions were better and the airship was taken from her shed into Cavendish Dock in the calm of the early morning.

What happened next is still to some extent obscure; most accounts state that when almost clear of the hangar a sudden squall caught the airship with her bows still inside, swinging her round against the building and breaking her back. This generally accepted report is not confirmed by one most reliable eyewitness, the man from the London *Times*, who covered the event at length for that august paper. The story was printed the following day, Monday, 25 September 1911:

> She came out of the shed stern first, a steam pinnace being used to tow her out. But as soon as the pinnace took hold of the airship's bows the vessel began to heel over to leeward .... whereupon the men [from the cruiser HMS *Hermione*], acting under orders released the check ropes that were used to guide the vessel out of the shed, and she then assumed an even keel; almost immediately, however, she began to bulge in the middle, while at the same time she heeled over once more. Then the afterpart of the airship started to sink in the water .... the gondola underneath the vessel, in which were Lts. Usborne and Talbot, became partially submerged ....
> The officers held to their post while one or two men in the towing boats .... dived into the water and made for a dinghy.

One can hardly blame them, for at that moment "The bulge which had been noticed at the centre of the vessel . . . . suddenly became larger, and there was a sound of tearing and cracking as if the airship was breaking in two". It was. The crowd of sightseers, mindful of the 640,000 cubic feet of hydrogen and the presence of the steam pinnace, quickly dispersed. As the *Mayfly* settled into the waters of the dock, the two sections, held only by the silk envelope, were ignominiously dragged back into the hangar and the doors firmly shut.

A Court of Inquiry into the loss of His Majesty's Airship No. 1 was convened. Admiral Sturdee, the President, summed up the whole affair as 'the work of a lunatic'. The embarrassment of the Navy can be judged from the fact that Winston Churchill, as First Lord of the Admiralty at the time the loss was investigated, would not permit the transcript of the Court to be published; later it became 'lost'.

The Times report, however, survives for all to see; it ends: "It is difficult to ascertain how the mishap occurred. The morning was an ideal one for the launch, the velocity of the wind being only 9 miles per hour. Moreover the area through which the ship passed in coming out of the shed was fully protected with wind screens . . . ." The story of the sudden squall would appear to be a cover which has become accepted with time; there was no squall. It would seem likely that, in a desperate attempt to increase its lift, the airship was simply lightened to the extent of compromising its structural integrity.

Captain D.C. Murray, writing of the *Mayfly* disaster in 1921,[1] confirms *The Times* report:

> On the 24th the ship was . . . . brought from her shed, but the structure of the hull was not strong enough to bear her weight and the ship collapsed. The gas bags were .7 tons over the estimated weight and the engines weighed 5.6 tons instead of 3. Various accessories also greatly exceeded the estimated weights.

With the loss of the *Mayfly* — to say nothing of the £35,000 (nearer £1 million in today's money) which went with it — aviation in general and airships in particular were out of favour with the Admiralty.

In the spring of 1911, fortunately before the *Mayfly* disaster, Mr Francis McClean, a member of the Royal Aero Club and a private aircraft owner, who flew from the club's airfield at Eastchurch, on the Isle of Sheppey, offered to lend the Admiralty, free of charge, two of his Short biplanes for the purpose of training four naval officers to fly. G. B. Cockburn, a fellow member of the Royal Aero Club, volunteered to act, without fee, as an instructor. Magnanimously the offer was accepted and on 1 March 1911 four young officers reported to the airfield to commence their pilot's training. They were Lieutenants R. Gregory, C.R. Samson, and A.M. Longmore and a Marine Lieutenant, E.L. Gerrard. As the four began their instruction, the first American naval officer to learn to fly, Lt Theodore Ellyson, was well on his way to qualifying for his Naval Aviator's Certificate, which he was awarded on 1 July 1911.[2]

Formal flying training to a fixed syllabus, as is now universal, was unknown in 1911; many of the pioneers taught themselves — Eugene Ely, for example. The two Short S27 biplanes which Francis McClean lent to the Admiralty had no dual control; the 'pupil', lashed to a strut behind the pilot in the open framework of the aircraft, simply reached over the instructor's shoulder, placed a hand on the stick and followed the movement of the controls as best he could. All the flying had to be done in calm air — usually only to be found just after dawn or just before sunset. This was because of the ultra light wing-loading of the early machines, typically 3lb per square foot,[3] which would make them virtually unmanageable in any sort of wind — certainly in the hands of an inexperienced pilot. On the other hand, the maximum speed of the Shorts would be unlikely to exceed 40mph and they probably stalled at under 30, so that any training accident would be more likely survivable than not.

However primitive the instruction, the four pupils at Eastchurch made good progress, and two of them, Lt Charles Samson — who was to make a distinguished contribution to naval flying — and Lt Arthur Longmore,[4] went solo in about two hours and were awarded their Royal Aero Club Certificates on 24 April 1911, after only two months' instruction. The other two officers qualified shortly afterwards.

During their training the pilots had been confined to the immediate vicinity of Eastchurch; now that they had their 'wings', Samson and Longmore decided in June to embark on a cross-country flight to the Surrey motor racing track at Brooklands, then one of the centres of British civil aviation. The journey was not unduly long — about 50 miles — but it took the two men three days to accomplish, though that includes at least two forced landings. Aircraft were still a long way from practical, reliable transport. They were not even particularly fast; indeed the world air speed record in 1911 stood at only 82.73mph.[5]

The original detachment to Eastchurch was to have been for six months, and the four naval pilots spent the remaining time taking up officers from the nearby naval dockyard at Sheerness for joyrides, which helped to spread an enthusiasm for aviation. The six months ended in September, to coincide with the *Mayfly* disaster and the virtual closing down of the Air Department.

Fortunately, in comparison with the airship fiasco, the Eastchurch experiment had been a success and, more importantly, had cost the Royal Navy nothing. Furthermore, Charles Samson, a forceful character and a natural pilot, though only a lieutenant, was on friendly terms with Captain Murray Sueter and the First Lord of the Admiralty, Winston Churchill, who was personally keenly interested in aircraft and had attended an important aviation meeting at Hendon, on the outskirts of London, in May 1911. The next year Churchill made his first flight and of it wrote:

> I am bound to confess that my imagination supplied me at every moment with the most realistic anticipation of a crash .... However we descended in due course with perfect safety .... Having been thoroughly bitten, I continued to fly on every possible occasion when my duties permitted.

With Churchill in a position of influence and authority, it was hardly surprising that the Air Department was reprieved: the Admiralty acquired the Eastchurch field, which became the first Royal Naval Air Station; and McClean's two Short S27s were bought outright, together with a third machine (another Short) which must have been in poor condition since it was quickly nicknamed 'The Dud'. Twelve naval ratings were posted to Eastchurch to be trained to maintain the aircraft and four more officers, selected from hundreds of applicants, arrived to be taught flying from the original trainees. Naval aviation, like it or not, had been established — very largely through the generosity and patriotism of McClean and Cockburn, and the enthusiasm of a very small band of men whom many senior naval officers regarded quite simply as insane.

Having secured a somewhat tenuous foothold for naval aviation, the next task of the aviators was to demonstrate the practical use to which their frail aircraft of the day could be put. In this matter the Royal Navy was perhaps a little behind the times, for elsewhere 1911 was to prove a seminal year for the development of military applications for aircraft. At Reims, in France, for example, there had been held — during September and October — a 'Concours Militaire' at which several military aircraft were displayed. In October also the first active service flight of an aircraft was made when an Italian, Captain Piazza, flying a French Blériot, made a reconnaissance flight from Tripoli to observe and report on the Turkish army dispositions at Azizia.

The US Navy, which, as we have seen, had pioneered naval aviation with Eugene Ely's historic flights, capitalised on them when Glenn Curtiss, in January 1911, demonstrated the first practical 'seaplane', then called a 'hydroaeroplane' (the word 'seaplane' was coined in 1913 by

Winston Churchill). The hydroaeroplane was an adaptation of Curtiss's standard pusher landplane but with the tricycle undercarriage replaced by a single central wooden float and with the addition of small cylindrical floats beneath each lower wingtip.

Glenn Curtiss must be acknowledged as the first pilot fully to master the technique of waterborne take-off and landing.[6] His most important flight was made on 17 February 1911, when he flew his machine from its North Island base out to the USS *Pennsylvania*, which was then at anchor in San Diego Bay. The aircraft was successfully hoisted aboard the cruiser, then lowered back on to the sea to make a water take-off and safe flight back to base. This demonstration of the ability of an aircraft to operate independently of a shore base so impressed the US Navy that on 8 May, as briefly noted earlier, Curtiss was given a contract to build an amphibian for the Navy. The funding of the order was due to the Secretary of the Navy, George L. von Meyer, supporting the case for naval aviation at a meeting of the House Naval Affairs Committee, when he requested and received 25,000 dollars — then about £6,000 — for the development of aeronautics. By naval estimates standards it was not an unduly large amount, but it sufficed to order the naval aircraft and to train additional pilots. The result of the contract was the Curtiss A-1, or Triad — so-called because it could operate with floats or wheels, or a combined wheel/float arrangement. The A-1 flew initially powered by a 50hp Curtiss engine driving a pusher airscrew. It was a biplane with 37 foot equal span wings; the length of the totally open framework fuselage was 28 feet, and when fitted with a later 75hp water-cooled Curtiss V8 engine, it had the very creditable maximum speed of 60mph.

The Triad was first flown (with the 50hp engine) by Glenn Curtiss himself at Hammondsport, N.Y., later flights being made by Lt T.G. Ellyson, the first United States Navy pilot. With the 75hp engine fitted, the A-1 could carry a passenger seated beside the pilot and reach an altitude of nearly 1,000 feet, with a range of around 60 miles, thus greatly extending the horizon of any ship to which it could be attached.[7] In fact the first A-1 was exclusively used for development flying in Chesapeake Bay in the hands of Ellyson and the Navy's third pilot, Lt Towers. The trials were marred by engine trouble, which resulted in Curtiss providing a rebuilt motor "without charge to the Government". The historic machine, the first on US Navy charge, was damaged and rebuilt many times, being struck off charge "except for motor" following a final crash at Annapolis in October 1912. It had served its experimental purpose well, demonstrating a practical use for spotting and reconnaissance from an aircraft operating with the fleet.[8]

Back in Britain, or more precisely Eastchurch, the small band of Royal Naval aviators were exploring the possibilities of their three Short aircraft. Fortunately, Eastchurch was also the base of the Short brothers, the manufacturers of the Royal Navy's first aircraft. Horace Short instructed the Navy men in the theory of flight, maintenance of engines and developments in aviation generally. Whether news of the Curtiss 'hydroaeroplane' flights reached Eastchurch is uncertain; on balance one is inclined to think it probably did: aviation was a small world in those days and new techniques were soon reported. In any event, Lt Longmore and another of the Short brothers, Oswald, became interested in the possibility of at least landing safely on water — a prudent expedient for naval aviators in the light of the unreliability of the early engines.

After some discussion, in November Short and Longmore fitted simple rubberised flotation airbags around the wheeled undercarriage and were gratified and doubtless relieved when Longmore made a successful, though damp, landing on the River Medway. To complete the experiment, the Short was towed ashore to the Isle of Grain, where the pilot dried out the engine, started it and flew safely back to Eastchurch.

While the Royal Navy's Short was being flown from the River Medway, in the north at Barrow-in-Furness another aircraft was being flown, rather intermittently, by a naval officer, Commander Schwann (later Captain of the seaplane carrier HMS *Campania*). The aircraft, an Avro hydroplane, was the officer's personal property, having been built for him by A.V. Roe; although he therefore owned the machine, he did not have any other qualification to fly it. Oliver Schwann, who had been assistant to Murray Sueter during the *Mayfly* building, was one of those men who firmly believed that an officer in His Majesty's Royal Navy could operate any mechanical device — aeroplanes included. On 18 November 1911 he managed to get the Avro into the air, thus being the first Royal Navy 'pilot' to make a waterborne take-off: the ensuing flight was short, the Avro crashing into the sea on landing. Neither the Avro nor its intrepid pilot seem to have been seriously damaged, for the aircraft was repaired and later successfully flown to and from water by Major S.V. Sipps.

Several published accounts of Schwann's flight maintain that airbags — paid for by passing a hat round the wardroom — enabled the Avro to make its waterborne take-off. The only known picture of Schwann's Avro shows it as a twin-float seaplane. It seems improbable that it was originally a landplane simply adapted with the addition of airbags; such an arrangement would allow a water landing of sorts, but to take off from water would require robust, rigid floats. It is possible, on the other hand, that the photograph, which is undated, depicts the Avro in its later, rebuilt form.

However Schwann achieved his 'first', the next official milestone was passed on 10 January 1912, when the S27 No.38 (T2) which Longmore had landed on the Medway was flown by Samson from a temporary platform constructed over the bows of an old cruiser, HMS *Africa*, while the ship was at anchor off Sheerness. The *Sheerness Guardian* covered the event:

> .... the pilot gave the word to "Let go all", and with his engine working perfectly he shot down the sloping rails clear of the ship's stern and was borne upon the air with the grace of some winged creature .... Cheer after cheer sounded from the great ship from the moment of the launch.[9]

(The *Guardian*'s man was slightly carried away by his own hyperbole; the aircraft flew over the ship's bows — not the stern.)

Officially that January flight was the first one made from a ship by a British pilot, but it has been said that Samson made an unofficial trial take-off from HMS *Africa* the previous month. An examination of HMS *Africa*'s log for 1911 reveals no such flight.[10]

The Admiralty was not apparently unduly impressed by Samson's flight from HMS *Africa*: after all, Eugene Ely had done it two years before. Possibly the fact that the *Africa*'s foreturret was out of action as a consequence of the take-off platform damned the experiment in the eyes of any right-thinking naval officer. In the event, Samson was preaching to the converted; for neither the fact that aeroplanes interfered with ships' guns nor that they might 'frighten the horses' — as the cavalry feared — prevented the formation in May 1912 of the Royal Flying Corps. Flying was now to be an integral part of the armed services.

On formation, the Royal Flying Corps (RFC) was to consist of two major Wings: a Military Wing, primarily concerned with the needs of the Army, and a Naval Wing. A memorandum[11] was published on 11 April 1912 and the relevant naval section reads, in part, as follows:

> It is impossible to overestimate the importance of experiments for the development of Hydroaeroplanes [seaplanes], and in flying from and alighting on board ship and in the waters under varying conditions.
>
> Until such experiments have proven conclusively how such operations are practicable it is impossible to forecast what the role of aeroplanes will be in naval warfare.

The above opening paragraph is interesting, for, when it was published, no one in Britain had landed on board a ship. Either Ely's flights were known or Samson had convinced the authors that such a feat could be demonstrated, for the memorandum continues: "Steps have been taken for purchase of 12 aeroplanes for first requirements." As far as airships were concerned, however, it seems that the *Mayfly* debacle was still casting its shadow: "The prospect of the successful employment of the rigid type of airship is not sufficiently favourable to justify the great cost, and it is therefore recommended that naval experiments should be confined to the development of aeroplanes and hydro-aeroplanes." Doubtless with a sideways glance at the rapid progress that Germany was making with Zeppelins, the memorandum concludes on a cautionary note: "The utmost vigilance will be taken, however, in watching foreign developments of the airship."

Many writers have maintained that one of the first public demonstrations of the newly constituted Naval Wing of the Royal Flying Corps was Commander Samson's celebrated take-off from HMS *Hibernia* whilst she was under way in Weymouth Bay. The occasion was said to have been graced by the presence of His Majesty King George V, reviewing his fleet from the Royal Yacht *Victoria and Albert*; the date is usually given as 9 May 1912.

There are good reasons for questioning the accepted version. The *Hibernia*'s logs for 1912 survive in the archives of the Public Record Office, London.[12] The first reference in the log to aircraft was made on 30 April 1912, when, while the ship was at Sheerness dockyard, aeroplanes were 'hoisted in'. The following day another aeroplane was taken on board with 'stores' — presumably fuel. The cruiser sailed for Portland that day (at 1100 hours). The next day, 2 May, at 1627 hours, the ship was off Portland making 5 knots; at 1630 she altered course to a north-westerly heading (N 56°W) to bring the vessel into wind, which was recorded as W by N Force 2 (a light breeze). By 1740 hours she had stopped and at 1755 the following entry is made: "Aeroplane T2 left ship, Commander Samson as pilot, and flew to Lodmoor." The log then records *Hibernia* as "At rest awaiting orders".

The Royal Yacht at that moment, according to her log,[13] was anchored at Portsmouth, some 50 nautical miles away. The King did not board her until 7 May; she sailed for Portland at 2145 hours.

The *Hibernia*'s logs undoubtedly clear up one point — the date of the flight. What they do not confirm is whether or not the ship was actually under way at the time of take-off. She was stopped at 1740 hours, though she could have got under way by the time of the take-off at 1755. On balance, one is inclined to think that she was under way at the time of the take-off, and a contemporary photograph seems to lend evidence to support this.

There was no flying from *Hibernia* on 9 May, but on 10 May we find "Hands refitting and hoisting in Hydroaeroplane", which Samson, according to some sources, flew over the Royal Yacht. The *Victoria and Albert*'s log does not confirm this.

There was more flying on 12 May, for the log records: "Stopped as required. Hoisted out Hydroaeroplane." This must have then taken off from alongside, for the ship was soon under way at 8 knots, but two hours later, "Turned back to Westgate to attend Hydroplane", which must have force-landed. (Some sources state that the aircraft was then towed back to Sheerness by a destroyer.)

The *Hibernia* completed her historic role on 13 May when, at Sheerness, the third hydroaeroplane was "hoisted out". This must have been a reserve machine, almost certainly in a dismantled state, for it would have been difficult to have stored three complete aeroplanes on board a ship and have only one visible in a contemporary photograph.

Two months after the *Hibernia* flight, the same Short S27 (T2) was successfully flown from another cruiser, HMS *London*, while she was steaming from Sheerness to Portsmouth. There is no doubt that on this occasion the ship was under way at 11 knots; her log records the event: "5.45 abeam Horse Fort, 60 revs. [11 knots]. Aeroplane T2 left ship."[14] The pilot is believed to have been Lt L'E. Malone.

The success of the flights from the ships led the Admiralty to convert an old cruiser, HMS *Hermes*, into a seaplane carrier. She had a tracked take-off platform built over her forecastle from which an 80hp French-built Caudron amphibian made several flights while *Hermes* was under way. Later she was used simply as a seaplane carrier with three Short 'Folders' — so called because they had folding wings to enable them to be housed in a canvas hangar on the forward deck. HMS *Hermes* was commissioned in May 1913 and can be considered the first aircraft carrier, though her seaplanes had to be put over the side and recovered for each flight, the ship obviously having to stop to launch and recover.

As these and other experiments continued, the possible role of naval aviation became clearer. In September 1913 a Short biplane made the first night flight from Eastchurch: there was a full moon though the night was misty, but the pilot had no difficulty in landing back at base. There were tests to spot submarines from the air and Samson had managed to lift and drop a 100lb bomb from his aircraft — no mean achievement with only 50hp or so (rather unreliably) available. Weight limitations were also to prove a handicap when the first tentative steps were taken to provide Wireless Telegraphy (W/T) communication between air and ship. The spark transmitters of the day were the Rouzet sets of 400 watts output, powered by a large generator which was clutched to the engine: the all-up weight of the equipment was over 100lb. Lt L.R. Fitzmaurice was appointed to develop the equipment and the techniques which were to prove promising, though the ranges were very short, owing to the insensitivity of the early receivers.

The difficulties facing the developing Naval Wing were much more acute than those of the Military Wing: the army simply had to fly its aircraft from a reasonably level field, with hangars and other facilities. The Royal Navy had to develop seaplanes and also find means of operating them and their aeroplanes with and from ships, which at that time had never been designed with such a requirement in view.

There were other purely naval problems, for although the army was to receive its aircraft ready for service use from Government factories — Farnborough, for example — the Navy did not. A contemporary history of the early days defines the difficulties:

> . . . . the Naval Wing from the outset obtained practically all their aeroplanes, seaplanes, engines and equipment from private firms by contract (which was the usual Admiralty procedure). This necessitated the Air Department being developed not only to embrace Administration Staff work but specialist technical officers and civilian experts had to be employed to deal with questions relating to the design of the machines, engines, etc., for which contracts were placed. [15]

To be fair, this state of affairs was to some extent of the Admiralty's own making; the original concept of the Royal Flying Corps was that of a single force with the two 'Wings' being complementary and to some extent interchangeable. The Admiralty, however, took the view that the function of the naval aviator was so different and specialised as compared with that of the Army that training and equipment must be a purely naval concern. The Navy simply ignored the unified aspect of the RFC and continued in its own way. For example, although Upavon was designated the main RFC pilot training base for both Wings, the Navy kept Eastchurch as its principal training station, with training courses of six officers and six ratings every three months.

The Isle of Grain, on the Thames estuary, was also retained by the Senior Service and used from early 1913 for research into float designs and W/T experiments. Calshot, on the Solent, became (in March 1913) a seaplane training base — though inhibited for a time by a lack of suitable aircraft. Several other seaplane bases were to be established round the east coast of Britain to cover the North Sea, Felixstowe (April 1913) being an early example, followed by other sites which formed a chain so that planes could fly from one station to another.

Perhaps the richest plum that fell into Admiralty hands was
Farnborough: the whole airship branch of the army based there was
handed over lock stock and barrel to the Navy on 1 January 1914. The
Navy, whose prejudiced view of airships was being changed by the rapid
development of the German Zeppelins, thus inherited four airships
(*Beta, Eta, Delta* and *Gamma*), two airship sheds and the aerodrome; and
several of the army officers and men who were experts in airship matters
willingly transferred to the Naval Wing. By January 1914 Farnborough
was training sixteen officers and sixty ratings a year in airship work.

The airship branch of the Royal Navy was destined to play a
significant role in the coming war; but for the moment the Naval Wing
was mostly concerned with aeroplanes. By early 1914 the role of aircraft
in naval service was being clearly defined as a result of the unremitting
enthusiasm and faith of the pioneers:

(a) Scouting from ships at sea with the Fleet.
(b) Offensive and defensive operation against hostile aircraft.
(c) Attacking vulnerable points on hostile territory.
(d) Protection of . . . . vulnerable points.
(e) Carrying out patrol work along our coasts and working in conjunction
with Patrol Flotillas and submarines.[16]

To achieve these objectives the Navy had, at the beginning of 1914, forty
land-based aeroplanes, with a further forty on order; fifty-five seaplanes,
with a further forty-five ordered; and 217 trained pilots.

The basic training of pilots was undertaken by the Naval Wing, but
officers who had gained their Royal Aero Club Certificate could claim up
to £75 reimbursement of the cost of their civilian training. However, on
commissioning as Sub-Lieutenants, they still had to obtain their
watchkeeping and engine-room certificates, and Marine officers had to
have first completed their military training before joining the Air
Service. Ratings who qualified as pilots received 4 shillings (20p) a day
flying pay if they had a First Class Certificate, or 2 shillings (10p) if only
Second Class. The naval aviators' ranks in 1914 were as follows:

Wing Captain	— equivalent to Captain
Wing Commander	— equivalent to Commander
Squadron Commander	— equivalent to Lt-Commander
Flight Commander	— equivalent to Lieutenant with over four years seniority
Flight Lt	— equivalent to Flight Lt
Flight Sub-Lt	— equivalent to Sub-Lt
Warrant Officer 1	— equivalent to WO1
Warrant Officer 2	— equivalent to WO2

The two senior ranks were the subject of objections from the army, which complained that it altered the meaning of 'Wing' as applied to the RFC. However, the Navy by now had dropped the naval 'Wing' title and was calling itself the 'Royal Naval Air Service' (RNAS), though that title would not be formalised until later in 1914.

During the long hot days of June and July 1914, Europe stood on the brink of what was to become World War I. Events moved swiftly: on 28 June 1914 the Austrian Archduke Franz Ferdinand was assassinated by a Serbian student at Sarajevo; Churchill ordered the mobilisation of the fleet, which assembled off Spithead on the Solent for the last peacetime review of the Royal Navy. It was a spectacular show of force: twenty-four new Dreadnoughts, thirty-five pre-Dreadnought battleships, twenty-five heavy cruisers, twenty-four light cruisers, seventy-eight destroyers and scores of lesser vessels — over 200 ships crewed by 80,000 officers and men were drawn up in lines for a total distance of 40 miles.

In addition to the grey hulls of the warships, in the lee of Gilkicker Point the new arm of the Navy — twenty seaplanes — was moored in five flights. Later they were to take off, each machine (according to *The Times*) "with a distinguished naval officer as passenger". The seaplanes were supplemented by a further eight land-based aeroplanes and four airships, all of which flew several times over the fleet (see page 369 for list of participating aircraft). "An imposing display of aircraft", headlined *The Times* on 17 July, and on 20 July the paper, in a leader commenting on the review, reported: "the seaplanes, wheeling hawklike here and there on every hand . . . . drew the gaze upward into the air in a way no Naval gathering has done before". Considering the gravity of the times and the presence of the most powerful navy in the world, that was no small achievement.

Prophetically, a month before, Admiral Sir Percy Scott had written: "Submarines and aeroplanes have entirely revolutionised naval warfare, no fleet can hide itself from the aeroplane-eye, and the submarine can deliver an attack in broad daylight . . . . Under these circumstances I can see no use for battleships and very little chance of employment for fast cruisers."[17] Ironically, the submarine, despised by surface sailors — "dirty little men in dirty little boats" — would twice nearly defeat Britain and would itself be defeated in another distant war by aircraft. No one could foresee that as King George V reviewed his fleet from the deck of the Dreadnought *Iron Duke*; but many could foresee the war with Germany that was to be declared in just seventeen days' time.

2

World War I

The Admiralty in Whitehall, London, unlike the nearby War Office, has always been an operational headquarters, and it was from the wireless aerials, high above the building, that the message "Commence hostilities against Germany" was transmitted at 2300 hours on 4 August 1914. It was relayed from base to base, ship to ship, across the world.

World War I was about to begin. As far as the Royal Navy was concerned, it was considered essentially to be a struggle against the German High Seas Fleet; battleship against battleship in set-piece naval actions which had changed little from the days of Nelson. The eight airships and eighty or so aircraft of the RNAS hardly affected the balance of naval power at all.

The contribution made by the RNAS in the first days of the war was confined to patrols over the North Sea, though on 25 August the Eastchurch Wing (C-in-C Wing Commander Samson) was ordered overseas to Ostend in Belgium to make preparations to cover a possible withdrawal of the British Expeditionary Force from Dunkirk — a distinct possibility in view of the rapid advance of the German Army. The advance, however, was halted in September and the long trench war of attrition began. The Dunkirk evacuation was to wait for twenty-six years and another war.

While the main battle fleets cautiously watched and waited on either side of the North Sea, one of the first engagements between the Royal and German Navies occurred a long way from home. It was a bizarre and long-drawn-out affair, and is of interest if only because it was the first naval action in which aircraft were to play a vital role.[1]

Prior to the outbreak of the war, the German navy had a number of cruisers deployed in distant waters to protect German colonial interests; one of these ships — the light cruiser SMS *Königsberg* — was off the East African coast on 29 September when she surprised and sank the cruiser HMS *Pegasus*, which was boiler-cleaning at Zanzibar. *Königsberg* was then pursued by HMS *Chatham* and took refuge in the Delta of the Rufigi river, Tanganyika (present-day Tanzania), in a maze of 100 miles of confusing channels bounded by mangrove swamps and innumerable islands, all rife with malaria, infested with crocodiles and covered by impenetrable jungle. Such charts as there were were inaccurate, owing to the constantly shifting mud banks and sand bars. Into this maze of shoals the *Königsberg*, taking advantage of a flood tide and up-to-date local knowledge (Tanganyika was then a German colony), disappeared.

Because of her greater draught it was not possible for the *Chatham* to pursue the German cruiser; equally it was impossible simply to leave the scene — the German ship epitomised the threat of 'the fleet in being'. At any time she could slip out into the Indian Ocean to menace Allied shipping: she would either have to be destroyed or a disproportionately large naval force would be tied down indefinitely, patrolling the Delta against the possibility of the *Königsberg* escaping. As an initial precaution, *Chatham* sank two colliers to block one of the main channels out of the Delta, but many remained clear. The *Königsberg*'s position up river was not even approximately known; she could not be observed from the entrance to the Delta. Clearly it was imperative to establish exactly where the German vessel was before any bombardment by *Chatham*'s gunnery could be undertaken. Land reconnaissance was out of the question; the banks of the river and Delta were in enemy hands, and were to some extent fortified and patrolled by Askaris led by German officers, to say nothing of the lack of maps and the hostile nature of the terrain.

The solution to the impasse was at that moment in Durban, South Africa, in the shape of a civilian pilot, H.D. Cutler, who had been giving exhibition flights in a dilapidated 90hp Curtiss flying boat, one of two that were owned by a mining engineer named Hudson. A latter-day naval press gang met the airman, commandeered one of the Curtisses and commissioned Cutler as a temporary Flight Sub-Lieutenant in His Majesty's Royal Naval Air Service.

The aircraft and its pilot were sent to Simonstown, where arrangements were made to ship them to the Rufigi Delta aboard the unarmed merchant ship HMS *Kinfauns Castle*. While waiting for the ship, there was a proposal that the Curtiss should undertake a bombing raid on the German Wireless Station at Windhoek; Cutler managed to persuade the Navy that Windhoek was too far inland for the Curtiss, which was undoubtedly true, but one suspects the real reason was that the Curtiss was in such a state of disrepair that it would have been quite incapable of lifting any sort of bomb, apart from its lacking even that most rudimentary navigational aid — a magnetic compass.

On 6 November the Curtiss was shipped aboard the *Kinfauns Castle*;

unfortunately, whilst it was lashed to the open deck, both ailerons were damaged by a heavy sea. The ship telegraphed Durban for the ailerons from the second machine to be sent to Niororo Island, where the Curtiss was eventually unloaded.

Cutler, assisted by Midshipman Gallehawke of the *Kinfauns Castle*, fitted the replacement parts and generally patched up his battered aeroplane. The hull was found to be leaking, as was the radiator; these were caulked but more trouble manifested itself — the humid tropical heat upset the temperamental engine's carburation. After two days of tinkering, the machine was ready for a test flight, but it was obvious that the engine was developing a good deal less than its nominal 90hp and the flying boat could barely lift itself and Cutler off the water. There was no question of taking off with an observer as had been planned. There was only one alternative: Cutler would have to undertake the flight on his own.

The tired flying boat staggered off on its first war reconnaissance at 0700 hours on 22 November 1914. Cutler, still lacking a compass, later wrote of the epic flight:

> I headed for the mainland but ran into a heavy rainstorm and low cloud coming down to 450′. I found myself over the land but could find no trace of the Rufigi. Having no compass I had reached the land several miles to the south of the Delta, though I was under the impression that I was north of it. I flew several miles in each direction and then turned inland but could find nothing by which I could locate my position, so as petrol was running short I turned out to sea .... [2]

One can imagine the feelings of the unfortunate pilot, flying over dense jungle, all of it in enemy hands, running out of petrol and hoplessly lost. He flew over the sea hoping to find a British ship; eventually he sighted a group of three uninhabited islands and, landing by one, he taxied ashore where amazingly he was rescued that afternoon by *Kinfauns Castle*. By sheer chance the ship had stopped an Arab dhow whose crew had seen the Curtiss flying to the south of the Delta and heading for the islands.

The long-suffering Curtiss was found to have holed its hull in the landing off the island; this was repaired in two days and once again Cutler took off in search of the *Königsberg*. Fortune favours the brave and on this, his second flight, Cutler found the German cruiser; she was 12 miles up river, moored at the apex of a bend in the main channel so that her guns could fire a broadside on any ship approaching up river towards her. Unfortunately Cutler's report, which was transmitted to HMS *Chatham* at Mombasa, was rejected on the grounds that it was impossible to get a ship of *Königsberg*'s size (3,400 tons) that far up river, which Royal Naval charts indicated was only navigable for 4 miles. A signal was sent by *Chatham* to the Admiralty: "Difficult to definitely fix [*Königsberg*'s] position until flight can be made .... with a trained observer if possible."

The rider"if possible" was a reference to the inability of the leaking Curtiss to lift two men. To try to improve its failing performance, the second machine at Durban was cannibalised for its hull, which duly arrived at Niororo Island; after reassembly, the Curtiss took off with Captain Crampton observing, and he confirmed the original position as reported by Cutler. *Königsberg*, doubtless disturbed by the aerial activity, then shifted position 1½ miles down river; this movement was noted on a third reconnaissance with Commander Fitzmaurice as observer. This time the Curtiss, which was flying at only 600 feet, was fired on by *Königsberg*, without any hits being made.

Bad weather stopped any flying until 10 December, on which day Cutler took off alone to see if the German ship had shifted her position yet again. The engine proved difficult to start and, at last airborne, Cutler was only a mile or so up river when "The [petrol] pressure failed .... I turned round but the engine cut out almost at once and I came down just in the mouth of the river. The Germans opened fire at once ...."

The machine had grounded on a sand bar some 50 yards from the shore. Cutler, under rifle fire, calmly tried to correct the cause of his engine failure: "I took the pressure valve down and pumped petrol into the top [wing] tank with the hand pump and started to crank the engine ...." As he was swinging the propeller, the Germans arrived on the shore 50 yards away, and bullets were ricocheting off the engine. Cutler took cover behind it and, having no means of setting fire to his machine, dumped his charts overboard. By that time a German officer and fifteen Askaris had waded out to the Curtiss, forcing Cutler at gunpoint to surrender, ending what was surely one of the shortest — though colourful — service careers in the history of the Royal Navy.

The German, extending the courtesy due to a captured fellow officer, took Cutler to his nearby hut and offered him a whisky and soda. Unfortunately, while the Prussian was engaged in observing protocol, the Royal Navy, having watched the Curtiss force-land, sent into the river an armed tug which, on finding no trace of the pilot, sank the Curtiss by gunfire. Cutler was to be a prisoner of war in East Africa until November 1917.[2]

The loss of the Curtiss, which for all its limitations was the only aircraft available, delayed the Rufigi operation. Since it was not possible to attack the *Königsberg* without a precise fixing of her position, and since the German ship had no intention of sailing out of the Delta into the guns of the Royal Navy — the *Chatham* had been joined by HMS *Fox*, *Kinfauns Castle* and two armed tugs — a stalemate was reached. Meanwhile in the South Pacific the Royal Navy lost the Battle of Coronel (1 November) and in the South Atlantic won the Battle of the Falkland Islands (8 December), which virtually left the *Königsberg* as the only surface warship of the German Navy outside European waters.[3]

The C-in-C, East Indies, was advised of the difficulties in attacking the *Königsberg* and he requested the Admiralty in London to send to Zanzibar a shallow-draught river gunboat, or failing that a seaplane which could bomb the German cruiser. Their Lordships reacted fairly promptly and on 7 January 1915 ordered Flt Lt J.T. Cull to lead a special flight to destroy the enemy ship by bombing. Cull's command consisted of another pilot, Flt Lt H.E. Watkins and eighteen ratings. Their first task was to take charge of two Type 807 Sopwith Seaplanes which were at Calshot and arrange for them to be shipped aboard the steamer SS *Persia* at Tilbury. These seaplanes did not greatly appeal to Cull, for in the official account of the operation which exists in the Public Record Office, London, there is the following marginal note which he pencilled: "Shorts were originally assigned but unfortunately could not be got ready in time; the only others available were these new and untried Sopwiths which soon after we left were discarded as being dud in England".

Hardly an auspicious start. The two suspect aircraft and the personnel sailed for Bombay, where the machines were re-embarked on HMS *Kinfauns Castle*, which was equipped to launch and recover seaplanes, finally arriving on 21 February 1915 off Niororo Island, about 100 miles south of Zanzibar.

The *Königsberg* had now been safely hidden up river for almost five months.

The Sopwiths were erected and test-flown far from prying eyes, but the engines and wooden airframes quickly proved to be unsuitable for the tropical climate. One of the seaplanes was almost immediately written off in a crash, the other gave continuous trouble. Cull noted in the margin of his official account: "The supply of spare parts was found to be quite inadequate and replacements were not forthcoming, the fact that the local works department of Zanzibar had to try their hand at making obturator springs — mainly of silver .... speaks for the primitive conditions."

The Admiralty was advised that the trials with the Sopwith were unsatisfactory and it was now considered that bombing or reconnaissance could not be carried out until replacement aircraft could be provided. Three Shorts (Type 827 "and very old ones too", Cull gloomily noted) were sent out from England, arriving at Niororo Island on 23 April 1915. Flight trails were made and on 25 April Cull, with Leading Air Mechanic Boggis as observer, took off to find the *Königsberg*. Cull wrote:

> The machine during this reconnaissance was not climbing well and a height of only 1,200 feet could be reached . . . . having arrived at the head of the Delta, turned and flew north . . . . The 'Konigsberg' was observed lying in the most western of the numerous channels . . . . she looked as though she had been newly painted, awnings were spread, smoke was issuing from her funnels, and in general she was looking very spick and span.

Cull then flew abreast of the German cruiser and calmly took several photographs with a heavy 7 x 5 Goetz Anschutz plate camera which was his personal property, there being no service cameras available in East Africa. He then turned his machine for home, but "as she turned, the 'Königsberg' opened fire, and for a first attempt made some very pretty shooting . . . .". The Short was in fact hit and the engine began to misfire, stopping altogether some miles from the anchorage of the British ships; however, Cull landed safely on the water and was towed home, where it was found that a rifle bullet had fractured the main oil pipe, causing the engine to seize up.

Apart from the accurate shooting of the *Königsberg*, the inability of the Short 827, with its 150hp engine, to lift a worthwhile bomb load sufficiently high in the tropics made any prospect of a successful aerial attack on the German cruiser a very doubtful proposition. In the event, the Admiralty had already decided to attack the enemy ship with monitors; three of these specialised shallow-draught warships had been ordered from British yards for the Brazilian navy before the outbreak of war and earlier in 1914 had been commandeered for the Royal Navy. Two of them, HMS *Severn* and *Mersey*, were dispatched from England for the Rufigi Delta.

The monitors were ideal for the task, having been designed for river work; for their size they were heavily armed with three 6 inch and two 4.7 inch guns. Their length was 266 feet with 49 foot beam, and though they displaced 1,260 tons, their draught was only 5 feet 8 inches. While the two monitors were making their steady way to East Africa, other units were closing in; indeed what had begun with a single cruiser was now becoming a sizeable naval force.

The ship which had chased the *Königsberg* into the Rufigi, HMS *Chatham*, incidentally, had to withdraw from the scene as she was ordered on 13 May 1915 to the Dardanelles, being relieved by the two monitors, which had reached Aden by 15 May. Further aircraft were also being shipped; these were landplanes which, equipped with wireless telegraphy (W/T) transmitters, would be able to signal the fall of shot for the two monitors, it being considered unlikely that the monitors would be able to make visual contact with the enemy cruiser.

The use of the far more efficient landplanes instead of seaplanes was made possible by the capture of Mafia Island, just outside the Delta, from the Germans on 10 January 1915. The possession of the island made the provision of an airstrip possible; however, the only level site on Mafia was overgrown with small trees and scrub and was bounded by swamp. The military governor set the local inhabitants to the task of clearing the area and preparing a landing strip; the work took several days but was completed by 18 June as HMS *Laurentic*, an armed liner, arrived off the island and unloaded the aircraft, stores and personnel, including naval pilots and observers.

The aircraft were French; two Henri Farmans and two G111 Caudrons, the latter single-engined biplanes with a 43 foot span, the two-man crew being accommodated in a curious boat-like structure, the tail being carried on long booms. In spite of its unconventional appearance, the Caudron could climb to 10,000 feet and carry a reasonably offensive load of bombs. Its maximum speed with a 100hp Anzani engine was 70mph.

The Henri Farmans were F27s, an improved version of an earlier Mark, though still hopelessly underpowered; their single 80hp Gnome rotary engine pushed the open-framed biplane at 60mph, taking 18½ minutes to climb to 3,000 feet. The sole virtue the Farmans possessed was the fact that the basic structure was of steel tubing, a great advantage in the torrid humidity of the Rufigi Delta. Nevertheless, when the cases containing the aircraft were opened, it was discovered that most of the wooden propellers, including the spares, had succumbed to the climate and were warped and useless, there being only sufficient serviceable airscrews to fly two of the aircraft. The shortage of propellers was almost immediately rendered less acute by two of the machines — one Caudron and a Farman — being written off in crashes during their test flights.

By 30 June 1915 the C-in-C reported to the Admiralty that the attack on the *Königsberg* was to take place on 6 July. HMS *Severn* and *Mersey* arrived off the Delta at dawn the day before, and the two monitors had quite a large fleet to back them up: a 5,200 ton Town class cruiser, HMS *Weymouth*, flying the Admiral's flag (Vice Admiral King-Hall); another

cruiser, HMS *Pyramus*, to support the monitors by engaging any enemy
fire which might come from the river banks; two further cruisers,
Hyacinth and *Pioneer*, to patrol offshore and bombard enemy positions
as a diversionary tactic; HMS *Laconia*, to patrol the southern entrance to
the Delta; and *Laurentic* to show the flag off the German-held port of
Dar-es-Salaam.

By 5 July all the forces were in position and ready. A reconnaissance
flight established the *Königsberg*'s position to the west of Kontoni Island,
8 miles up river.

The aircraft on Mafia Island had the following orders:

1. Spotting for the monitors.
2. Bombing *Königsberg* or any force which might be attacking the
   monitors with machine guns, rifles or field artillery.

Because of the unfortunate crashes on Mafia Island, only two aircraft
were available; the Farman was detailed for the spotting, the Caudron
for bombing. The spotting aircraft's wireless was tuned to 'D' wavelength,
as were those of the *Severn*, *Mersey* and the flagship, HMS *Weymouth*.
Instructions were issued that all W/T traffic on other wavelengths was
to be kept down to a minimum. (For the codes used during the
operation, see page 373.)

The stage was now set for one of the most unusual naval engagements
to date. It was still dark at 0400 hours, when the stillness of the Delta was
broken by the two monitors weighing anchor and slowly moving up the
dank, sullen river, edging past a large sandbank and into the main
channel of the Delta. The sound of the ships' machinery alerted
watchers on the banks, for at 0520 hours, while still dark, a field gun to
the north of the river entrance opened fire on the shadowy outlines of the
monitors. After ten months the Battle of the Rufigi Delta had begun.

The German ground forces were under the command of a naval
officer, Korvettenkapitän Schoenfeld, his 'Delta' detachment consisting
of naval ratings, European reservists and Askaris, armed with about 150
rifles, a few Maxim guns and some light field artillery pieces. Despite
their small numbers, the defenders had the advantage of local knowledge
and prepared positions; furthermore it was later to become apparent that
they were very well informed about the movements of the monitors.

The first three shots fired from the German field guns were blanks,
probably a prearranged alarm signal, for they were quickly followed by
live rounds. The monitors were not hit, however, passing out of artillery
range by 0540 hours, when they came under rifle and machine gun fire
from the west bank; this soon petered out when the ships returned the

fire. In the silence that followed distant explosions were heard at 0600 hours, caused by bombs that were dropped from 6,000 feet on *Königsberg* by Lt Watkins, who had taken off in the Caudron from Mafia Island at first light. Two exploding bombs were observed by the pilot very close to the ship. As the Caudron flew back to base, the monitors were seen to be just dropping anchor 10,800 yards down river from *Königsberg*.

By 0623 hours both monitors were anchored bow and stern, though *Severn* dragged her stern anchor in the flood tide, which delayed the start of operations. This was unfortunate, for Flt Lt Cull, with Sub-Lt Arnold as his observer in the Farman, had taken off from Mafia at 0540 hours and had been circling *Königsberg* signalling that they were ready to spot from 0617 hours. Because of the *Severn*'s anchoring difficulties, it was not until 0648 that the monitors opened fire with their 6 inch guns. Arnold in the Farman signalled that the first salvo was 200 yards short and to the left. *Königsberg* returned fire at 0700 with rapid and extremely accurate salvoes from at least four 105mm (4.1 inch) guns. From the first salvo both British ships were straddled, shells continually falling within 10 yards of the vessels, some just over, others just short. At 0740 hours *Mersey* had been hit twice, one shell exploding on the port side of the forward 6 inch gun shield, killing the trainer, sight-setter, cartridge number and CPO in charge; the remainder of the gun's crew, including the guns officer, were knocked unconscious by concussion. Another shell from *Königsberg*'s well-aimed salvoes hit a motor boat which was moored alongside *Mersey*; the boat sank but absorbed the explosion, which nevertheless bent a bulkhead and dented the ship's plating below the waterline (in a post-action report, the monitor's Captain considered that the motor boat saved his ship from being sunk).

After the two hits, *Mersey* pulled out of line and temporarily retired from the action to attend to her wounded and to enable damage control parties to make an assessment. As *Mersey* left her billet, another salvo exploded on the exact anchorage she had just vacated.

Eleven minutes after *Mersey* had been hit, *Severn*, which had continued firing, received from Cull's spotting aircraft a 'hit' signal and for the next twenty minutes the monitor's shot was accurate and many shells from her 6 inch guns were reported as exploding on *Königsberg*. *Mersey* rejoined the battle shortly after 0800 hours, by which time Cull and Arnold were relieved by the Caudron flown by Flt Lt Blackburn, with Assistant Paymaster Badger as his observer, who continued the spotting for the monitors. Though repeatedly hit, *Königsberg* was far from defeated and her return fire was as accurate and as heavy as before. *Severn* was now straddled repeatedly and her Captain (Captain H.T.

42   WORLD WAR I

Fullerton) decided to shift his position. As he did so, the uncanny accuracy of the German gunners was explained: four men were seen high up a tree on an adjacent island, where they were undoubtedly spotting for the cruiser and signalling the fall of shot by field telephone. The tree was felled by the *Severn*'s 3 pounder and a salvo of three 6 inch lyddite (high explosive blast) shells was fired into the position, which effectively silenced the spotters. After that the accuracy of *Königsberg*'s fire fell off considerably for the remainder of the day.

At 0945 hours *Severn* anchored near *Mersey* and both ships opened fire at 0950 at a range of 11,300 yards. Unfortunately the W/T link from the spotting Farman failed. There was then a lull in the firing from both sides as the two monitors were moved up river to close the range and endeavour to sight the *Königsberg*'s mast tops over the intervening jungle. This proved impossible and *Severn* went aground, which delayed the reopening of the attack, which was not resumed until 1430 hours.

Results as reported by the spotting aircraft were disappointing, most shots falling short despite signals to that effect. *Königsberg* continued to return fire, though now only from two guns; her former accuracy seemed lost, for although some shells fell in the river near the British ships, many crashed into the trees that lined the south-west bank. At 1525 hours Cull was relieved by Blackburn, whose engine began to misfire and forced him to return to Mafia Island. At 1545 the monitors withdrew for the day, anchoring outside the Delta bar, out of range of both *Königsberg* and the two field guns on the mainland, which had again fired on the gunboats as they passed, but without scoring any hits.

On analysing the day's shooting, it was found that *Severn* and *Mersey* had fired 635 rounds, and 78 spotting corrections had been logged, but the melancholy fact from the British point of view was that *Königsberg* was still afloat and battleworthy, despite numerous hits. The two aircraft had spent no less than 15 hours 19 minutes in the air, with four pilots and two observers under constant fire from the cruiser's secondary armament. *Königsberg* did not, of course, possess true anti-aircraft guns; nevertheless, the frail RNAS machines were in machine-gun and rifle range for much of the time, and their engines and fuel tanks were exposed and unprotected. Enemy action apart, simply to keep the aircraft in the air for a whole day in the tropics with rudimentary field service was itself a major feat; had they been forced down, either from hits or 'normal' engine trouble, the only possible landing would have been in the river, teeming with crocodiles, hippos and predatory fish — not an inviting prospect for the crews of flimsy landplanes which would break up and sink on impact.

After additional aerial reconnaissance, Vice Admiral King-Hall decided that a further action, which it was hoped would prove decisive, would take place the following day. However, bad weather delayed the start until Sunday, 11 July 1915.

From experience gained during the first day's battle, the tactics were somewhat changed for this final assault on the enemy cruiser. It was decided that only one monitor would fire at a time, it being considered that confusion had been caused by the two ships firing simultaneously and the spotting aircraft's observer therefore being unable to identify the source of the shots.

Severn and *Mersey* were ordered to cross the Kikuaja bar into the Delta at 1100 hours, *Severn* to anchor slightly further up stream than previously, *Mersey* to remain under way north of *Severn*, which was to be given half an hour to try to knock out the German ship. If *Königsberg* was still afloat after that time, then *Mersey* was to proceed up river to close to within 6,000 yards and endeavour to finish off the cruiser. The C-in-C reported to the Admiralty: "Although I think the chance of success fairly good it is very possible monitors may while disabling *Königsberg* be so damaged that they cannot return."

The monitors entered the Kikuaja channel of the Delta at 1145 hours; as they did so *Mersey* was hit by two shells from a field gun, one exploding on the Captain's cabin and wounding three members of the after 6 inch gun crew. The second shell hit sandbags round the after capstan, without causing any casualties or significant damage. After passing out of the field gun's range, the ship encountered and returned small arms fire.

Königsberg again had observers reporting the fall of shot. According to one account, she relied on field telephone messages "from a German officer sitting in a tub sunk in the mud, 30 yards from where the *Severn* was anchored — a situation, one might think, well calculated to encourage accurate reports".[4]

At 1212 hours the German cruiser opened fire; the first four gun salvoes went over the monitors but corrections were signalled, fire being then concentrated on *Severn* with accurate 105mm salvoes straddling the ship. Two shells were near misses, pitching within 10 feet of the *Severn*'s stern, shaking the ship and flooding her decks. At 1233 *Severn* returned fire at a range of 10,000 yards with her 4.7 inch guns.

The fall of shot of the opening salvo was not spotted as the Farman was midway between the German and British ships. As soon as the aircraft arrived over the enemy, it began to report by W/T hits from the eighth salvo; indeed from 1242 onwards the observer, Sub-Lt Arnold, was reporting a succession of direct hits on the German ship. However,

the spotting aircraft was itself in trouble and Cull, its pilot, was having
difficulty in controlling the machine, owing to severe turbulence, which
caused the under-powered Farman to lose height and drop down to only
2,500 feet, at which point, according to the pilot's post-action report:

> There was a violent bump which must have been caused by a lucky shot with
> high explosive shell .... this was followed by the engine completely
> seizing up .... and we started planing down towards *Mersey*, though the
> *Severn* was nearer I did not want to interfere with her fire just [as] all her shots
> were falling before the forebridge of *Königsberg*. On our way down Flt. Sub
> Lt. Arnold continued very coolly sending corrections and gave one very
> important one; bringing the *Severn*'s shots forward to amidships and we had
> the satisfaction of seeing shells falling on the middle of the *Königsberg* before
> we lost sight of her. He also informed *Mersey* we were hit and asked them to
> send a boat.

The Farman touched down in the river about 100 yards from *Mersey*; it
at once somersaulted and broke up. Arnold, the observer, was thrown
clear, but Cull was trapped in the sinking wreckage and nearly drowned
before he struggled clear and, with Arnold, was picked up by the boat
from *Mersey*.

In view of the way in which Cull and Arnold had remained over the
enemy as long as possible, even though hit, and had then continued
coolly to transmit gunnery corrections to the *Severn* as they glided down
to a certain crash landing, the C-in-C recommended both officers for the
Victoria Cross. They got DSOs.

Severn continued to pound the *Königsberg* and by 1345 hours had fired
forty-two two-gun salvoes; eight heavy explosions were then heard and a
large cloud of black and yellow smoke rose over the jungle from the
stricken German ship. *Mersey* was now ordered to pass *Severn* and close
to within 7,000 yards of *Königsberg*. *Mersey* fired twenty-eight salvoes,
hits being reported from the third shot by the spotting aircraft — the
Caudron — whose pilot, Flt Lt Watkins and observer, Lt Bishop (Royal
Marines), reported that the cruiser was on fire aft and her middle funnel
gone; shells were hitting her continuously and many explosions were
seen.

By 1420 hours return fire had ceased, and it was considered that
Königsberg was a wreck and further firing by the British ships was a
waste of ammunition. *Severn* then began to proceed up river towards the
German ship to ascertain the extent of her damage, but the C-in-C
signalled by W/T the order for the two monitors to retire, which they
did, continually firing into the river banks, supported by HMS *Pyramus*,
which had also engaged the shore artillery to cover the monitors'
withdrawal; no hits, however, were made by either side.

Above the Delta, Watkins, the pilot of the Caudron, seeing the monitors retiring, flew back to Mafia Island; however, in his excitement to report the news of the successful outcome of the battle, he rather misjudged his landing, running into the swamp, which overturned the aircraft. He got out unscathed and ran towards the crowd of mechanics to tell them the news, forgetting his unfortunate marine observer, who was trapped underneath the wreckage, trying to undo his safety harness with his head in the swamp. His cries were heard and he was hauled out, in the words of the official report, "not much the worse for wear".

Since the only two available aircraft had now been lost, it was not possible to carry out a final reconnaissance over *Königsberg*. However, from previous reports it was obvious that the ship was a total wreck, though in the shallow river she would not sink.

The RNAS unit on Mafia Island embarked its stores on HMS *Laconia* and returned to Zanzibar with the fleet, where a celebration was held. From Zanzibar the *Laconia* sailed for Mombasa, Kenya, where the RNAS men were landed to operate two Caudrons and three Shorts which had arrived from England. With replacement aircraft available, Vice Admiral King-Hall decided that a flight over the *Königsberg* be made to confirm that the ship was indeed a total loss. A Caudron was shipped to Mafia Island from Mombasa with a party of RNAS mechanics to erect it.

Bad weather delayed things but on 5 August 1915 the Caudron, piloted by Flt Lt Blackburn, with Flt Commander Cull as observer, flew over the German cruiser. *Königsberg* had a list to starboard of 15°; her starboard battery was under water and her central funnel was lying across the rusty decks; her masts were gone and there was no doubt she would never sail again. The Germans had come to the same conclusion, for the two airmen noted that a lighter was alongside the listing wreck, de-ammunitioning the ship and removing stores. The possibility of this happening had been feared by the British but with the Delta still in German hands, short of mounting another costly full-scale naval and air operation, there was little that could be done to prevent the salvage.

General von Lettow-Vorbeck, the German C-in-C, in his report of the action stated:

On 11 July, *Königsberg* suffered severely. The guns crews were put out of action. The wounded Captain had the breech blocks thrown overboard and the ship blown up. The loss of the ship .... had at least an advantage for the land campaign, in that her crew and stores were now at the disposal of the Protection forces.

Korvettenkapitän Schoenfeld then set about salvaging the cruiser's guns; the breech blocks were raised from the river and the ten 105mm guns were removed from the hulk and placed on improvised trucks, then, incredibly, manhandled miles overland through the bush. Five were mounted at Dar-es-Salaam and others up country. These guns were to give the German forces in Tanganyika valuable service until the end of the war.

The RNAS detachment that fought the battle of the Rufigi Delta split up after the action. On 12 August 1915 Squadron Commander Gordon with three Short seaplanes and naval mechanics embarked on HMS Laconia for Mesopotamia. On 8 September, twelve months after the action against *Königsberg* had commenced, Flt Commander Cull, with Flt Lt Watkins and their mechanics, left Mombasa by train for British East Africa to support the land operations against the German forces in Tanganyika.

So ended one of the most unusual of naval battles. There is no doubt that without the aircraft the action would have been far more costly, if not impossible. It is most unlikely that the monitors alone could have successfully engaged the *Königsberg*; they would have come under heavy, rapid and accurate fire at point blank range the moment they appeared round the last bend in the river. The only alternative would have been a costly continuous blockade outside the Delta, tying down several warships, perhaps for the remainder of the war.

When the protracted campaign against the *Königsberg* began in September 1914, Wing Commander Samson's 'Eastchurch' wing, No 1 RNAS, was preparing to cover a possible withdrawal of the British army from France. There was something traditional about this; most of the wars fought by the British — and they are many — begin with a series of crushing defeats which somehow eventually turn to victory. So it was in September 1914; the German army's advance was contained and the western front stabilised. By the time that had happened, Samson's wing — a collection of ten or so very miscellaneous aircraft — had retreated from Ostend to Dunkirk.

Before the retreat, four of the wing's aircraft, on 22 September 1914, had made the first British air raid over Germany. It was not a success; in thick mist, only one aircraft, that piloted by Flt Lt Colet, found the target — the Zeppelin sheds at Düsseldorf. Unfortunately his two 20lb bombs failed to explode, probably due to their being dropped from too low an altitude.

In October, after the retreat to Dunkirk, it was decided to make another attack on the Zeppelin sheds at Düsseldorf and also to bomb an airship base at Cologne. The raids were to be made on 8 October, using two of only three Sopwith Tabloids the RNAS possessed. The Tabloids were the outstanding aircraft of their day, among the first 'modern' aircraft, with a streamlined cowling over the rotary engine and a clean fabric-covered structure. Harry Hawker had designed the small biplane — its span was only 25 feet 6 inches — for speed and, when first publicly flown at Hendon in November 1913, it had caused a sensation with its 90 mph maximum and its rate of climb of 1,200 feet per minute. The prototype was a two-seater, and a seaplane version was fast enough to win the Schneider Trophy at Monaco when flown in that event by Howard Pixton, just before the outbreak of the war.

A military version was immediately ordered; it was a single-seater powered by a 100hp Gnome Monosoupape, which offered a 92mph maximum speed and an endurance of 3½ hours. It could lift two 20lb bombs.

By October 1914 the RNAS had three military Tabloids on its strength: two of them — Nos 167 and 168, with No. 1 RNAS — were detailed to bomb the German airship sheds. The distance of the target precluded the use of Dunkirk as a base, so the two aircraft were flown to Wilryck, on the outskirts of Antwerp, on 8 October. Unfortunately the airfield was already under German artillery fire; however, after refuelling, the two Tabloid pilots, Squadron Commander Spenser-Grey and Flt Lt R.L.G. Marix, took off from the shell-pitted field, Spenser-Grey setting course for Cologne 125 miles away, and Marix a similar distance for Düsseldorf. Spenser-Grey was unlucky; the ground mist which had foiled the earlier raid again shrouded the airship sheds but the main railway station was visible and this was bombed as an alternative target.

Twenty-five miles to the north, Marix found conditions to be ideal; he spotted the large airship sheds without difficulty and dived to within 600 feet, at which height he released his two 20lb bombs. While he did so, his aircraft came under heavy machine-gun fire from the ground; as the Tabloid climbed away at full throttle, Marix turned to see the shed erupt into an enormous 500 foot fireball. His two small bombs had exploded inside the shed which housed the brand new Zeppelin Z9. Three-quarters of a million cubic feet of hydrogen ignited and burnt out the airship and its shed. Z9 was the first German Zeppelin to be destroyed by a British aircraft.

As Marix climbed away from the blazing target, his elation was somewhat tempered when he tried to turn the aircraft but found that the rudder-bar was no longer connected to the rudder; machine-gun bullets had severed the control lines. This was unfortunate, for the one 'dated'

design feature of the Tabloid was in the use of wing warping instead of
the more modern movable ailerons, which were much more efficient and
which would have banked the aircraft round without difficulty. As it
was, Marix now found himself using up his rapidly dwindling petrol
supply flying deeper into Germany. Very slowly, by using maximum
warp, he managed to turn his Tabloid back towards Antwerp. The wind
had veered and freshened and, in failing light and down to the last drop
of fuel, the Sopwith was landed safely in a field near a railway line some
distance north of Antwerp. A passing Belgian policeman confirmed
Marix's position and volunteered the information that a light locomotive
was due to go into Antwerp that evening to bring out a train load of
refugees. The engine duly appeared and the driver was flagged down;
Marix rode on the footplate to within 5 miles of the city, got hold of a
bicycle and peddled to Wilryck, which was still under heavy German
artillery fire. The resourceful aviator had arrived just in time to rejoin
No. 1 Wing in the retreat to Dunkirk. The Tabloid was never seen again
by the RNAS, but its loss was a small price to pay for the destruction of
Z9.

Both Marix and Spenser-Grey (who had flown back directly to
Wilryck) were awarded the DSO.

The raid which destroyed the Z9 was the first successful British air
bombing attack on German territory. The loss to the enemy of the Z9
was particularly welcome to the Royal Navy, which was becoming
increasingly wary of the German Zeppelins; they could, and indeed did,
shadow the British fleet and the slow climbing, cumbersome seaplanes
could do nothing about them. Also the Navy was responsible for the
defence of the British Isles: the RFC, being now principally an adjunct
to the army, had largely departed with the British Expeditionary Force
to Belgium and France. The Admiralty took the view that at any
moment the Germans might use their airships to raid ports and military
targets in Britain. (It is doubtful if anyone in Britain at that time
seriously thought that the bombing of civilians in towns and cities would
take place.)

Because of the near impossibility of intercepting the night-flying
Zeppelins, the Admiralty, encouraged by the success of the Düsseldorf
raid, decided that it would mount what would now be termed a pre-
emptive strike on the main Zeppelin base at Friedrichshafen on Lake
Constance. Friedrichshafen was the birthplace of the Zeppelin; it was
here that the man who gave the German airships his name — Count
Ferdinand von Zeppelin — in 1896 formulated his plans for the first
airship at the Hotel Kurgarten, and the first trials took place from Lake

Constance. By 1914 the area contained not only the massive erecting sheds and hydrogen-gas-producing plants, but also extensive workshops and a large floating dock for the Zeppelins: it was clearly an ideal target.

The raid, which was the first ever strategic bombing attack, was planned by the Admiralty with care and in the greatest secrecy. The aircraft selected were Avro 504s, one of the truly great designs, which are now mainly remembered as trainers, a role they were to fill to perfection in RAF service for over fifteen years. In 1914 they were thoroughly modern military aircraft; biplanes it is true, but with a simple, sturdy structure they weighed only 924lb, enabling the 504 to lift about 100lb of bombs in addition to a single Lewis machine gun which could be carried. The 504's speed with an 80hp Gnome rotary engine was 82mph at sea level, and it could remain airborne for four and a half hours. Perhaps the greatest feature of A.V. Roe's design was in its forgiving flying character- istics; it was viceless, a 'gentleman's aeroplane'. The prototype first flew at Brooklands in July 1913 and attracted immediate attention from the military, the Admiralty ordering seven in early 1914.

For the Friedrichshafen operation, a special flight of three brand new 504s, plus the original Avro on naval charge, No. 179, was formed at Manchester, where the Avro factory produced the machines. In October 1914 Squadron Commander P. Shepherd was given the command of the Friedrichshafen attack, which it had been decided would be launched from Belfort, a few miles from the frontier between France and Switzerland and some 120 miles from the target. Because of the surrounding mountains, the only possible route was up the valley of the Rhine.

The four Avros — 179, 873, 874 and 875 — were each equipped to carry 4 x 20lb bombs, a heavy offensive load in 1914. They were test flown, then dismantled and shipped to Le Havre on board the appropriately named *Manchester Importer* on 13 November. Still crated, the four aircraft were loaded on to railway flatcars, the train being shunted into a disused factory siding until nightfall; then, under cover of darkness, they were transferred to army lorries and driven to an airship shed at Belfort.

By next afternoon the Avros inside the large hangar were assembled and rigged, engines run up and tested, fuel tanks filled and all four bombed up, each with four 20lb bombs. Because of the need to preserve secrecy, it was decided not to test-fly the Avros — a tribute to the reputation of the design. Next morning the pilots arrived, but bad weather delayed the operation. On 17 November there was a slight improvement and Commander Shepherd decided after all to test-fly his aircraft, 179; unfortunately he had what was described as "a taxiing accident", which must have been spectacular for it smashed the propeller, buckled a wheel and crushed a wingtip. The damage was repaired by the air mechanics in the airship shed, but the work took some days and it was not until 20 November that all four aircraft were available and the raid possible. By that time Shepherd had been taken ill and his machine was to be flown by a reserve pilot, Flt Sub-Lt Cannon.

The temperature on the morning of 20 November was -7°C. To facilitate starting, in the biting cold, the mechanics drained all the castor oil from the aircraft and warmed it up before returning it to the oil tanks, which had been lagged in layers of red flannel.

The doors of the airship hangar opened and the four Avros were wheeled out. The engines, with their warm oil, started easily and the four biplanes taxied for take off. It was 0930 hours exactly as Avro 873, piloted by Squadron Commander Featherstone-Briggs, took off; he was followed by Flt Commander Babington at 0950 in 875, and 5 minutes later Flt Lt Sippe lifted off 874; as he circled overhead 'old 179' was beginning its take-off run. But Flt Sub-Lt Cannon found his engine was short of 100rpm and the heavily laden Avro showed little sign of becoming airborne, so Cannon cut the throttle and taxied back down wind to try again; in his anxiety to get 179 off in time to join the others, he ground looped and broke the tail skid. Since repairs would take too long, the machine was wheeled back to the hangar and dismantled.

The three remaining Avros were by then well on their way to target. The most lucid account of the operation is to be found in Flt Commander Babington's log:

> 0950: Left Belfort. Engine revs 1,050rpm.
> 0952: Shaped course for Basle astern of Briggs in 873.
> 1010: 873 circling.

The aircraft skirted Basle in neutral Switzerland and soon were:

> 1020: Over the Rhine near Basle. Shaped course Rhine valley at between 4,000/5,000ft.
> 1130: Over clouds between Schaffhausen and Konstanz, observed 874 3 miles ahead enter clouds.
> 1147: Sighted Zeppelin sheds, shaped course for them at 3,500ft.
> 1153: Abeam empty [Zeppelin] shed. 874 seen low over lake, shrapnel bursting over him.
> 1155: Steep dive from 4,000ft, straight down. [Released] bombs at 950ft. Heavy fire, machine nearly vertical down over shed, felt shock of bombs.

Under intense anti-aircraft fire, having released his bombs into the Zeppelin shed, Babington pulled his aircraft from its near terminal velocity dive and zoomed up and turned towards the sun to confuse the German gunners. As he did so he saw Sippe attack, dropping three bombs (the fourth hung up) from 700 feet. The effect was spectacular beyond expectation; a hydrogen generating plant exploded with tremendous violence, flames leaping high into the sky. Workshops were also hit and inside the sheds the Zeppelin Z7 was badly damaged.

Squadron Commander Featherstone-Briggs had his Avro shot down by heavy ground fire over the target; he landed his damaged machine safely but was taken prisoner. He was the only casualty. Sippe, still carrying a hung-up bomb, was the first to return, landing 874 safely back at Belfort at 1350 hours, nearly out of petrol. Flt Commander Babington overflew the Belfort airfield, which was shrouded in mist, until on the limit of his Avro's endurance, landing at an RFC field near Largerwells after a flight of four and a half hours in an open cockpit in sub-zero weather.

For the loss of one aircraft, the RNAS had achieved a significant victory. In a tribute to the operation, Winston Churchill told the House of Commons on 23 November: "This flight of 250 miles, which penetrated 150 miles into Germany, across mountainous country, in difficult weather conditions, constitutes, with the attack a fine feat of arms."

The successful raid on Friedrichshafen, though mounted by the RNAS, had been made by landplanes flown from an airfield. The seaborne aspect of the RNAS was at that time confined to seaplane reconnaissance patrols, mainly over the North Sea from the chain of bases around the British Isles. The aircrafts' short endurance, however, limited their range: to enable seaplanes to operate further afield, carriers were required. The sole 'carrier' was HMS *Hermes*, which embarked three seaplanes, but she was torpedoed by a U-boat in the Channel on 31 October 1914. *Hermes* was followed into service by HMS *Ark Royal*, which had originally been laid down as a tramp steamer and bought by the Admiralty in 1913 while still on the stocks. She was redesigned with her machinery space, funnel and bridge aft, like a tanker, giving a clear 'flying-off' deck of 130 feet; unfortunately, by the time she was commissioned in 1915, her top speed of only 10 knots was too slow for the relatively heavy aircraft then in service to take off directly from her deck, and she was therefore operated as a simple seaplane tender, hoisting her aircraft in and out of the water by crane.

In August 1914 the Admiralty had converted three fast cross-Channel ferry boats — *Engardine*, *Riviera* and *Empress* — as seaplane carriers; each was able to operate at least three seaplanes and, though little more than navigable hangars, they had a relatively high speed of 21 knots, which enabled them to keep up with the fleet. In November 1914 it was decided that these three ships would launch a seaborne air attack on the German Zeppelin base at Nordholtz, south of Cuxhaven. This was to be the first true naval air operation, with the attacking aircraft being launched from seaplane carriers at sea, the target being far out of range of any land-based machine.

The Cuxhaven operation was planned with great care and attention to detail. It had a dual purpose: first, to try to destroy Zeppelins and their base, and, secondly, to try to get at least part of the German fleet out of Wilhelmshaven and drawn into a naval action. The latter was seen at the Admiralty as the primary reason for the whole operation, the seaplane raid being "incidental and subsidiary, though very important in itself . . . ." The seaplanes were to be launched at first light on Christmas Day, 1914.

The attacking force of *Engardine*, *Empress* and *Riviera*, escorted by two cruisers and ten destroyers, slipped out of Harwich at 1700 hours on 24 December. Ten submarines, which were to screen the surface vessels and also act as rescue ships should any of the seaplanes have to ditch, were already at sea.

The task force crossed the North Sea under cover of darkness without incident until 0430 hours, when HMS *Arethusa* passed a number of trawlers, one of which transmitted a W/T signal. Half an hour later the wireless room on the cruiser intercepted a very powerful signal coming from Heligoland; this was coded but the transmission contained the known German Navy 'urgent' prefix. However, no hostile force appeared. At 0600 the three seaplane carriers had arrived at the designated flying-off position: lat. 54°27'N, long. 8°00'E — 40 nautical miles due north of the Frisian Island of Wangerooge. It was bitterly cold in the pre-dawn darkness but the sea was calm with little wind, and visibility when dawn broke, judging by the brilliant night sky, would be unlimited.

As the fleet hove to, the pilots, who had been in readiness since 0500 hours in the crew rooms, were getting their final briefing and checking the fairly comprehensive survival kit that each man was required to carry on the flight: "1 service revolver and 6 packets of ammunition, 1 Very light pistol and 6 rounds, a lifebelt, two torches, boxes of matches, a knife, first aid kit, provisions for 48 hours, maps and charts." There was also a long inventory of tools, including ignition wire and spare sparkplugs and a copper drift hammer. Clearly the staff at the Admiralty were taking the raid seriously.

In addition to the primary target, the Cuxhaven Zeppelins, the pilots were also ordered to report on several specific points. Was the Elbe Light

Vessel No. 1 in position? Where were the ships in the Schelling Roads anchored? What were numbers and classes of ships inside the basin at Wilhelmshaven? Information was also required on boom defences and the buoyed channels through the minefields, in addition to likely targets in the various German naval anchorages which British ships could attack.

The pilots and observers climbed aboard their seaplanes at first light, 0700 hours. *Engardine*, the flotilla leader, signalled the other ships to hoist out their aircraft; the only sound across the slow dark swell was the clanking of the steam winches as the nine seaplanes were lowered into the water. (See page 374 for aircraft and pilot allocations.) A few minutes after 0700 *Engardine* gave the 'start engines' signal and, in the pale light of that freezing Christmas dawn, the unfortunate air mechanics began the difficult task of starting the stone cold engines, swinging the heavy wooden propellers from precarious footholds on the narrow floats — at times awash. In getting the reluctant engines to fire, two mechanics ended up in the icy sea. Nevertheless, within fifteen minutes seven of the nine seaplanes were taxiing away from the ships into wind to begin their take-off run, which was to be long, owing to the light wind and the smooth sea. Two machines (122 and 812) could not be started and were ignominiously hauled back on board their ships.

By 0730 hours the seven seaplanes, marked under their wings with a large red roundel, in addition to the then usual Union Jack, were airborne and steering for Cuxhaven, 40 nautical miles away. At 0735, as the sound of their engines faded, the German Zeppelin L6 was sighted approaching from Heligoland; she made for *Empress* and dropped several bombs near her. HMS *Undaunted* then opened fire with her 6 inch guns and several shells burst near the L6, which moved off out of range. An enemy seaplane also briefly appeared but soon left the scene, flying towards land.

In the open cockpits of the British seaplanes it was very cold; so cold the engines were running roughly and misfiring, in part because the low octane fuel was not vapourising and possibly also because of carburettor ice. All the machines were, by any standards, underpowered and slow; the fastest — the 'Folders' — had a maximum speed of only 78mph, and the slightest drop in engine revs made it very difficult for the pilots to maintain height.

Although the visibility at sea had been excellent, as the aircraft flew over the East Frisian Islands, long fingers of mist were seen forming below. The seaplanes dropped lower and lower until pilots and observers could see only a confusing landscape of sand dunes and saltings passing a few feet beneath; the mist thickened and, as the target area was approached, the pilots were trying desperately to climb above a dense fog bank which completely blanketed the ground. Not one of the seven attackers so much as glimpsed the Nordholtz Zeppelin sheds.

Turning south-west over the Heligoland Bight to look for secondary targets, the seaplanes found Wilhelmshaven clear. Flying over the German naval basin, their observers were able to note the details of the anchored fleet, identifying individual ships; several bombs were dropped on a cruiser and a seaplane base in the face of heavy anti-aircraft fire that was encountered. The ground fire was accurate and most of the attacking seaplanes were hit, although miraculously none of the crews were injured. More warships were also sighted in the Schelling Roads: two pilots, Flt Commander Kilner and Flt Lt Edmonds, reported ships raising steam and noted seven battleships, eight cruisers (two under way) and a number of destroyers. Edmonds dive-bombed one of the cruisers. As the attackers, having dropped their bombs, made for the recovery position off Borkum, the effects of the German gunfire took its toll: four seaplanes with leaking fuel tanks or damaged engines had to ditch. At 0930 hours Flt Commander Ross put 119 down near HMS *Lurcher*, which took him on board and put a towline to his aircraft, Ross reporting that he had bombed a German U-boat off the island of Wangerooge.

Ross was not the only pilot picked up that day. The most spectacular rescue was that performed by the submarine E11, whose Captain, Lt Commander Nasmith, saw through his periscope Flt Lt Miley's Short 'Folder' alight. E11 surfaced, causing some concern to the crew of the seaplane until they noticed the broad band of red and white checkerboard which had been painted as a recognition signal on all the British submarines taking part in the operation.

Miley was lucky; he had no idea the submarine was on hand when he ditched, out of petrol and 13 miles short of the fleet. E11 embarked the 'Folder's' crew and took the seaplane itself in tow; soon afterwards, at 1000 hours, two more Shorts — 814 and 815 — sighted the E11 and alighted alongside, the pilots hailing the submarine's bridge with the information that they too were out of fuel. At that moment another submarine was sighted diving towards the E11; simultaneously a Zeppelin appeared and raked the submarine and seaplanes with heavy machine-gun fire, while manoeuvring to bomb the E11. Her Captain, menaced from above and below, cast off the 'Folder' which was in tow and ran alongside the drifting Short, 815. The other seaplane, 814, was now sinking as a result of the Zeppelin's fire and both the crews swam towards E11, hoping the submarine would remain surfaced long enough for them to scramble aboard. Fortunately the Zeppelin, possibly because of the gunfire which the E11 was directing at her, had drawn away but was only waiting for the submarine to clear her decks to dive. As the last airman was hauled aboard, Lt Commander Nasmith, though expecting to be torpedoed by the other submarine at any moment, crash-dived and, with only 40 feet of water above his vessel, the Zeppelin's bombs burst on the surface overhead. No torpedoes struck the E11; the 'enemy' submarine was in fact the British D6, which had prudently dived on first sighting the Zeppelin.

At the same time as the dramatic rescue by submarine, two further seaplanes, 136 and 811, were recovered by *Riviera* — the only two of the original seven raiders to rejoin the fleet. No. 136, flown by Flt Commander Kilner from *Riviera*, carried as her observer Lt Erskine Childers, the author of *The Riddle of the Sands* — a story of espionage set in the area over which he had just flown.

No. 135, another of *Riviera's* aircraft, flown by Flt Commander Hewlett, was now the only seaplane unaccounted for; it had last been sighted by another crew flying 8 miles west of Heligoland. It was later learnt that Hewlett's engine had seized up owing to an oil pipe being fractured by gunfire and he too had alighted on the sea alongside a Dutch trawler, *Maria van Hutten.* He was to be interned, though was later released from Holland, then a neutral country.

Seen in purely military terms, the attack, apart from the reconnaissance of the German fleet, had been a failure, but in point of fact it achieved a good deal more than was at first apparent. In the first place, the sudden appearance of a force of British aircraft over a major enemy naval base, supposedly far out of range of any possible airborne attack, caused a good deal of consternation. As a consequence the Germans were forced to strengthen their anti-aircraft defences, not only at Cuxhaven and Wilhelmshaven, but at all vulnerable military installations, thus tying down a sizeable force of men and guns which could have been disposed along the western front.

The Cuxhaven raid's real value lay in its very execution; true it was flawed, but it demonstrated that an airborne force could operate with the fleet. The operation mounted on Christmas Day 1914 pointed the way to a totally new concept of naval warfare which would prove devastating twenty-seven years later at Taranto, Pearl Harbor, and the Battles of the Coral Sea and Midway, when the aircraft carrier became the capital ship of the world's navies. An American naval historian said of the Cuxhaven raid: "If any single action gave birth to the concept of aircraft carrier operations .... this raid would qualify ...."[5]

The man who planned the Cuxhaven raid, Squadron Commander Cecil l'Estrange, had grasped the fundamentals, for he wrote in his official report:

> I look upon the events which took place on 25 December as a visible proof of the probable line of developments of the principles of naval strategy. One can imagine what might have been done had our seaplanes, or those sent to attack us, carried torpedoes instead of light bombs. Several of the ships in Schelling Roads would have been torpedoed, and some of our force might have been sunk as well.

The fact that fog had shrouded the sheds at Nordholtz and thus prevented any of the seaborne attackers from bombing the Zeppelins was to have almost immediate repercussions. Just fifteen days later, the Kaiser, who had qualms about using airships to bomb England, finally made up his mind and authorised raids on targets within the British Isles with, however, the reservation that such raids were "expressly restricted to military shipyards, arsenals, docks and, in general, military establishments .... London itself was not to be bombed."[6]

The Imperial German Navy had three Zeppelins immediately available — L3, L4 and L6, all of which were of the 'L3' class, and 520 feet long with a hydrogen capacity in excess of 800,000 cubic feet. Though these airships were slow, with a maximum speed of only 50mph, they had an operational ceiling of 10,000 feet and could remain airborne for over thirty hours, which must have seemed a very long time indeed to the crews in the open gondolas and gun positions.

An airship properly trimmed, unlike an aeroplane, has a high degree of inherent stability which enables it to fly blind in cloud or at night. Furthermore, their long endurance permitted the German airships to take off from their bases in daylight, arrive over a target in Britain under cover of darkness, then fly back to Germany to a daylight landfall.

In those distant pre-radar days the only hope of interception by defending aircraft lay in a visual contact, which at night was extremely unlikely. However, airships did suffer from one operational disadvantage: they were difficult to handle on the ground in a wind. High winds delayed any prospect of a raid over the British Isles until 13 January 1915, when the three Zeppelins started out to attack English targets, but deteriorating weather over the North Sea forced them to return to Germany without crossing the enemy coast.

Six days later, the weather having improved, L3, commanded by Kapitänleutnant Hans Fritz, together with L4, Kapitänleutnant Graf von Platen-Hallermund, were eased out of the large double shed at

Fuhlsbüttel, near Hamburg, and set course for the Humber, on the British east coast, 300 miles to the west. As the L3 and L4 crossed the River Elbe at Cuxhaven, the Nordholtz shed, which could be rotated, had been turned into wind and the third Zeppelin, L6, emerged. Her Captain, Oberleutnant Horst Freiherr Teusch von Buttlar-Brandenfels, had on board a distinguished passenger — the Commander of the German Naval Airship division, Korvettenkapitän Peter Strasser, who had for some time pressed the Kaiser to sanction a Zeppelin raid over Britain and who had planned the present attack. L6 rose from her base, heading across the North Sea for the Thames estuary.

The Zeppelin L6 was sighted over the Dutch Frisian Island of Terschelling at 1330 hours, followed soon afterwards by two others (L3 and L4). To the intense disappointment of Strasser and her crew, L6 then suffered engine trouble over the North Sea and was forced to return to Nordholtz without even sighting the coast of England.

The two remaining Zeppelins approached England in darkness and remained undetected; the coast was crossed at 2000 hours. There was no blackout in force at that time in England and towns could clearly be seen. The navigators soon established their position over Norfolk, 85 nautical miles south of their intended landfall. The two airships seemed then to separate; L3, flying at 5,000 feet in sleet and snow showers, dropped several bombs on the fishing port of Great Yarmouth at 2030. The Kaiser's strictures about military targets were, it would seem, not heeded; the bombs (not all of which exploded) destroyed a number of dwelling houses, killing two civilians and injuring several others.

Meanwhile L4 flew northwards over East Anglia, passing a seaside resort, Hunstanton, at about 2200 hours, then following the Great Eastern Railway line to Heacham, a small village on the Wash, where it dropped a single bomb which exploded on open ground, causing no damage. The Zeppelin then passed directly over Sandringham, 6 miles away, where King George V was in residence, giving rise later to unfounded stories that the whole object of the raid was to kill His Majesty, a most unlikely intention; after all, the King was Kaiser Wilhelm's cousin. L4, unaware of the royal presence below, flew on until it sighted the lights of Kings Lynn, dropping the remainder of its bombs at 2300 hours on that town. They demolished a row of cottages in Bentinck Street, killing a woman and a youth, and injuring his four-year-old sister and twelve other people.

Throughout the spring and summer of 1915 the Zeppelins continued sporadic raids on the British Isles: there were attacks during April on Blyth in the north-east, Lowestoft, Malden in Essex, Ipswich and Bury

St Edmunds. None of these bombing raids could be described as heavy, though people were killed and injured. In the following month Ramsgate and Southend were the targets; the East End of London was bombed on 31 May, the Kaiser having lifted his ban on attacks on the capital, with the proviso that the objectives were to be the docks and that no bombs were to be dropped west of the Tower (probably for fear of hitting Buckingham Palace). The worst raid to date came on the night of 6 June when sixty-four people were killed in Kingston-upon-Hull; this raid caused a wave of anti-German feeling and shops with German-sounding names had their windows smashed and were looted.

The RNAS, which was at that time responsible for the defence of the British Isles, could do little. Anti-aircraft fire was ineffective and the fighter aircraft available could only stand by until a Zeppelin was reported, then take off at night (without blind-flying instruments) in the hope of making a lucky visual interception. Very few such interceptions were made; the pilots who did sight an airship quickly found that a Zeppelin dropping its water ballast could climb at 1,000 feet a minute, far faster than a contemporary fighter, few of which could do better than 300 feet a minute. Until June 1915 not a single Zeppelin had been shot down.

The only possible defence against the German raiders was the same tactic as that which would be used against the V2 rockets twenty-nine years later: attacking the bases from which they flew. This strategy, as we have seen, had already destroyed and damaged Zeppelins and, seeking further success, four aircraft of No.1 Squadron RNAS, based at Dunkirk, had taken off on 7 June 1915 to attack Zeppelin bases at Bercham Ste Agathe, near Bruges, and at Evere, near Brussels. Unknown to the Navy pilots, three of the airships from those bases had that night set out to raid London; they were the army's LZ37, 38 and 39, all three of which suffered engine trouble soon after take-off and were forced to return. While on its homeward flight, LZ37 was sighted over Ostend by Flt Sub-Lt R.A. Warneford, flying a Morane-Saulnier type 'L' monoplane, No. 3253, of No. 1 Squadron.

The aircraft, one of the four heading for the airship sheds, was laden with six 20lb bombs slung under the fuselage, which reduced its already marginal rate of climb to the point where, should it be sighted by the Zeppelin, the latter would have no difficulty in outclimbing the small Morane.

Flying at 8,000 feet, Warneford stalked the dark grey shape of the Zeppelin for forty minutes when, near Bruges, the airship began to descend towards its base. Warneford, about 2,000 feet above, pushed the

Morane into a steep dive; a German air gunner, positioned on the top of the Zeppelin, sighted the aircraft and immediately opened fire. Other gunners also fired and Warneford had to break away but kept his precious height advantage, pushing the Morane at maximum power to its ceiling of 11,000 feet.

Alarmed by the sudden attack, the Zeppelin's Captain decided to make a run for home and the big airship began to descend, the air gunners ceasing fire, possibly because the ship was venting hydrogen. It was a fatal decision. Half a mile astern and with a 2,000 foot height advantage, Warneford cut his engine and silently glided down over the Zeppelin, which was now at about 7,000 feet and descending rapidly. Coolly Warneford positioned his 34 ft span monoplane 150 feet over the airship and pulled the wire toggle to release his six bombs; five missed but the sixth was a direct hit. Instantly there was a tremendous explosion; LZ37 was doomed.

The shock of the exploding hydrogen threw the Morane on to its back and Warneford, hanging from his straps, watched, upside down, the Zeppelin falling like an enormous firebrand with a trail of burning fabric and wreckage. The blazing hulk of LZ37 crashed into a convent outside Ghent, sadly causing casualties among the nuns and orphan children. Warneford righted his aircraft but the engine cut out and refused to restart; after a protracted glide in the dark, he made a forced landing in a field 30 miles behind the German front line. Miraculously the Morane was undamaged, and by torchlight Warneford discovered the cause of the engine failure: a leaking petrol union. This he repaired and, thirty-five minutes after force-landing, took off for Dunkirk.

Within thirty-six hours Warneford received, by telegram from the King, his country's highest award: the Victoria Cross. Sadly, ten days later, when flying a Farman over Paris with an American journalist, Harry Needham, as passenger, his aircraft rolled over during a steep turn and both men, who were not strapped in, fell to their deaths.

The raid on Hull was the first to cause a large-scale wave of public indignation; the next attack on London on 13 October killed 149 people in the Strand area. The public outcry was immediate: the Germans were called 'baby killers', cartoons in the popular press depicted Germans as monsters bayoneting children, and the German word *Schrecklichkeit* — frightfulness — was quoted to describe the behaviour of the Zeppelins' crews. It is difficult now, after the heavy air raids of World War II, in almost any of which more civilians were killed in a single night than throughout the entire Zeppelin campaign against England, to imagine the feelings of revulsion that were felt in 1915.

Wars up to that time had been fought by soldiers and largely, as far as Britain was concerned, on foreign soil; to attack 'innocent' civilians asleep in their beds was just not done. But it had been done. The Zeppelin raids were a military watershed, the first shots fired in a new concept of war — Total War, in which the entire populations of the belligerents were to become as legitimate a target as the soldiers in the field, with terror as the weapon.

The Royal Navy, though concerned with the consequences of the Zeppelin raids on the towns and cities of Britain, was in reality far more worried about another and, militarily, more important aspect of the German airships: their constant shadowing of the British Fleet and the reporting by wireless on the movement of naval units. As Admiral Fisher complained in a memorandum written early in 1916: "The Grand Fleet cannot stir out of Scapa Flow for any surprise operation, as the German Zeppelins report by W/T the British Fleet's every movement."[7] The search for a counter to this humiliating challenge to the Royal Navy was to lead eventually to the true aircraft carrier. There were, however, to be several false dawns before that sunrise.

The seaplane had seemed in the early formative days of the Naval Wing an obvious and logical choice, and for reconnaissance the type still had its uses, but for a given engine power seaplanes with their massive floats, which were both heavy and drag-producing, would always be out-performed in every aspect of flight when compared with landplanes, apart from their ability to take off and alight on water. Even then, seaplanes could only operate if the surface of the sea was neither too rough nor too calm; the ships which carried them had to heave to to launch and recover, a procedure which did not appeal greatly to their captains once the German U-boats began to be a serious menace. Even when they became airborne, the seaplanes could only carry a very small offensive load, and their operational ceiling, when bombed up, was around 3,000 feet and often a good deal less, bringing them into the range of machine-gun and small arms fire.

It is perhaps not realised today that the majority of aircraft in service up to 1916 were powered by engines which, even if they had been reliable, were in a power range that would now be considered marginal for a light aircraft. Yet seaplanes powered by 100hp radial engines were required to lift two men, bombs, machine guns, wireless, survival kits, and at least two floats, which, when the seaplane was flying, contributed nothing but drag. Seaplanes, because of the rough treatment to which they were inevitably subjected in being hauled in and out of the water and in taking off and alighting from the open sea, had of necessity to be more strongly constructed than landplanes, which of course made them heavier. It is difficult to obtain exact comparisons of landplane versions of the early seaplanes, but the Fairey Flycatcher, a biplane naval fighter of 1923, was available in both landplane and seaplane configuration. The performance figures[8] are illuminating:

	Landplane	Seaplane
Loaded weight	2,979lb	3,579lb
Maximum speed	133mph	126mph
Ceiling	19,000 feet	14,000 feet

The Flycatcher of 1923 was, when compared with the 1916 aircraft, highly developed and powered by a 400hp engine, yet the seaplane version had a markedly inferior performance and was far less manoeuvrable — a vital consideration for a fighter.

It is of significance that all the successful RNAS bombing operations — the raid on Friedrichshafen, Warneford's destruction of LZ37 and the protracted operation in the Rufigi Delta — had used landplanes, which were more efficient flying machines. To intercept Zeppelins and attack them in the air, any form of seaplane would be unable to get within effective range. The seaplane, except possibly when used for reconnaissance, was by 1916 at a dead end. Nowhere was this more evident than in the disastrous Dardanelles campaign — usually now simply referred to as 'Gallipoli', a place name which has for British arms much the same sad aura of defeat as Dunkirk.

In essence, the object of the operation in the Dardanelles, as conceived by Winston Churchill, was to defeat the Turks and force their Austrian and German allies to deploy a large force in the area, thus relieving the pressure on the western front. Had it succeeded, the Balkan states would probably have joined the Allies, the Russians could have marched east, and the war might well have ended in 1915. The Dardanelles operation, which began in February 1915, was a seaborne action; HMS *Ark Royal*, the converted tramp steamer, was part of the naval force and carried six seaplanes, which were required to act as spotters for the battleships. Unfortunately, during the battle the sea was nearly always unsuitable for launching seaplanes and, although *Ark Royal* had a 130 foot flying-off deck, her slow speed of 10 knots was, as already noted, insufficient to launch landplanes and the deck was never used; therefore *Ark Royal's* contribution was strictly limited.

The Dardanelles action began badly. The attacking ships ran into a Turkish minefield laid in the Straits and three battleships — *Inflexible*, *Irresistible* and *Ocean* — were sunk and the attack broken off despite urgent telegrams from Churchill insisting the battle be resumed.

The next phase of the ill-starred venture was an amphibious landing at Gallipoli, which was in scale overshadowed only by the Normandy landing thirty years later. That latter assault took three years of detailed planning and training; the Gallipoli landing was barely planned at all. It was to end in disaster and cost the lives of 50,000 men.

Number 3 Wing, under the colourful Commander Samson, arrived from Dunkirk to support the Gallipoli landing. It operated a motley collection of landplanes from a makeshift airstrip at Tenedos; short of stores, spares and wireless equipment, the main contribution it made was photographic reconnaissance, which supplemented the vague and inaccurate maps of the area. Commander Samson himself flew his BE2a No. 50, which had seen action over France and Belgium, using it to bomb the Turks at every opportunity. As he wrote later: "I started bombing the Turks, I didn't hit anything, but it was good practice, and I felt it was time the Turks realised that Eastchurch [the original RNAS wing] had arrived on the scene."[9]

A single German submarine, U21, joined the battle, torpedoing HMS *Triumph* and *Majestic;* at this point *Ark Royal,* which was too slow to hope to evade torpedoes, anchored in the harbour at Mudros. She was relieved by another seaplane carrier — a converted Isle of Man packet, the *Ben-my-Chree,* a fast ship which possessed a flying-off deck, though even when later commanded by Samson, this was never used; seaplanes were still launched and recovered over the side, though Samson managed to achieve this manoeuvre without actually stopping the ship.

Among the seaplanes embarked aboard were the first and second prototype Short 184s. It was in one of these aircraft, on 12 August 1915, that Flt Commander C.H.K. Edmonds made an historic flight. His seaplane was armed with an 810lb, 14 inch Whitehead naval torpedo slung between the floats, and though barely able to get off the water (to do so Edmonds had to fly solo and with only sufficient fuel for 45 minutes' flight) he spotted a 5,000 ton Turkish supply ship off Injeh Burnu. The ship was not under way when Edmonds glided down from his enforced ceiling of 800 feet to within 15 feet of the water and, at a range of 350 yards, let go the torpedo, which ran true, exploding against the ship's hull abaft the mainmast.

This was the first time a ship had been attacked by a torpedo dropped from the air. The feat was only slightly marred by the fact that the Turkish vessel was stationary, for the very good reason that she was fast aground in shallow water, having been sunk some days previously by the British submarine E14. However, five days later Edmonds torpedoed an undamaged Turkish supply ship, which he sank. Later that day, 17

August, the other Short 184, piloted by Flt Lt G.B. Dacre, torpedoed a large tug in False Bay. These were the only successful torpedo attacks by Allied aircraft during the Great War. The official *War in the Air* explains why this was so:

> Unhappily, the torpedo-loaded Short seaplane [the 184] could only be made to get off the water and fly under ideal conditions. A calm sea with a slight breeze was essential and the engine had to be running perfectly.
>
> Further, the weight of the torpedo so restricted the amount of petrol which could be carried that a flight of much more than three-quarters of an hour was not possible. So it came about that while a number of torpedo attacks were attempted, only three were successfully concluded.[10]

In Britain, seaplanes were still the backbone of the Coastal Naval Air Stations, of which there were nearly fifty by the end of 1915, but the seaplanes based on them, though not lacking a certain spindly elegance, were of little use against that *bête noire* of the Royal Navy, the Zeppelin. It is possible to argue that the Navy overestimated the menace to the fleet which airships posed and, as far as the defence of Great Britain against them was concerned, that became the responsibility of the RFC from February 1916, leaving the Navy free to concentrate on purely maritime aspects of the Germans' prying flights. In a sense the problem confronting the RNAS was a good deal more straightforward than countering the night raiders, for the fleet-spotting Zeppelins clearly had to operate in daylight, so that visual interception was possible. It was obvious, however, that to make an air-to-air attack on the intruders, fast-climbing fighter aircraft were required.

There was one promising fighter already on naval charge: the Bristol Scout. This small (24 feet 7 inches span) single-seater biplane, designed by Frank Barnwell, first flew in February 1914 and entered service with the RNAS a year later. The 'Type C' — the version first delivered to the Navy — was for its time a very fast aircraft, with a top speed of 93mph, which it achieved with only an 80hp Gnome rotary engine — a tribute to the clean lines of the airframe. It had an initial rate of climb of 1,000 feet per minute, with a service ceiling of 15,000 feet. The Bristol Scout was a fighter that could attack Zeppelins with at least an even chance of success, and about sixty were on naval charge by the spring of 1916. The problem was simply that the Scouts were landplanes and had a limited endurance of two and a half hours, restricting their radius of action. To overcome this deficiency, two Bristol Scouts were embarked on yet another seaplane carrier, a converted Isle of Man packet SS *Viking* which became HMS *Vindex*, with a 64 foot flying-off deck.

On 3 November 1915 Flt Lt H.F. Towler flew Bristol Scout No. 1255 from the ship at Harwich. This was the first take-off by a landplane from an aircraft carrier. Not that there was any question of landing back on deck! Towler landed ashore, probably at the nearby RFC airfield at Martlesham Heath. It was, however, a small step nearer to the true aircraft carrier, though six years after Eugene Ely's landing on the *Pennsylvania* and three years after Samson had flown from HMS *Hibernia*. Samson, incidentally, had some Bristol Scouts in his wing at Gallipoli and why these were not flown from the *Ben-my-Chree*, which he commanded, is something of a mystery.

The Navy now at last had the means to intercept any Zeppelins that were shadowing the fleet. HMS *Vindex* could make 20 knots — fast enough both to keep station with the fleet and launch fighters, though it is true that unless they could reach land there was no alternative open to the pilots, after flying off, other than ditching alongside a friendly ship.

RNAS Bristol Scouts were not normally armed with machine guns, though they carried twenty-four Ranken darts — small incendiary missiles which were dropped from two containers in the cockpit floor. Flt Lt Freeman, flying from *Vindex*, attacked the Zeppelin L17 over the North Sea, but, unfortunately, though some of the darts hit the airship, he had to break off the attack when a fuel line fractured, cutting his engine, before the L17 received a fatal hit. Freeman ditched and was rescued by a Belgian ship and interned for a day or two in Holland.

Neither the Ranken dart nor a later anti-Zeppelin weapon, the Le Prieur rocket, five of which were carried on each of the outer interplane struts, was really effective against airships. It was not until the introduction of incendiary Buckingham ammunition, fired from twin machine guns and synchronised to pass through the propeller arc, that the intercepting fighters had a telling weapon with which to attack the Zeppelins.

In their search to get short-range fighters within sight of the marauding Zeppelins, the most ambitious proposal tested by the Navy was to carry a Scout on the top of a Porte Baby flying boat. (Just why it was called the Porte 'Baby' is difficult to imagine; it was, for its day, very large indeed, with a 124 foot wingspan and three engines — two tractors, one pusher.) The flying boat, designed by Squadron Commander Porte, flew from Felixstowe RNAS base piloted by Porte himself, with a Bristol Scout from *Vindex* (No. 3028) perched on the top wing. The composite aircraft climbed up to 1,000 feet above Harwich harbour when the Scout's pilot, Flt Lt Day of *Vindex*, opened up his engine and successfully separated, climbing away to land at Martlesham. The date of that notable but strangely neglected 'first' was 17 May 1915. Flt Lt Day was killed in France soon afterwards and the experiment was not repeated.

(*Above*) The only known picture of an historic moment: 15.16 hours, 11 November 1910 when Eugene Ely, a civilian pilot, flew his 60hp Curtiss from a temporary wooden deck, only 83 feet long, on the forepeak of USS *Birmingham* (*US Naval Photographic Centre*); (*below*) the USS *Pennsylvania* alongside Mare Island, San Francisco Navy Yard: the finishing touches are being made to the wooden platform for Eugene Ely's landing. The maximum available length was only 120 feet and a sloping canvas screen abaft the mast was erected in case the crude arrester system failed (*US Navy*)

(*Above*) The 'Old Glory' flying from the after mast of the *Pennsylvania* shows that Eugene Ely had to make a downwind landing with a considerable cross-wind (*US Navy*); (*below*) Ely's judgment was good: he has caught the 9th, 10th, 11th and 12th rope and the Curtiss, about 10° and three feet from the centre line, comes to a standstill on the *Pennsylvania* (*US Navy*)

(*Above*) Ely stands by his Curtiss prior to taking off from USS *Pennsylvania* (*US Navy*); (*below*) Commander Samson walks away from S27 on the field at Lodmore, having just landed after his flight from HMS *Hibernia* (*Flight*)

(*Above*) *Königsberg*, after the action, sunk in the Rufigi (East Africa). The central funnel has gone and her hull is riddled with shrapnel (*Archiv BfZ*); (*below*) the ten 10.5cm guns from *Königsberg* were salvaged from the sunken ship and, under Korvettenkapitän Schoenfeld, were manhandled across country through the bush on improvised carriages, to serve the German ground troops throughout the remainder of World War I (*Archiv BfZ*)

(*Above left*) The original HMS *Ark Royal*, a 6,900-ton tramp steamer bought in 1913 by the Admiralty in frame and converted into a seaplane carrier — the first ship to be commissioned exclusively in that role (*FAAM*); (*above right*) HMS *Hermes*, the Royal Navy's sole 'carrier', sinking in the Channel after being torpedoed by a German U-boat on 31 October 1914 (*FAAM*); (*below*) HMS *Engardine* hoisting aboard a Short 184 in the Firth of Forth. *Engardine*, together with her near sisters, *Riviera* and *Empress* — all converted from cross-channel steamers — took part in the attack on the Zeppelin base at Nordholtz: the first naval air operation, mounted on Christmas Day, 1914 (*IWM film still*)

(*Above*) A Short 184 — one of the most successful designs of World War I naval aircraft. Even by the standards of the day their performance was not electrifying, the maximum speed being 88½mph and they took over 8 minutes to climb to 2,000ft (*IWM film still*); (*below*) the Sopwith Type 9700, always known as the '1½ Strutter', was the first British aircraft to enter service with a synchronised Sopwith-Kauper interrupter gear, enabling its single Vickers .303in gun to fire through the airscrew arc (*British Aerospace*)

The 'flying-off' deck of HMS *Furious*: the Sopwith Pup is being hoisted out of the 'hangar' below decks, still in a rather primitive way. The tight fit of the small fighter is obvious; one wonders what this operation was like in a seaway (*British Aerospace*)

(*Above*) 2 August 1917 and an unknown photographer records a milestone in naval aviation as Sqdn Cmdr Dunning lands his Sopwith Pup aboard HMS *Furious*: the first deck-landing on a ship under way (*IWM*); (*below*) this photograph of a Bristol Boxkite illustrates perfectly the nature of aviation in its formative years (*Moore Collection, FAAM*)

The relative approach speed of the Sopwith biplane was such that the main problem was not 'arresting' aircraft but retaining the light machines on the deck; even such experienced pilots as Rutland had difficulty in controlling their aircraft in the turbulence caused by the hot gases from the funnel and the ship's structure, as can be seen in these still-frames from Rutland's accident (*IWM film stills*)

(*Above*) The C (Coastal) airship, of which 26 saw service. These non-rigid airships had a duration of 11 hours at 45mph; hard-going in open cockpits over the North Sea in winter (*IWM film still*); (*below*) the car of a slightly more sophisticated airship — the SSZ — as used in 1916 (*IWM film still*)

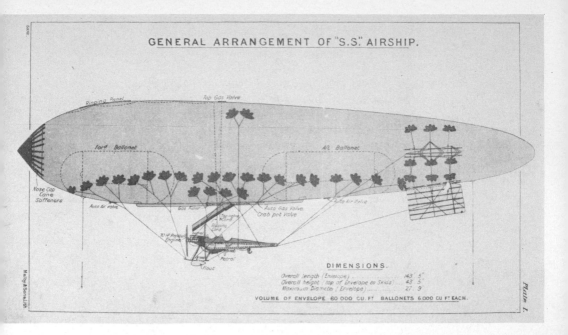

GENERAL ARRANGEMENT OF "S.S." AIRSHIP.

(*Above*) During World War I, the RNAS used a number of types of airships to patrol coastal waters on anti-U-boat patrols. The early SS (Sea Scout) used the fuselage of a BE2c, propelled by a 70hp Renault engine (*Public Record Office, London*); (*below*) the method of attacking a sighted U-boat had the merit of simplicity: the small bombs were simply dropped over the side. U-boats respected the blimps enough to crash-dive as soon as they were sighted and no ship was lost to a U-boat when escorted by them (*IWM*)

(*Above*) HMS *Eagle*, 24,550 tons (1922), the first 'island' carrier and a prototype for generations of fast Fleet carriers. *Eagle* had a flight deck 625ft long with an 'island' offset to port, leaving a 96ft-wide runway. Two lifts gave access to a hangar, 400 x 33ft, with a minimum headroom of 20ft 6in, which could embark some 18 aircraft. *Eagle* remained the only 'island' carrier to have twin funnels (*MOD (RN) Crown Copyright*); (*below*) HMS *Hermes*, commissioned in 1923, has the distinction of being the first aircraft carrier to be built as such. She was not a large vessel — 10,850 tons — and could embark only 15 aircraft. She was sunk by Japanese bombers in April 1942 (*RAF Museum*)

(*Above*) HMS *Furious* in one of her many forms: as she appeared during the 1930s as a flush-decked carrier (she was to have a vestigial port island during the war years). The photograph shows the short 'slip flight' deck which *Furious* had in common with HMS *Glorious* and *Courageous* (*Flight*); (*below*) the 'Covered Wagon' — CV-1, USS *Langley*. The first US Navy carrier, *Langley* had an open hangar, topped by a simple wooden flight deck, which was a common feature of American carriers, unlike British Fleet carriers, in which the hangars were in the form of an armoured box (*US Navy*)

HMS *Courageous*, just prior to the outbreak of war. Her port island arrangement is well illustrated, as is the entrance, or exit, to the hangar deck from which the 'Slip Flight' emerged. (Presumably the flag staff was removed when flying was in prospect) (*Flight*)

(*Above*) A close-up of the five hooks on the undercarriage of Lt Pride's Aeromarine engaging the fore and aft wires (*US Navy*); (*below*) a Gloster Sparrowhawk, a fighter built in 1921 solely for the Japanese Navy, at Hucclecote, Gloucester. These aircraft were manufactured in substantial numbers by the Japanese Naval Arsenal. Speed: 120mph, when powered by a 200hp Bentley BR rotary engine (*Gloster Aircraft*)

By far the largest seaplane carrier to enter service with the RNAS was the *Campania*, an old (1893) Cunard liner which had been on the Liverpool—New York run pre-war; the 20,000-ton ship could achieve 21 knots and was bought by the Admiralty following a request made to the Director of Transports on 6 September 1914: "C-in-C Home Fleet requires a large ship to carry seaplanes, can one be provided capable of steaming at 20 knots, and with sufficient coal to keep up with the Fleet."[11] Several ships were available, including two German prizes; *Campania* was initially offered on charter at £13,000 a month, any structural alterations that were effected having to be made good after the charter. After consideration the ship was bought outright at a cost of £180,000 and a very substantial conversion put in hand. She emerged with a 120ft flying-off deck and accommodation in her hangar deck for ten large seaplanes. Eight 14 inch Whitehead torpedoes were to be placed in her magazines for "discharge from seaplane".[12] The old ship was, in addition, given eight 4.7 inch quick-firing guns, one 3 inch anti-aircraft gun and four 24 inch searchlights. She was a coal-burning ship, and it is of interest to note that she required no less than 480 tons of it per day: stokers worked hard in those days.

Campania embarked her seaplanes and sailed from Cammell Laird's yard, Birkenhead, on 1 May 1915, commanded by Oliver Schwann, the intrepid self-taught pilot who had tested floats on his Avro at Barrow. Fortune was not to smile on his ship. She worked up, launching and recovering her seaplanes over the side, which caused Admiral Jellicoe to minute the Admiralty in July 1915:

> No seaplane [using a wheeled trolley] has yet succeeded in flying off *Campania* .... we will not only be powerless to carry out aerial spotting but .... we shall be unable to prevent the Germans doing so by means of their Zeppelins .... I regret that I am unable to propose any means of meeting this menace, unless it can be done by the use of aeroplanes rising from the deck of *Campania*, capable of climbing above the Zeppelins .... [13]

Stung by this criticism, one of *Campania*'s seaplanes, notably a Short 184, took off from the deck using a wheeled trolley but, even with the ship steaming at 17 knots, the lumbering Short used every available inch of the 120 foot deck to become airborne. In the light of this trial the *Campania* went back to Cammell Laird's yard to have the flying-off deck extended to 200 feet.

While *Campania* was refitting, *Vindex* and *Engardine* made, in March 1916, an unsuccessful attack with seaplanes on Zeppelin sheds thought to be at Hoyer; they were in fact at Tondern, so the first attack, not surprisingly, found no target. The second raid, six weeks later, consisted of eleven Sopwith Baby seaplanes, though for various reasons only three actually became airborne and of the three only one found the target, which the pilot missed.

It must have been apparent to the most obtuse officers at the Admiralty that operating small seaplanes, heavily loaded with fuel and bombs on the open sea, was not perhaps the best use of seaborne airpower, but still the Navy persevered with seaplanes in spite of the fact that *Campania* now had a 200ft deck, long enough to operate almost any existing landplane in service. When she rejoined the fleet in April 1916, however, she still had only seaplanes embarked — three Short 184s and seven Sopwiths. The Shorts were now at last able to take off from the lengthened deck without difficulty, as one of their pilots, Richard Bell-Davis, wrote in his book *Sailor in the Sky*:

> At sea, aircraft [seaplanes] took off from the launching deck, being lowered on to a wheeled axle, held in place by guides on the floats and by a wire span attached to a quick release gear in the pilot's cockpit. The tail float had a fitting which slid into a split tube guide set at a height which kept the machine in her flying attitude.
> After take-off, the pilot released the wheels and axle which fell into the sea and could be recovered by attendant destroyers.

Campania joined the Grand Fleet at Scapa Flow, that desolate northern anchorage in the Orkney Islands, just in time to be included in the biggest, and last, naval action of World War I: the Battle of Jutland.

The battle was complex; indeed, certain aspects of it remain the subject of controversy among naval historians. Reduced to its essentials, it was to be a decisive action between the German High Seas Fleet and the British Grand Fleet, which had been sparring across the North Sea since the outbreak of the war. German cruisers had made hit and run raids on the English east coast, bombarding Whitby, Scarborough and Hartlepool; the RNAS had raided Wilhelmshaven; and in January 1915 an action between British and German battle-cruisers had been fought inconclusively off the Dogger Bank.

Both the British and the German naval staffs had planned operations which it was hoped would entice the two great Battle Fleets into a decisive battle which could shorten the war for the victor. The British plan, 'Operation M', was to be implemented on 1 June 1916; the German plan drawn up by Admiral Scheer, which actually precipitated the battle, was to be activated a day earlier, on 31 May. It called for a group of five battle-cruisers and four light cruisers, commanded by Admiral Hipper, to sail for a position off the Norwegian coast, in the hope that the neutral but pro-British Norwegians would inform London by telegraph of the presence of the German naval force. The plan relied on the fact that the British Grand Fleet was split up: the main battleship force, under the Commander-in-Chief, Admiral Sir John Jellicoe, was at Scapa Flow, with the second battle squadron, under Vice Admiral Sir Martyn Jerran, some distance down the Scottish coast in the Moray Firth, and the fifth battle squadron, under Vice Admiral Sir David Beatty, at Rosyth on the Firth of Forth, 185 nautical miles south of Scapa. Scheer had hoped that Beatty's battle-cruisers, on receipt of information from the Norwegians, would steam out of Rosyth for Hipper's force, while Scheer himself would lead the High Seas battle squadrons 50 miles behind Hipper's and trap the British cruisers between the two German forces, before Jellicoe could intervene from Scapa Flow.

Unfortunately for the Germans, a wireless message that was the signal to start the operation was intercepted and partially decoded in the legendary 'Room 40' at the Admiralty, the code-breakers there gleaning sufficient information to be able to warn of possible large-scale German naval movements in the North Sea. Acting on the information from Room 40, the Admiralty ordered the entire Grand Fleet to put to sea as soon as possible and to assemble some 100 miles north-west off the Horn Reefs on Denmark's western coast.

At 1735 hours on the evening of 30 May, all the ships in the anchorage at Scapa Flow were alerted to be ready to sail. *Campania*, because of her position in the anchorage 5 miles from the boom, was to be the last vessel to leave, following HMS *Blanche*. During the twilight of the summer evening the ships raised steam and prepared for sea; as they did so the fleet had to observe strict wireless silence, for it was assumed that German direction-finding stations would be listening for any signals which could indicate unusual activity from the known anchorages of the Grand Fleet, which had been reconnoitred by Zeppelins.

At 2100 hours, as darkness fell, the order 'Weigh' was flashed from the flagship — the Dreadnought, *Iron Duke*. One by one the great grey battleships, with their attendant cruisers and destroyers, left the sheltered waters of Scapa Flow, steaming through Hoxa Sound and out into the North Sea. As they did so, from Rosyth Vice Admiral Sir David Beatty's Fifth Battle Squadron was passing under the great railway bridge spanning the Firth of Forth; among those ships was the seaplane carrier *Engardine*.

During that summer evening and into nightfall the long lines of battleships headed across the dark waters of the North Sea: it was an impressive armada, no fewer than 150 warships carrying among them 104 guns of between 12 and 15 inch calibre, and manned by 60,000 officers and men. At least there should have been 150 ships: in point of fact there were only 149, a deficiency not realised until, at 2255 hours, a signal lamp blinked an order from C-in-C to *Campania*. There was neither acknowledgement nor reply, for the very good reason that she was still moored to her buoy in Scapa Flow. The seaplane carrier had never received her final sailing order; it was not until the boom defence officer came alongside to inform her Captain that he was about to close the anti-submarine boom defences that *Campania*'s commander realised that the fleet had sailed.

Immediately *Campania* steamed after the fleet with a double watch in the stokehold, pushing the old ship as fast as she could go. By 0200 hours the C-in-C, Admiral Jellicoe, was informed that the *Campania* had at last sailed but she was then four hours behind the fleet and unescorted; it was feared that she could be torpedoed by one of several U-boats that were reported to be in her vicinity. Jellicoe made a decision and at 0437 he ordered *Campania* to return to Scapa Flow. She arrived back at the deserted anchorage at 0900 hours on 31 May.

One can imagine the feelings of her Captain and crew; she had been out in the North Sea the previous day, exercising her aircraft, and had returned to the furthermost point of her anchorage near to Scapa airfield to embark three additional machines, from which position she could not see the dim station-keeping lights of the other ships as the fleet sailed. Just why she received no positioning and timing signals, and had not been missed until 2255 is a mystery which has never been solved.

While *Campania*'s disconsolate crew stared at the seagulls perched on the vacant mooring buoys, across the North Sea the two fleets were closing towards each other. At 1428 hours HMS *Galatea* and the German light cruiser *Elbing* were both investigating a Dutch steamer, *N.J. Fiord*; they sighted each other, simultaneously opening fire. The Battle of Jutland had begun.

Beatty's and Hipper's fleets were now racing towards each other at 22 knots, with dense clouds of smoke pouring from the funnels of the coal-fired ships, making them visible below the horizon. At 1445 hours Beatty ordered *Engardine* to launch a seaplane to reconnoitre the German fleet. It took about fifteen minutes for *Engardine* to heave to and swing out into the sea a Short 184 (No. 8359), piloted by Flt Lt F.J. Rutland, with Assistant Paymaster S.S. Trewin as observer. The weather was poor with a 900 foot ceiling; however, Rutland, scouting to the north-east, sighted Hipper's fleet and, to make certain of the identity of the enemy warships, he flew to within 3,000 yards, under heavy fire from the German gunners; even so, the observer was able to transmit a clear report of the position, course and composition of the enemy fleet. Unfortunately, Rutland had to terminate the reconnaissance when a petrol pipe fractured, allowing him just sufficient time to fly out of range of the enemy fire. He then force-landed in the sea, repaired the faulty fuel line and took off for *Engardine*, being hoisted aboard at 1600 hours, only to learn that his reports had not been received by the C-in-C. That was not, by any means, to be the only failure of communications that day.

Deteriorating weather prevented any further launching of *Engardine*'s seaplanes, though there was no lack of pilots prepared to fly; her participation in the Jutland battle ended with her towing HMS *Warrior*, until that badly damaged cruiser heeled over and sank.

The battle raged on, with Beatty's ships engaging Hipper's cruisers and Scheer's High Seas Fleet racing from the south towards the battle, while the ships of the British Grand Fleet tore down from the north. It was a confusing running action with Jellicoe left in ignorance of the position of the enemy, owing to a lamentable failure of communications. In essence, had Jellicoe known the exact position of the enemy, there is no doubt that the High Seas Fleet would have been destroyed or at least gravely weakened; as it was, when the presence of the Grand Fleet, which outgunned his own, was revealed to Admiral Scheer, he skilfully broke off the action and narrowly escaped being trapped.

Before that, the battle with Beatty's squadron had starkly revealed shortcomings in the design of many British ships, especially in armour and the vulnerability of their magazines. The gunnery of some of the cruisers was poor, and the armour-piercing shells were found to shatter against German armour; many shells which were direct hits simply failed to explode.

When the action ended with the German fleet slipping below the horizon as darkness fell on 31 May, the North Sea off Jutland Bank was littered with the flotsam of battle and the bodies of nearly 9,000 sailors. The Germans had lost the battleship *Pommern*, the battle-cruiser *Lutzow*, the light cruisers *Wiesbaden*, *Rostock*, *Frauenlob* and *Elbing*, together with five destroyers; 2,551 officers and men died with their ships. The British lost the three battle-cruisers *Queen Mary*, *Indefatigable* and *Invincible*; the three cruisers *Warrior*, *Defence* and *Black Prince*, and eight destroyers, together with 6,087 sailors.

The Battle of Jutland was to end as a victory for the Royal Navy, but it was a Pyrrhic victory. The British had sustained almost double the losses of the Germans, and it was a victory only in the sense that the German High Seas Fleet never sailed again during the Great War. The High Seas

Fleet, however, had not been destroyed; it remained 'a fleet in being' which had to be watched and countered until the last day of the war, tying down the British Grand Fleet.

The most apt comment on the whole battle from the British point of view was the ultimate in understatements made by Beatty to his Flag Captain, as HMS *Queen Mary* disintegrated with exploding magazines before his eyes: "There's something wrong with our bloody ships today, Chatfield." And not, one might add, only with the ships . . . .

Had *Campania* sailed as planned, it is not too fanciful to argue that her aircraft could have played a decisive role. The Germans had no aircraft at Jutland and the low clouds prevented Zeppelins from participating; *Campania* could have given Jellicoe the vital information on the enemy fleet's position and course — intelligence he never received during the critical phase of the action, when, by the smallest of margins, Scheer's fleet escaped being trapped. One seaplane with a good observer and W/T could so easily have turned the battle from a virtual stalemate into a great victory.

The Battle of Jutland was the last great classical naval engagement fought by gunnery with ships in visual contact: in future naval battles, aircraft would play first a significant and later a decisive role. Rutland's solitary 184, buffeted by the concussion of the battleships' gunfire, was a storm petrel driven before a gale-force wind of change.

Along the corridors of power in Whitehall, the aftermath of the Battle of Jutland must have shaken faith in the assumed invincibility of the Royal Navy. The shortcomings of many of the British ships, the failure of the armour-piercing shells, to say nothing of glaring errors of judgement made by some senior officers, posed problems enough for the naval staff; it is unlikely that the failure of the RNAS to provide adequate reconnaissance during the action figured very largely in the post-mortem. However, one point was manifestly clear: seaplanes taking off from and alighting on the open sea, particularly the North Sea in winter, was, to say the least, a marginal operation; also obvious, though not as a result of the Jutland action, was the inadequate performance, particularly against Zeppelins, of the seaplanes then in service.

In many ways the Admiralty, with its tradition of direct procurement from manufacturers, was far better placed than the Army in the question of new types of aircraft, particularly fighters, the development of which, during the war years, was spectacular. Thus, while the RFC was to some extent forced to operate outmoded machines, such as the lumbering BE2c, the Navy was under no such obligation, and its freedom of choice was to have a profound effect, not only on the search for a shipborne fighter, but also in the critical air battles over the Somme.

In the early months of 1916 the Germans began to shoot down large numbers of the slow two-seater RFC reconnaissance aircraft. This sudden and dramatic increase in losses was due to a small single-seater fighter designed by the Germans, or rather their Dutch employee Anthony Fokker. It was the Eindecker monoplane and not, by the standards of the day, an outstanding machine, but it enjoyed one important innovation: it was the first warplane in service to have a machine gun synchronised to fire through the propeller arc. This in airfighting was a great advantage; the pilot simply had to fly his aircraft straight at his opponent and, when in range, open fire. The Fokker was in fact a flying gun and, as such, the prototype of all fighters up to the age of the air-to-air guided missile. The concept was not new, indeed it was not even German: a French pilot, Roland Garros, had used the idea in a crude form a year earlier, when he had simply clamped a machine gun on the engine cowling of his Morane-Saulnier monoplane and fitted steel plates to the wooden propeller blades to deflect any bullets which struck them. A sufficient number cleared the airscrew to enable Garros to enjoy considerable success before, perhaps not surprisingly, engine trouble forced him down behind the German lines. His Morane landed intact, was examined with interest and the advantage of the idea immediately grasped; the Germans soon fitted similar deflectors to an aircraft and began tests. However, as soon as the pilot fired his machine gun, the propeller blades disintegrated; the reason was that, unlike the French, who used copper-jacketed lead bullets, the thorough Germans used chrome steel jackets which, on hitting the deflector plates, did not deform but shot the blades off. The result was that Anthony Fokker, a clever engineer and a very skilful pilot, was asked to devise a system that would enable the German Air Force to redress the balance of the advantage the French undoubtedly held.

Within 48 hours Fokker, aided by his two engineers, Leinberser and Lübbe, had designed a system of cams and levers which allowed the machine gun to fire only when the propeller blades were clear. It can be argued that others had already designed similar systems, notably the German Franz Schneider and the British Edwards brothers; nevertheless, the Fokker gear was the first to be used in action. It was fitted to the Fokker Eindecker E1-M5K/MG — the 'MG' standing for Maschinengewehr (machine gun), which was typically a 7.92mm Parabellum, placed on the pilot's eyeline, slightly offset to the right. In the hands of such pilots as Oswald Boelcke and Max Immelmann (who invented the climbing turn, which still bears his name, to exploit the new armament) it was lethal; by early 1916 each of these pilots had shot down eight Allied aircraft, had been awarded the coveted 'Pour le Mérite' — the legendary Blue Max — and had become national heroes. Less gifted pilots also found the lumbering RFC spotter aircraft easy meat, so much so that the period became known in the British press as the 'Fokker Scourge'. It was taken seriously by the technical journals too; in May 1916, for example, the influential British magazine *The Aeroplane* carried a full page advertisement soliciting financial support to build a 'Fokker Fighter'.

In mechanised war between powers of roughly the same industrial development any innovation is either independently arrived at, copied, or countered, almost as soon as it appears. The Fokker with its interrupter gear was no exception. That the RFC needed a fighter which could take on the Eindecker was only too obvious to the pilots and squadron commanders in the field, and the aircraft manufacturers, too, were aware of the need to keep producing new and better designs, particularly of fighters. The success of the German fighters simply served to convince the politicians and generals that new Allied types already existing in prototype form must be placed in service.

Ironically, the two aircraft that would do much to redress the balance for the RFC on the western front were already in production — for the Admiralty — at the Kingston-on-Thames factory of the Sopwith Aviation Company. One was the Type 9700, always known to the Royal Navy as the '1½ Strutter' — a curious name given, it is believed, because of the 'W' arrangement of its centre section struts. The second aircraft, a single-seater fighter biplane, was given the official designation 'Sopwith Type 9901', but in spite of, or because of, official displeasure, it was always known as the 'Sopwith Pup': it was small and seemed to the pilots to be the 'pup' of the 1½ Strutter.

Chronologically, the first of the two types was the 1½ Strutter. It had, as noted, initially been ordered by the Admiralty, the company being one of the Navy's contractors; their Sopwith Schneider and its derivatives were by far the best single-seater seaplanes available. Now, with the realisation that landplanes would be the only types to have any chance of intercepting the fast-climbing Zeppelins, it was logical that it should be to the Kingston-on-Thames company that the Navy turned.

The designer of the '1½' was Herbert Smith, and he had produced a neat biplane which was the first British aircraft to have a synchronised machine gun firing through the propeller arc; the gun was a Vickers .303 inch, with either a Scarff-Dibovski or Sopwith-Kauper interrupter gear. The '1½' was constructed in two forms: as a single-seater bomber with provision for 65lb of bombs carried internally, thus reducing drag;

or as a two-seater fighter/bomber which was the main variant of the 550 supplied to the RNAS. The observer sat in the rear seat and was armed with a single .303 Lewis gun mounted on a rotatable Scarff ring, which gave an excellent steady field of fire above, abeam and, most importantly, aft. This position of the gunner in the rear cockpit, which now seems the most obvious place, was in fact a new concept; the early RFC reconnaissance planes and pusher fighters placed the observer with his machine gun in the front seat, making it impossible to defend those aircraft when attacked from the rear by the agile, forward-firing Fokkers. The two-seater fighter was thus established with the 1½ Strutter, to be perfected later in the war over the western front by the RAF with the Bristol Fighter.

The new Sopwiths entered RNAS service with No. 5 Wing in April 1916 at Coudekerque, where they made successful bombing raids on Zeppelin sheds at Evere, Berchem Ste Agathe and Cognelée, in addition to attacking enemy troop concentrations and airfields. Further equipping of other naval wings in France was delayed by the preparations for what was to become the Battle of the Somme; the RFC was in desperate need of aircraft for that vital battle and many 1½ Strutters were released by the Royal Navy to the army. The contribution the type made to shipborne operations, as will be related, was thus delayed.

The Sopwith Pup, though in production for the Navy, specifically for shipboard use, also attracted the attention of the RFC. It was powered only by an 80hp rotary Le Rhône; nevertheless, it could achieve 111 mph, and not only could it climb to 17,000 feet, it could fight at that altitude; its handling was such that pilots generally considered it to be the most pleasant to fly aircraft of all produced by British constructors during the Great War. It could outfly such formidable opponents as the German Albatros, which, with twice the Pup's horsepower, had replaced the Fokker E1 as the RFC's 'Scourge'. The celebrated 'ace' McCudden wrote of the Pup: ".... when it came to manoeuvring, the Sopwith would turn twice to an Albatros' once .... at 16,000ft the Albatros began to find its ceiling just where the Pup was still speedy and controllable...."[14]

Though the Pup was used extensively by the RFC, naval pilots first established the fighter's reputation. The first deliveries had been made to the RNAS in September 1916 to No. 1 (Naval) Wing. The first victory the Pup achieved was a week or so later on 24 September. A German LVG had just bombed the Dunkirk airfield of No. 1 Wing; Flt Sub-Lt S.J. Goble took off in a Pup and, within two minutes, chased the two-seater and shot it down in flames over Ghistelles. It was the first of many

victories. No. 1 Wing, however, was not the only naval unit to operate
Pups on the western front.

The bitter fighting on the Somme, which had started so confidently
on 1 July 1916, was still unresolved by late autumn; the RFC was
suffering very heavy loses and was desperately short of aircraft and
pilots. The Admiralty was asked to help and quickly formed No. 8
(Naval) Squadron. 'Naval Eight' was to become one of the most famous
RNAS units; it was to be based at Vert Galand and went into action
within days of arrival in October 1917, with three flights of six Nieuport
Scouts, six 1½ Strutters and six Pups. Very soon after, the squadron was
entirely equipped with Pups and, by the New Year, had destroyed
twenty enemy aircraft at a time when the German Air Force was
dominant.

Before turning to the important contribution the 1½ Strutters and
Pups were to make as shipborne fighters, brief mention must be made of
another Sopwith fighter which was used by the RNAS over the western
front exclusively: the Sopwith Triplane. Though it had an operational
career of only seven months — February to September 1917 — it fought
in the great aerial battles with three Naval Squadrons: 1, 8 and 10. The
triplanes of the Canadian 'Black Flight' — B Flight of Naval 10 — were
named *Black Death*, *Black Maria*, *Black Prince*, *Black Rodger* and *Black
Sheep*, and they were led by Flt Sub-Lt Raymond Collishaw. They shot
down among them eighty-seven German aircraft between May and July,
Collishaw being credited personally with shooting down sixteen in
twenty-seven days. He went on to score sixty victories and became the
third highest ace with the British forces, behind such illustrious names
as Mannock (seventy-three) and Bishop (seventy-two).

For Britain, 1917 was to be the critical year of World War I. It was the
first year during which the meaning of Total War was brought home to
the civil populations of the belligerents; with shortages of food and
consumer goods, and long hours of work in munition factories and
blacked-out cities, the appalling carnage on the western front brought
about a depressing realisation that the stalemate of trench warfare, now
in its third year, was likely to continue indefinitely.

The German Navy, its High Seas Fleet unwilling, after Jutland, to
enter upon another major surface engagement, commenced in February
an unrestricted U-boat campaign as a counter to the Royal Navy's
blockade, which was causing desperate shortages of food and raw
materials in Germany. Partly because of the U-boat offensive, the
United States entered the war on 6 April. Intensive U-boat warfare had

sunk so many British merchant ships — 881,000 tons in April alone — that food reserves were down to about six weeks' supply and, had the Germans been able to sustain the rate of sinking, there is no doubt that the Allies would have had to sue for peace. The U-boats were eventually contained by the formation of the convoy system, with a large contribution made by the RNAS anti-submarine patrols.

Admiral Beatty had replaced Sir John Jellicoe as Naval Commander-in-Chief, and in January 1917 the RNAS was reorganised to the extent that a Fifth Sea Lord was appointed to assume responsibility for the Naval Air Services. Beatty, possibly recognising — a little late in the day — that efficient air reconnaissance could have made the outcome of the Jutland action much more favourable to the Royal Navy, enquired of their Lordships at the Admiralty what the future naval policy for the aviation wing was going to be. He was informed vaguely that the RNAS was to be confined to naval duties, to provide reconnaissance and Zeppelin protection for the Grand Fleet, and to police the shipping lanes on anti-submarine patrols. To achieve that, they added, "we must have [ships with] flying-off decks and alighting decks, to allow fighters maximum endurance".

Beatty, whatever his other failings might have been, was most emphatically not an organisation man; he did not wait for the slow wheels of officialdom to revolve but at once set up his own Grand Fleet Aeronautical Committee. It was a move long overdue. The Committee came to conclusions: it was not interested in naval units from land bases fighting over France to bail the RFC out of its self-created problems of obsolete aircraft; it was not concerned with the bombing of distant strategic targets or even with defending Great Britain from air attacks; its recommendation was first and foremost the necessity of operating modern, high-performance aircraft with the Grand Fleet, exclusively for the Grand Fleet.

To implement these proposals, the Committee submitted in February 1917 that the seaplane carrier *Campania* and another converted packet, *Manxman*, together with *Nairana* (formerly a 3,000 ton Australian mail steamer) and *Pegasus* (the erstwhile Great Eastern passenger steamer *Stockholm*, also of 3,000 tons) should embark Sopwith Pups to replace their seaplanes, and that these landplanes should incorporate flotation bags, which would enable them to be safely ditched and later recovered. The Committee considered that twenty Sopwith Pups would be required as fighters, in addition to twenty two-seaters for reconnaissance duties.

The carriers for the aircraft, not the aircraft themselves, presented the problem: taking off was not thought to be difficult; landing on the 'alighting deck' was the snag. *Campania* was too old and was not considered to be worth rebuilding; the others were too slow. In fact there was a suitable ship available — the Italian liner *Conte Rosso*. She had been commandeered by the Admiralty while still in frame at a British shipyard in 1916 and was to be completed with two decks: a forward deck for take-off and another aft for alighting. The two decks were, in reality, continuous, but were to be divided by a 20 foot high bridge across the ship carrying the wheel and charthouse, much as on some car ferries today. Although this design was nearly an aircraft carrier, the proposal, when first mooted in 1916, was that only the front deck be used for take-off; the rear deck was to be for ranging seaplanes, then hoisting them over the side. Eventually the *Conte Rosso* would be completed with a flush flight deck but that event was, in 1917, still distant. While she was being built, or rather rebuilt, it was envisaged that on completion she would be used solely as an 'alighting-on deck' for landplanes which had taken off from other vessels.

With benefit of hindsight, it seems incredible that the — to us — so obvious solution of the flush-deck carrier was so reluctantly embraced. Yet the idea had been put forward even before the war, and H.A. Williamson, who had served as Senior Flying Officer on board *Ark Royal* at Gallipoli, had submitted detailed proposals for an island carrier (the island on the starboard side) with arrester gear to the Director of Naval Construction as early as September 1915; but the suggestion had been rejected as impracticable, possibly an early example of what, during World War II, was known as the 'Not Invented Here' syndrome.

It seems unbelievable that men who could provide ships which had to heave-to in action to lower and recover seaplanes from the open sea, with the ever-present danger of submarine attacks, could decide that a flush-deck carrier was 'impracticable', yet such was the case. Perhaps the most cogent explanation is to be found in the official history, *War in the Air*:

> Most of the newly appointed administrative officers had no previous knowledge of aircraft or aircraft operations; what they were chosen for was their power of organisation, their strict sense of discipline, their untiring energy, and their pride in the ancient service to which they belonged.
> The senior naval officer who was inexperienced in the air was promoted over the heads of the pioneers of naval aviation.[15]

Beatty's committee had, if anything, gone too far in the opposite direction. It was proposed that the enemy High Seas Fleet be attacked by waves of torpedo aircraft, flying at dawn to the naval bases, operating from fast converted merchant ships with flying-off decks, though the only one actually under construction was the *Conte Rosso* and she would not be ready for a year. Beatty was not prepared to wait that long and a suggestion was made to adapt HMS *Furious*, one of three 'hush-hush' or mystery ships which had been instigated by Admiral Fisher at the outbreak of war. *Furious*, and her sisters, *Glorious* and *Courageous*, were strange warships. They displaced 22,500 tons, were lightly armoured, very fast (31 knots), and mounted two gun turrets fore and aft. The guns on *Furious* were of no less than 18 inch calibre, capable of hurling a 3,320lb shell for 30 miles. The three warships had been envisaged by their creator as shore bombardment vessels.

Furious had her forward turret removed; a hangar was built on the forecastle and a flying-off deck, measuring 228ft x 50ft, constructed on the top. A large hatch in this deck permitted access to the hangar, a derrick being available to haul out the aircraft stored below. The loss of the forward turret perturbed Beatty, and the after turret was retained. It is doubtful if anyone considered how such a ship could use both the gun and the aircraft or what the effect on the latter would be when the former was fired. W.G. Moore, a pilot on board, wrote of the gun:

> My cabin was immediately beneath it [the after turret] and the *Furious* was built in a very light way, certainly not strong enough to carry a gun like that. Every time she fired, it was like a snowstorm in my cabin, only instead of snowflakes sheared rivet-heads would come down from the deckhead and partitions.[16]

HMS *Furious* was commissioned in June 1917, with, originally, three Short Type 184 seaplanes that had folding wings and five Sopwith Pups that did not, which made them tricky to hoist out of the hangar. With her high speed there was no difficulty in aircraft taking off, but landing was still either by ditching (the Pups) or being hoisted aboard (the seaplanes). *Furious*, in spite of the experience gained with *Campania* and the other seaplane carriers, simply represented improved means to unimproved ends.

The Pups, which had by this time been supplemented on the western front by the fiery Sopwith Camel, were considered more or less expendable; they certainly cost no more than one of *Furious'* monstrous 18-inch shells, but not unnaturally the pilots rather objected to ditching after each flight away from land bases. One of them, Flt Sub-Lt W.F. Dickson, later to become Marshal of the RAF Sir William Dickson, recalled:

> I was one of her fighter pilots .... Now when *Furious* was in harbour at Scapa Flow we used to practise flying from a small airfield called Smoogroo and when we flew alongside our ship and waved to our friends aboard we had this idea: couldn't we possibly somehow land back on this rather splendid

flying-off deck; we explored the possibilities of it; of coming up alongside and rolling our wheels along the deck. [17]

This of course was when *Furious* was at anchor, and as she lacked any form of arrester gear, the deck was not long enough even for the lightly loaded Pup to land on. However, the pilots considered that with the vessel steaming at 25 knots or so into a wind, an actual landing should be possible. It was agreed that a trial should be made and a date was fixed: 2 August 1917. That day there was a steady wind of 21 knots into which *Furious* steamed at about 26 knots, there being insufficient space in the Scapa anchorage for her to work up to her maximum of 31 knots.

It was something of an occasion. Moore recalls: "We were weighed down with brass hats and gold braid that day, as there were many Admirals and even Generals on board to see the experiment."[18]

With a 'felt' wind of some 47 knots blowing down the deck, the relative speed differential between the Pup, with its light wing-loading (5lb per square foot) and the ship would be around 3 knots, if that. It was arranged that the aircraft would sideslip over the deck with a group of seven pilots standing by. Sir William Dickson was one of the seven: "The role of us pilots on the deck was to run under the aircraft. There were rope toggles, one under each wingtip; our job was to seize the rope toggles and then as the pilot cut his engine we would haul the aircraft down."[19]

Sir William, who in 1980 is the sole survivor of the party, was on the port wing toggle. The pilot for that historic test flight was his CO, Squadron Commander Dunning. W.G. Moore was also a member of the handling party:

> [Dunning] took off, and after a circuit of the ship made his approach along the port side, side-slipping in and centering up over the deck in the right position. We dashed out and grabbed our appointed toggles. He then cut his engine. We had some difficulty hauling the aircraft down squarely onto the deck and holding it in the wind.[20]

Which is hardly surprising as the light aircraft was in the equivalent of a full gale; nevertheless, the machine was secured and Dunning stepped out. It was an historic moment: Dunning's Sopwith Pup — its identity is unknown — was the first aircraft to land on a ship under way. It was, however, over six years after Eugene Ely had pointed the way by alighting on the anchored USS *Pennsylvania*, and he had used arrester gear.

The success of Dunning's landing caused something of a stir in the Navy; telegrams of congratulation were sent between the Admiralty and the C-in-C. But Dunning himself was not satisfied; he considered that the performance had been flawed because the aircraft was still airborne when the handling party had grabbed the toggles and hauled it down. He decided to make another landing before allowing his pilots to try.

This time he instructed the deck party not to take hold of the aircraft until the wheels touched down. The second attempt was made on 7 August: the touchdown was satisfactory but the handlers could not prevent the brakeless aircraft from being blown backwards and the elevators were damaged against the coaming of the hangar hatch. Dunning climbed out, went up to another Pup, No. 6452, in which Moore was sitting with the engine running in preparation for a take-off to try a landing. Dunning shouted: "Come out of that, Moore. I'm not satisfied with that run and I'm going again."[21]

Off went the Pup with the disconsolate Moore watching from the deck. Dunning flew a wide circuit round *Furious* and sideslipped over the deck, but, as Sir William Dickson witnessed, "he came in a little too high, he waved to us to get clear as he was going round again. He opened the engine up and choked it a little, the aircraft stalled and cartwheeled over the side."[22]

As the Sopwith fell into the sea, Dunning was knocked unconscious, probably by striking his head against the instrument panel. There had been no picket boat standing by and by the time the 22,000 ton ship had hove to and launched a boat, twenty minutes had elapsed. The Pup with an airbag in the tail was floating, but Dunning, still strapped in the cockpit, had drowned.

In the light of Dunning's crash, all further attempts at landing on the forward deck were forbidden; for the next three months when *Furious* flew her Sopwiths off they either alighted on land or ditched in the sea as before; this was clearly unsatisfactory and the suggestion was then made that the after gun turret be removed and a 'landing-on' deck provided. This was opposed by the Commander of *Furious*, Captain Wilmot Nicholson, who sent the following telegram:

To Admiralty   12.9.17.   Cypher 'K'
Captain of HMS Furious represents that a flying-on deck is most undesirable owing to following reasons:
(1) Eddy currents aft render landing more dangerous than forward.
(2) Great difficulty and delay in transporting machines from aft forward.
(3) Loss of offensive power owing to after guns having been removed.
This question is of vital importance and it is requested that Admiralty officers may be sent up [to Scapa Flow] to confer with Captain of HMS Furious in order that an immediate decision may be arrived at. There appears to be considerable divergence of opinion on this subject amongst flying officers.[23]

The telegram is interesting: on the one hand, it reveals a slightly Luddite attitude to the removal of the, surely useless, after gun turret; on the other hand, it contains a shrewd forecast that turbulence aft could prove troublesome. As to the statement that there was "considerable divergence of opinion" among the pilots, one can imagine that the deck, even with severe turbulence, was a rather more attractive prospect than ditching in those cold northern waters. In the event, the captain was overruled, the decision was made to construct the after flying deck, and *Furious* went to a yard in Newcastle for the work to be carried out.

While the *Furious* conversion was taking place, other experiments were being conducted into shipborne fighter operations. A light cruiser, HMS *Yarmouth*, had a 20 foot platform constructed on her forecastle; from this, in June 1917, F.J. Rutland, who had flown the solitary 184 at Jutland, successfully flew off in a Sopwith Pup. On 21 August Flt Sub-Lt B.A. Smart made his first take-off from *Yarmouth* in his Pup and shot down Zeppelin L23 at 8,000ft. He ditched just off the Danish coast and was picked up by HMS *Prince*; the Pup was abandoned.

L23 was the first Zeppelin shot down by a fighter that had taken off from a ship; the feat, however, had required *Yarmouth* to sail directly into wind. This was considered a disadvantage when working with the fleet; the wind direction could well be different from the course of the other ships. It was suggested by Lt Commander Gowan that the problem could be solved by a temporary platform being placed on a ship's forward gun turret, which could turn into the 'felt' wind without the ship itself having to alter course.

This was tested on 1 October 1917 when Squadron Commander Rutland flew a Pup from a 17 foot platform mounted on 'B' turret of HMS *Repulse*.[24] The wind was 230°, the ship's course 145° at 24 knots. The turret was trained 42° to starboard, giving a felt wind of 31.5 knots directed on to the turret platform, as indicated by a flag held by a rating. Rutland ran up his engine, a rope holding the Pup was slipped, and he took off with 6 feet to spare. Some days later the feat was repeated from *Repulse*'s after turret. In the light of these successful flights, on 17 October the Admiralty ordered that all light cruisers and battle-cruisers should carry fighters, provided that the guns were in no way interfered with.

The after flying deck being built on *Furious* was 284 feet long. This was considered to be on the short side, and an investigation into arrester gear was made at the Isle of Grain under Squadron Commander H. Busteed. A dummy deck was laid out on the airfield, and one of the prototype Pups — No. 9497 — was used in the trials. The early arrested landings were to utilise the system that Eugene Ely had pioneered when landing on USS *Pennsylvania* — transverse wires loaded with sandbags, which were engaged by a long hook suspended below the aircraft fuselage. Although successful with Ely and later to become the standard on carriers, it did not prove satisfactory during the experiments on the Isle of Grain. Several different hook positions were tried — one under the aircraft's centre of gravity. Other variations included sprung skids in place of a wheeled undercarriage, with a hook to engage transverse wires and horns to engage fore and aft wires; these were not, strictly speaking, 'arrester wires', but were designed to hold the aircraft on the deck. Possibly Dunning's going over the side influenced the provision of these.

After many trial landings on the dummy deck, it was decided to finalise on rigid twin skids, which had a greater retarding effect than wheels; each skid carried two pairs of horns to engage the fore and aft wires running the length of the flight deck and held up 9 inches from it. The Pups thus modified for deck-landing were designated 'Type 9901a'; they were ready to be embarked aboard *Furious* when she returned to Scapa Flow in March 1918.

The after gun turret had gone (to be fitted to a shore-bombardment monitor HMS *General Wolfe*); in its place was a long deck built like the forward one over a roomy hangar. To facilitate aircraft movement between the flying decks, two gangways skirted the ship's main superstructure to port and starboard. No arrester as such was fitted but the longitudinal wires tested at the Isle of Grain ran for most of the flight deck's length. At the extreme limit of the alighting deck a gantry suspended a number of thick vertical manilla ropes to act as a final barrier to prevent overshooting aircraft crashing into the superstructure amidships. The pilots coming aboard the refit viewed the alterations with dismay. Sir William Dickson said: "We were not very pleased, because it was pretty obvious that to ask one to land on that afterdeck abaft the superstructure, with the ship steaming fast to give us enough relative wind to land into, was going to set up a lot of turbulence."[25]

The foreboding of the pilots notwithstanding, HMS *Furious* embarked ten Pups and fourteen 1½ Strutters (some of the 1½s still retained their wheeled undercarriages), and, hoisting the flag of Rear Admiral P.F. Phillimore, the first Admiral Commanding Aircraft, set sail from Scapa to conduct trials of the new landing deck on 20 March. They were a disaster: even such a superbly controllable aircraft as the Sopwith Pup could not cope with the turbulence over that after deck. Hot funnel gases and areas of partial vacuum, alternating with violent eddies, put even the experienced Rutland over the side — fortunately without repeating

Dunning's sad fate. Of thirteen landings, only three Pups were put down undamaged; the rest, in the words of W.G. Moore, "just dropped on the deck like shot partridges . . . ."[26]

Rear Admiral Phillimore telegraphed the Admiralty on 7 April:

> Furious — Alighting on.
> . . . . further trials of landing aeroplanes on the after deck of HMS Furious have so far proved unsatisfactory.
>   In order to avoid delay and in anticipation of their Lordships' approval, I have instructed the Captain . . . . to proceed with experiments in alighting on the forecastle . . . .[27]

This was, in fact, still considered to be too dangerous, and the pilots reverted to ditching in the sea. Thus, apart from the large number of aircraft she embarked, *Furious* was still no better placed to recover her landplanes than the numerous battleships now equipped with gun-turret platforms. Even a relatively heavy 1½ Strutter, with a crew of two and full wireless equipment and armament, had taken off from HMAS *Australia*; and some forty others, together with fifteen Pups, were at sea embarked on various ships of the Grand Fleet, in addition to HMS *Furious*, *Campania* and several smaller seaplane carriers.

By 31 March 1918 the RNAS had on charge 2,949 aircraft and 103 airships, with a muster of 67,000 officers and men manning 126 naval air stations: it was to be the culmination of the hopes of the very small band of pioneers that had begun the service. The next day the entire strength was handed over to the control of the newly created 'Independent Air Force'. The Royal Flying Corps and the Royal Naval Air Service were now combined to form the Royal Air Force.

The change was to be to the advantage of the RFC; for the Navy, it represented the relinquishing of direct control of its naval wing. Yet, in many ways, the Navy had been hoist with its own petard. The Admiralty, as has been noted, established direct negotiations with the aircraft manufacturers in the same way that they had contracted for ships down the centuries. The result of the close working collaboration between maker and user had produced the Short seaplanes, which, whatever their shortcomings against Zeppelins, were satisfactory shore-based reconnais-sance machines. When the RNAS required a light, easily controlled fighter, able to operate from short take-off platforms and decks, Sopwith provided the Pup to meet their specifications. When the Sunbeam Company had difficulty delivering over 1,500 aero-engines to Admiralty contracts in 1917, the Navy bought Hispano-Suizas. The result of this direct procurement enabled the RNAS to maintain a surplus of aircraft held in store, at a time when the RFC was desperately short and the Navy had to supply machines to the army in France.

In contrast to the RNAS, the RFC, from its inception, had its aircraft largely supplied from the Royal Aircraft Factory, and, while this did eventually procure such an excellent fighter as the SE5a, it also produced the BE2c which, despite the appalling losses its lack of performance and armament caused, continued in production long after it should have been superseded. In short, many generals and senior officers felt that the RNAS was growing fat at the expense of the army. As far as the public in Britain was concerned, the Zeppelin, now largely defeated by the incendiary bullet, had been replaced by a new terror — the Giant Gotha bombers, which arrived over England in daylight in June 1917. Eighteen bombers attacked London, killing and injuring 500 people; although a large number of fighters took off to intercept, not one of the raiders was shot down.

The public image of the RFC in 1917 was that built up by the popular press, which to some extent had created the mystique of the fighter pilot taking on the German hordes in the great dogfights high over the western front. That such engagements, though undeniably colourful, were irrelevant to the trench war beneath them was not questioned as the young 'aces' were lionised — though, to be fair, that is not what most of them sought. When, however, 'Hun' bombers could fly unscathed in broad daylight over London, the outcry was immediate and prolonged. Why had no British bombers raided Berlin? Why could the Gothas raid with such impunity? There was a growing feeling that something was very wrong with the RFC.

In Britain the usual palliative to any disaster is for the Prime Minister of the day to appoint a committee. Soon after the Gotha raids, in July 1917, Lloyd George appointed a member of the War Cabinet, General Smuts, to head a Committee to look into two aspects of the conduct of the war in the air. One was to consider the defence of London; the other section of the brief was to examine the more far-reaching question of 'Air Organisation'. The upshot was a recommendation by the Committee to create an 'Independent Air Force' to be responsible for all aspects of the air war. The Navy had argued that putting forward a case for an independent air force did not necessarily preclude the need for an additional specialised Navy or Army air wing. The Admiralty in the end were acquiescent; one suspects that there were still senior officers who viewed the growing naval air arm with suspicion.

The Air Force Bill was drafted and became law on 1 April 1918. Perhaps it was just a coincidence that it was April Fools Day ....

In practical terms the event made little difference to the personnel of the two constituent services: all naval squadrons, including the illustrious Naval Eight and Ten, simply had 200 added to their numbers, to become 208 and 210, RAF; the uniforms remained the same, khaki or navy blue; ranks did change, but until 1920, when the present day RAF ranks were introduced, ex-RNAS officers took their equivalent army titles (after all, the Marines had been using army ranks for years). One portent, however, was significant: the post of Fifth Sea Lord, who had been responsible for naval aviation, was abolished after existing for only a year; naval aviation was now to be the concern of the new Air Ministry.

The RAF's first and last operation of the Great War mounted with shipborne aircraft was to be another attempt to raid the Zeppelin base now known to be at Tondern, on the Danish border.

During the late summer of 1917 a new fighter had made an appearance over the western front — the Sopwith Camel. This development of the earlier, slightly smaller, Sopwith Pup is generally considered to have been the best British fighter of the war; incontrovertibly it shot down the most enemy aircraft — 1,294, of which 386 fell to the twin synchronised guns of naval Camels. It was, incidentally, the fairing over the guns which gave the aircraft its nickname; officially it was 'F1'. The Camel was used by both the RFC and the RNAS, but a later version, the 2F.1 Camel, was supplied exclusively to the Navy, having been designed specifically for shipboard use. Although it did not have a wing-folding capability, the fuselage was detachable just behind the cockpit to facilitate stowage. The main purpose of the naval Sopwith Camel was to intercept Zeppelins over the North Sea, for they continued to fly reconnaissance patrols over the Grand Fleet. To this end, many Camels — 112 by October 1918 — were embarked to fly from the gun turrets of the warships of the Grand Fleet. Many larger battleships carried a Camel on the forward turret and a 1½ Strutter, used for spotting purposes, flying from the after guns.

Although the Camel was conceived primarily as a fighter, for the Tondern raid at least seven of them were fitted with bomb racks below the lower wings, to carry two 50lb bombs. They were to sail aboard *Furious* for the raid, but, before that, intensive practice runs in low level bombing were made at Turnhouse, on the Firth of Forth. After a false start, due to bad weather, the raid, 'Operation F7', was to take place on 18 July 1918.

Force 'A', comprising *Furious*, escorted by the 1st Light Cruiser Squadron and eight destroyers of the 13th Flotilla, slipped from Rosyth on the night of 16—17 July and arrived at the flying-off position, 56°03N, 7°35E, due east of Ringkøbing Fjord, Denmark, at 0300 hours on 18 July.[28]

The seven Camels were hoisted up from the hangars below to the forward deck; the engines were started and warmed up. *Furious* turned into wind and increased her speed. Just as dawn was breaking, Captain W.D. Jackson, who was to lead the first flight, opened the throttle of the 150hp Bentley rotary; the Camel ran along the deck and was airborne at 0314 hours. He was followed by Captain W.F. Dickson and Lt N.E. Williams. Within four minutes the first flight had formed up to starboard, and the second flight, led by Captain B.A. Smart, followed by Captain T.K. Thyne, Lt S. Dawson and Lt W.A. Yeulett — the last pilot to take off — left *Furious* at 0320. The second flight formed to port, then both set course south east for Tondern, 80 nautical miles away.

En route to the target, Captain Thyne suffered trouble and force-landed in the sea, but was rescued by a British destroyer; the other six flew on to Tondern. Visibility was only fair under an overcast sky, though this had the advantage of concealing the two flights from enemy observation. The planned track of the raiders kept them clear of Danish territory until the northern tip of Sylt was sighted, when the two flights turned inland across the wide bay formed by the inner spit of land which joins Sylt to the mainland. German fighters were known to be stationed on Sylt and many warships were sighted at anchor in the bay, but no enemy aircraft appeared as, according to the prearranged plan, the Camels split up to make individual attacks on the Zeppelin sheds from different directions.

Visibility had improved and the huge sheds were easily seen just to the north of the town of Tondern. The three Camels of the first flight started their bombing dive, attacking one of the sheds and scoring hits; there were heavy explosions of hydrogen, followed by an enormous fire. The second flight attacked the remaining shed, which also exploded; inside the blazing buildings two Zeppelins, L54 and L60, were gutted.

Most of the attacking aircraft came under rifle or machine-gun fire; of the first flight, the leader, Captain Jackson, ran out of petrol — probably due to a hit in his fuel tank — and landed near Esbjerg, Denmark. Williams also force-landed, owing to lack of petrol, near Scullinger, Denmark. The second flight's leader, Captain Smart, returned to the fleet, as did Dickson, who was hauled aboard the British destroyer HMS *Violent* after ditching ahead of her. Lt Dawson made a navigational error and landed at Ringkøbing, just 20 miles east of the fleet's position. One aircraft was missing, that of Lt W.A. Yeulett; his fate is a mystery. He attacked the target but neither he nor his aircraft was seen again, the assumption being that he crashed unseen into the sea.

The Tondern bombing raid, the first to be made by landplanes flown from a ship at sea, was a success. But *Furious* could not land her aircraft back on board and only one of the Camels was salvaged after ditching.

Although *Furious* did not take part in any further raids, a Sopwith Camel was to shoot down a Zeppelin over the North Sea, having taken off from a lighter towed at 30 knots behind a destroyer. This unusual 'carrier' was tested by Commander Samson, who had made an experimental flight in May 1918. The Camel had a skid undercarriage which fitted into two troughs running the length of the barge. The destroyer worked up to 32 knots and the lighter itself almost became airborne; then everything went wrong. As Samson opened up the throttle the Camel leaped out of the troughs into the air, then stalled, crashing into the sea just ahead of the lighter, which ran over it. Miraculously Samson bobbed up in the wake and, as he was hauled aboard a whaler, his first words were reported to be "I think it well worth trying again".[29]

It was tried again, but not by Samson; Lt S.D. Culley, on 31 July 1918, successfully flew a normal wheeled Camel off a towed lighter. A few days later, on 10 August, the lighter was being towed by a destroyer of the Harwich Flotilla when the Zeppelin L53 approached; Culley took off and from 19,000 feet shot down L53, the last Zeppelin to be destroyed in air combat. Rear Admiral Tyrwhitt made this signal to the fleet: "Flag general: your attention is called to Hymn 224, verse 7."

The Naval hymnal was thumbed in the charthouses of the fleet and the following decode emerged:

> O happy band of pilgrims,
> Look upward to the skies,
> Where such a light affliction
> Shall win you such a prize!

Culley's Camel survives in the Imperial War Museum, London.

Although Samson was the man who had seen the possibilities of fighters being launched from a towed lighter, the idea of towing aircraft had earlier been suggested by J.C. Porte, the Commanding Officer of the RNAS flying-boat base at Felixstowe and a great innovator to whom the Royal Navy owed a good deal. It was Porte who had proposed the composite aircraft earlier referred to; his Curtiss 'America' flying boats were being used to patrol the North Sea on anti-U-boat sorties and, in order to increase their radius of action, Porte had suggested that they be towed behind a destroyer to begin their patrols out to sea, rather than waste time and endurance flying to their stations. Another proposal was that such towed aircraft could be used to bomb the German coast.

John Porte, though English, had worked in America, where he had joined the American Curtiss Company in 1913, but returned to England as soon as war was declared and was commissioned as a Squadron Commander in the Naval Wing. He then talked the Admiralty into buying two Curtiss flying boats (Nos. 950 and 951), which were taken on charge in November 1914 to be evaluated at the seaplane base at Felixstowe. They must have impressed the Navy, for a further sixty-two were ordered (of which eight were licence-built in Britain). The boats were given the designation Curtiss H4.

In truth, as a type, the H4 had certain defects; in particular its wooden planing hull was lacking in seaworthiness. But after modification by Porte the hull shape was improved, conferring better take-off and alighting characteristics. The H4 saw a great deal of operational service with the RNAS and taught the Navy a good deal about flying-boat operations.

The H4s were followed in May 1917 by a larger boat, the Curtiss H12; this became known as the 'Large America', the remaining H4s then being called the 'Small America'. Seventy-one H12s were eventually delivered to the Royal Navy from the Curtiss plant at Hammondsport, Buffalo, New York. It was a big flying boat, with a wingspan of 92 feet 8½ inches, but, like the smaller H4, the planing bottom of the hull was prone to failure in any sort of seaway. The original engines, 160hp Curtiss, also proved disappointing, and were replaced by two Rolls-Royce Eagle Is of 275hp. (Later aircraft had 375hp Eagle VIIIs.) The H12 carried four air gunners armed with Lewis guns and could carry two under-wing 230lb bombs.

To a Curtiss H12, No. 8660, goes the honour of flying the first of the famous 'Spiders Web' anti-U-boat patrols on 13 May 1917. The Spiders Web was centred on the North Hinder Light vessel in the North Sea. As its name suggests, it was an octagonal pattern of patrol lines, 60 miles across, covering 4,000 square miles of sea, which could effectively be patrolled by four flying boats in five hours.

The Spiders Web patrols began when, with the increase of U-boat activities in early 1917, intercepted wireless traffic from the submarines enabled British direction-finding stations to discover the routes which the U-boats took when crossing the North Sea on the surface to their bases at Kiel and Wilhelmshaven. It took a surfaced U-boat about ten hours to cross the area covered by the Spiders Web, and after only seven days' of its operation, UC36 was sunk 10 miles east of the North Hinder light by Flt Sub-Lts Morrish and Boswell in their 'Large America'. Several other U-boats were claimed as sunk, though most remained unconfirmed.

The H12s could hardly be considered as fighters; nevertheless, in spite of their size and low speed — maximum 85mph — the flying boats also shot down two Zeppelins. The first was the L22, which Flt Lt Galpin, flying H12 (No. 8666), attacked 18 miles off Texal Island on 14 May 1917; the second was the L43, which was shot down by 8677, piloted by Flt Sub-Lt Hobbs on 14 June 1917.

The success of the American Curtiss boats led to the creation of the Felixstowe F2A, a flying boat in many ways the ancestor of the Sunderland, which was to provide the backbone to Coastal Command twenty-three years later. The F2A owed much to John Porte's work and embodied a hull to his design. The wings and tail were Curtiss, the actual H12 unit initially being used. The F2A had a very long endurance — six hours on standard tanks, and up to nine hours with additional fuel carried aboard in tins. It had a sturdy airframe, as was demonstrated on 4 June 1918, when a formation of four F2As and an H12, engaged on a Spiders Web patrol, had a running fight with no fewer than fourteen German seaplanes. One of the F2As (N4533) was forced down with a common trouble of the type — a faulty fuel line — before the fight began; another F2A (N4302) suffered similar trouble just as the action started. The remaining three flying boats — two F2As (N4295 and N4298) and the Curtiss H12 — fought the enemy seaplanes and shot down six of them without further loss: an incredible performance for aircraft weighing 5½ tons.

The obvious advantages demonstrated by the Felixstowe F2A led the Curtiss Company, no doubt after consultation with John Porte, to produce an improved version of its H12. This was the H16 'Large America', 160 of which, powered by two 345hp Rolls-Royce Eagles were ordered for the RAF. The war, however, came to an end before all were delivered and the last fifty were cancelled, leaving seventy on charge by October 1917. Another fifty H16s were being operated around the British coast by the United States Navy based at Killingholme. The US Navy's H16s were generally similar to those supplied to the RAF, though powered by 330hp Liberty engines. It is said that one American pilot actually succeeded in looping his H16, such was the confidence in its structure.

Anti-U-boat patrols were not solely flown by flying boats, and brief mention should also be made of the role played by the RNAS airships' anti-submarine patrols. The Navy's interest in rigid airships never quite recovered from the *Mayfly* affair, though, as related, several were acquired in the early years; but these were for the most part rigid airships which were difficult and expensive to build and maintain. It was only too plain, however, that the Germans had overcome the difficulty. Thus, when Admiral Fisher returned to the Navy as First Sea Lord in October 1914, he talked to all the airship officers serving in the Royal Navy to discover what part, if any, British airships could play in the war at sea. One result of the conference was the ordering, in February 1915, of thirty-eight small, non-rigid airships for patrol work. These were quickly and cheaply built (£5,000 each). The first type, the SS (Sea Scout), was little more than an elongated gas bag with the wingless and tailless fuselage of a BE2c suspended beneath. They were used principally for anti-submarine work in the English Channel and Irish Sea from May 1915. Subsequent developments culminated in the Coastal, which had a trefoil envelope of 17,000 cubic feet. This airship had a specially designed open gondola with two engines of 220 and 100hp, and could patrol at 45mph for 24 hours. Small airships were very effective as anti-submarine weapons, and when the shipping convoy system was introduced in 1917, only one ship was lost to a U-boat in convoys escorted by airships. They were also very reliable; one of the Coastal class ships logged no fewer than 66,000 miles in two years' continuous service.

Interesting and valuable as the contribution of the flying boats and airships was, the long-term future of Royal Naval flying lay with shipborne aircraft taking off and alighting on decks of ships at sea. *Furious* had demonstrated, with the Tondern raid, the feasibility of taking off, but operationally landing on was, in September 1918, still to be achieved.

After the failure of the *Furious* 'alighting' experiments, due to excessive turbulence, the ex-Italian liner *Conte Rosso* had been further modified and, in September 1918, the 15,750 ton ship emerged, renamed HMS *Argus*. Her gantry had gone and she had a flush flight deck from stem to stern, 567 feet long. The world's first true aircraft carrier, she had no island, indeed no superstructure of any description to cause turbulence over the flight deck; the uptakes from the boiler rooms vented from ducts over the stern. The ship was conned from a small charthouse which was raised hydraulically above the flight deck for navigation, being lowered flush when the carrier was flying off or landing on her aircraft. It is said that during her acceptance trials, under full speed in crowded anchorage, the movable wheelhouse suddenly descended, leaving the helmsman and officers of the watch stuck between floors, so to speak. As her Captain, H.H. Smith, put it: "There was *Argus*, steaming full speed and spinning under full helm in crowded waters, while all those in authority over her were disappearing below."[30] Fortunately *Argus* could be conned from two positions: the disappearing wheelhouse on the flight deck, used when making passage, and a bridge wing either side of the flight deck when flying was in progress.

With the *Argus* in service — her unconventional appearance caused her to be dubbed the 'Ditty Box' by sailors — the Royal Navy was within hailing distance of having an aircraft carrier at last. Her sea trials were soon over and she began to work up to her true function.

Commander Richard Bell Davis made history in October 1918 by flying off his Sopwith 1½ Strutter from the deck of *Argus* and safely alighting back on board. *Argus* had no arrester gear but used the longitudinal guide wires as fitted to *Furious*. The 1½ had a normal wheeled undercarriage and, with the ship's speed around 20 knots, heading into a reasonable wind, there was no difficulty in bringing the lightweight Sopwith to a standstill; nevertheless, in the absence of arrester wires, the ship's aircraft lift was lowered 15 inches and a 10 foot wooden ramp placed on top. The aircraft landing dropped into this 'trap' as it it was called, which was near the end of the aircraft's landing run, and the ramp brought the machine to a final stop, when it was secured by a deck party of men who were waiting in catwalks at either side of the flight deck. The war was to end before *Argus* was operational, but by the end of her working-up period she had achieved nine successful landings — six by 1½ Strutters and five by Camels. Even so, a contemporary report cautioned: "The operation of alighting on the deck of a ship remains and is likely to remain one of considerable risk, and to be undertaken only by experienced pilots . . . ."[31]

Argus eventually embarked Sopwith Cuckoos, which were the first aircraft designed as torpedo carriers, armed with Mk IX 18 inch torpedoes. However, these were not operational until after 1918, though had the war continued, Beatty's vision of hundreds of such aircraft attacking the German fleet could have become at least a partial reality.

When World War I finally ended on 11 November 1918, the Royal Navy had the world's only aircraft carrier in commission. Another, *Hermes*, under construction, was to be the first carrier to be built from the keel up as such, though the hull was based on that of a Hawkins class cruiser.

Furious, which was one of the ships to escort the surrendered German High Seas Fleet to Scapa Flow, then went to a yard to undergo her third, but by no means last, conversion — this time to a flush deck carrier, similar to *Argus*. The Admiralty, in early 1918, had requisitioned the

Chilean battleship *Almirante Cochrane*, which had been under construction on the Tyne at the outbreak of the war and began a rebuild which was to be completed in 1923 as the carrier HMS *Eagle*.

Of the existing seaplane carriers, *Campania*, as if to atone for missing the Battle of Jutland, dragged her anchor and rammed the battleship *Royal Oak*, then quietly sank into the Firth of Forth. The old *Ark Royal* was renamed *Pegasus* (in 1935), to plod on as a seaplane carrier and depot ship throughout World War II. The converted ferry boats were either scrapped or returned to their lawful occasions plying across the English Channel or the Irish Sea.

With the coming of peace, Britain thus had the prospect of four aircraft carriers, the only naval power in the world to be so equipped. Acquiring the carriers was one thing; the provision of the aircraft and the men to fly and service them, however, was now in the gift of the Royal Air Force — and that was to be altogether a different proposition.

From that day in March 1911, when the four young naval officers reported to Eastchurch to learn to fly, to 11 November 1918 was just over seven years. In that time the amalgamation of the RNAS with the RFC had created the greatest air force in the world: 22,000 aircraft in 188 squadrons, manned by 291,000 men. It would quickly shrink to a tiny fraction, of which the Navy's share would be miniscule. But that melancholy story belongs to a later chapter. For the moment, most of the men of the RNAS were being demobilised to pick up the threads of their pre-service lives in a much-altered world; the fate of naval flying was no longer their concern:

> Now there is nothing, not even our rank,
>   To witness what we have been;
> And I am returned to my Walworth Bank,
>   And you to your margarine.

RUDYARD KIPLING
('The Changelings')

3
The Years of Peace

World War I ended in a military sense on 11 November 1918. After the first heady days of peace; after the marches and the celebrations of the victors, the politicians began the protracted conferences to negotiate a peace treaty.

The meetings were, in many ways, difficult; the German forces had not suffered a catastrophic defeat, indeed they had not been defeated at all. In November 1918 the German army on the western front was fighting on foreign soil; at no point had the Allied ground forces crossed the frontier. What had caused the Germans to sue for peace was the collapse of their allies in the Balkans, the *prospect* of defeat in the west — an eventuality made probable by the entry of the United States into the war — and the cumulative effect, particularly on the civilian population, of the Royal Navy's ruthless blockade, perhaps the deciding factor.

When the armistice was rumoured in Germany, it was by no means universally welcomed; the Navy — or at least the Admirals — certainly did not want it. It is not entirely clear if the High Command wanted a final battle with the Royal Navy: a maritime *Götterdämmerung*, or if they simply intended to wreck the peace negotiations, which the German Navy considered were imposed on them by the politicians. Whatever the reason, the High Seas Fleet was ordered to raise steam and prepare for action. It was too much for the sailors, most of whom, since the Battle of Jutland, had been safely at anchor in harbour. The ratings mutinied and their unrest spread to the civil population of Kiel. Thus Germany, bereft of allies, with sections of the civil population in revolt and near starvation — the Allied blockade continued during the peace negotiations — could not contemplate a renewal of war in the west, with the probable outcome being a long retreat to inevitable defeat on German soil. The only alternative was an armistice followed by a negotiated peace.

After a great deal of wrangling among former Allies, a peace treaty was signed on 28 June 1919 in the Galerie des Glaces, in the Palace of Versailles. Ironically this was the very hall from which the German Empire had been proclaimed in 1871.

The Treaty of Versailles, as the instrument was to be called, was deeply resented by the Germans as punitive and unjust; it compelled them to surrender the High Seas Fleet (to the British), pay massive reparations and concede considerable territory in Europe and all the former African colonies. The Germans considered that they had been duped and signed under duress, fully intending to repudiate the treaty as soon as politically convenient; the bitterness engendered by Versailles was to nourish the seeds of World War II; but that was still, in 1919, twenty years distant. For the moment Europe was, nominally, at peace.

The four years of war had caused losses to the fighting powers in men and materials that were entirely without precedent; the physical damage was relatively slight and would soon be restored but the economic and political consequences of the struggle were to be felt for a great deal longer. A.J.P. Taylor, the historian, wrote: "The old financial stability was shaken, never to be restored. Depressed currencies, reparations, war debts, were the great shadows of the inter-war period . . . ."[1]

It was against that background that the nascent Royal Air Force fought for its very existence. The man in whose hands the destiny of the RAF was placed was the first Chief of Air Staff, Sir Hugh Trenchard, who was to earn his subsequent sobriquet 'Father of the Royal Air Force'.

Almost as soon as the fighting ceased, the RAF began to decline from the pre-eminent position it had held in November 1918: barely a year later, it had shrunk to a little measure indeed — 31,500 men, and twelve squadrons of 1918 vintage aircraft. The share grudgingly allocated to the Royal Navy was rather less than the equivalent of two squadrons.

Once it had become clear that the RAF was not just a wartime expediency but to be a permanent part of the British forces, a bitter internecine struggle developed between the RAF and the two older services. It was not entirely a question of airpower, though the army had been reluctant to lose control of the Royal Flying Corps and the Royal Navy clearly needed an air arm for its carriers and fleet protection; fundamentally it was a matter of money. The defence budgets in the immediate post-war years suffered swingeing cuts in the prevailing atmosphere of retrenchment, disarmament and growing pacifism: the cake had now been sliced thinly, three ways instead of two.

It was the Navy that was to prove the most intransigent opponent to the new service. Beatty, in particular, who had succeeded Sir Rosslyn Wemyss as First Sea Lord in November 1919, reiterated the view he had placed before the Smuts inquiry, that the existence of an Independent Air Force did not necessarily preclude the possibility of the Navy — or the army — having a specialised wing and, that said, the Admiralty should be allowed to include the cost of such a wing as part of the Naval Estimates, in the same way as the cost of the aircraft carriers.

When the formal plans for the peacetime organisation of the Royal Air Force were presented to the House of Commons in November 1919 by Winston Churchill, then Minister for War, he seemed to concede Beatty's proposals, saying: "In addition [to the independent RAF] there will be a small part of it specially trained for work with the Navy, and a small part specially trained for the Army, these two small portions probably becoming, in future, an arm of the older services." Trenchard fought to forestall that proposal; he firmly believed in what he called the 'Unity of the Air', which meant in practice that all service aircraft, together with the aircrews that flew them, the service and training echelons, were under the direct control of the RAF.

The Royal Navy never accepted that doctrine and fought unceasingly to regain Admiralty control over a Fleet Air Arm. That was to take nearly twenty years. The immediate effect of 'Unity of the Air' was the loss to the Navy of all the war-experienced airmen; the young officers who had created naval flying, pioneering the difficult art of carrier operation, were now told in effect that if they wished to continue service flying, they would have to leave the Navy and transfer to the RAF. Most of them, reluctantly, chose the latter course. The short-term effect was the loss of trained pilots and aircrew: the longer-term loss was to be far more serious, for it was to deprive the Navy of officers of Flag rank, with practical experience of flying and an understanding of naval airpower, at a critical time when, on the eve of World War II, the Royal Navy was desperately re-establishing the Fleet Air Arm, the total control of which it was eventually finally to wrest from the RAF in May 1939.

It was a long struggle with concessions reluctantly yielded by the Air Ministry. The first was in 1921, when it was agreed that naval officers could be trained as air observers. Then the Balfour Committee of 1923, whose terms of reference were to investigate 'The relations of the Navy and Air Force as regards the control of Fleet Air Work', recommended not only that all naval observers were to be naval officers, but that up to 70 per cent of the carrier pilots should be naval, with dual RAF and Navy rank. For reasons which remain obscure, the Air Ministry opposed the training of naval petty officers as pilots on the grounds that they were less 'suitable' than RAF sergeants.

By April 1924, the Admiralty had succeeded in getting the carrier branch of the Royal Navy officially called 'The Fleet Air Arm', though it still formed part of the RAF. Three carriers were by then in commission:

HMS *Argus*, *Hermes* and *Eagle*. *Furious* was undergoing yet another rebuild and her two sisters, *Glorious* and *Courageous*, were being converted. The development of these vessels will be considered later; for the moment the existing three had embarked the total muster of aircraft the RAF released for naval duty: seventy-eight.

Small and tenuous though the Fleet Air Arm was, naval officers could now pursue a flying career and still remain in their chosen service, though strangely such men did not seem to enjoy the same promotion prospects as 'normal' naval officers; still they wore naval uniforms and carried naval ranks with their distinctive wings on their sleeves. If certain admirals did not appreciate the need for the Fleet Air Arm and its carriers — and some did not — the attitude of Lord Trenchard, now the first Marshal of the Royal Air Force, was still as intransigent as ever over what he considered to be a breach of the 'Unity of the Air' which the Fleet Air Arm represented. He publicly let slip his attitude as he presented wings to newly qualified pilots on No. 4 Course, Flying Training School, Netheravon, when he said to a group of naval and marine pilots, in the voice which had earned him the nickname 'Boom': "I congratulate you on becoming pilots — but I'll be damned if I can understand the colour of your uniforms."

In view of the trouble that the Admiralty had in getting aircraft and crews from the RAF for service with the fleet, it is perhaps surprising that the far more costly carriers survived the orgy of scrapping that followed the conclusion of the war; in fact for a time only HMS *Argus* continued in commission. *Furious* was paid off and de-commissioned from 1919 to 1922, partly because the ideal arrangement for a flush-deck carrier had yet to be finalised. This was due in part to the less than satisfactory ducting of the boiler gases on *Argus*, which tended to obscure the final approach path of the landing aircraft; there were also subsidiary difficulties with the trunking overheating. So, pending further trials, the final configuration of both *Furious*, awaiting conversion, and *Hermes*, which had only been completed as a hull, was in abeyance. The ship that was not only to decide the final shape of these and later carriers, but was also to be the first true fleet carrier, was the incomplete ex-Chilean Dreadnought battleship *Almirante Cochrane*, requisitioned by the Admiralty and renamed HMS *Eagle* as a result of a request by Sir David Beatty in 1917 for a fast and well-armoured carrier to operate 'Scouts' (fighters) with the Grand Fleet.

The Director of Naval Construction, Sir Eustace Tennyson d'Eyncourt, proposed a design on the Chilean battleship's hull that would have offered a 400ft hangar, with two aircraft lifts connecting to a 640ft flush flight deck. The original layout was to have had the same amidships 'goalpost' arrangement envisaged for *Argus*, with two 'islands' port and starboard, each with uptakes from the boiler rooms and a connecting bridge with a wheelhouse some 20ft above the flight deck. The trials aboard *Furious* had indicated the danger of turbulence inherent in such a layout, and the goalpost idea was dropped; after consultation with the commander of *Furious*, Captain W.S. Nicholson, and pilots who had flown from that ship, it was decided to place a single narrow island containing the uptakes and navigational offices and wheelhouse bridge on the starboard side of the vessel. The decision to place the island on the starboard, or right-hand side of the carrier was made as it was discovered that most pilots tended naturally to turn left after take-off, or if attacked (a fact to be exploited by German fighter pilots stalking World War II bombers and even today most airfields have left-hand circuits or patterns). Another consideration for the adoption of the left-hand circuit is thought to have been the pronounced and dangerous tendency of the Sopwith Camel to drop its nose when turning to the right, owing to the gyroscopic effect of its large rotary engine, which had caused many fatal spins with the Camels in the hands of inexperienced pilots.

The concept of the 'island' carrier had been advanced earlier (see p. 92) and one could have been built from 1915 onwards, but possibly the use of high performance aircraft at sea was less obvious at that time.

The Director of Naval Construction's staff had little precedent to guide them; the after deck of *Furious* was known to be unusable, owing to turbulence of the amidships structure. *Argus*, the only operational carrier at that time, had no structures whatever on her deck, which, although it was generally satisfactory from a flying point of view, was less than ideal when operating the carrier from the ship-handling point of view. Some compromise was therefore desirable.

A great deal of testing and experiment went into the finalised design of *Eagle*. A model was extensively tested in a wind tunnel at the National Physical Laboratory (NPL) to determine the position and shape of the island; one subsidiary result of the tests was the characteristic 'rounddown' on the entry of the flight deck of British carriers.

After the tests it was decided to complete *Eagle* as an 'island' carrier, but work had barely started at the Walker Navy Yard when construction was halted in October 1919 on orders from the Admiralty, owing to serious reductions in the Naval Estimates. As a consequence *Eagle* was very nearly abandoned as a carrier altogether and rebuilt as a battleship, to fulfil the original Chilean order; however, it was pointed out that the money already spent in conversion and the additional cost of rebuilding would involve an overall loss on the contract that would be of the order of £1 million. Another consideration in favour of completing the vessel as a carrier was that she was the prototype 'island carrier', and both the rebuilding of *Furious* and the completion of *Hermes*, and indeed of any future carriers, depended on the outcome of practical flying trials to and from her deck at sea. By November 1919 the decision to complete as a carrier was given.

When finally commissioned, *Eagle* would be unique in being the only British fleet aircraft carrier ever built with twin funnels, but for her limited flight trials she was to use only two boilers with one operational funnel; the second was lashed to her flight deck when she arrived at Portsmouth in April 1920.

With a wooden hut as a wheelhouse on the island and with jury-rigged lighting in her red-leaded accommodation, she began her trials. No arrester as such was fitted but 190 foot longitudinal wires, as on *Furious* and *Argus*, were installed on her 625 x 96 feet flight deck. For the trials the hangar was not used; the lift wells were plated over.

An 'Eagle' flight was trained ashore at Gosport; then in May 1920 a 2F.1 Camel and a Parnell Panther were winched aboard and taxiing trials, with *Eagle* alongside in Portsmouth harbour, were made on the flight deck to test the wires. The Panther, incidently, is of interest in that it was the first aircraft to be designed from the outset for carrier operations, and it started a tradition of ugliness which most of the early Royal Navy fleet aircraft maintained.

Eagle put to sea and the Camel, Panther and a Bristol Fighter made 'touch and go' landings as she steamed through the English Channel. Little turbulence seems to have been experienced and on 1 June 1920 Flying Officer W.F. Dickson — now in the RAF, but the same pilot who had tested the after deck of *Furious* and flown as a naval officer on the Tondern raid — landed a 2F.1 Camel, N8134, safely aboard *Eagle*; the Panther and Bristol Fighter followed. Within the next fortnight these three aircraft and a Sopwith Cuckoo safely made fifty-nine take-offs and landings without incident.

After further trials *Eagle* paid off at Devonport on 16 November 1920, pending completion. Additional wind-tunnel tests were made at the NPL, as a result of which the shape of the island was modified. The tests had actually suggested that the island should be shorter and wider than in fact it was, but it was decided to retain the existing structure, slightly altered in shape.

The knowledge gained from the experimental flights, sea trials and laboratory wind-tunnel work was to be incorporated in the next British carrier, HMS *Hermes*, then on the stocks but to be commissioned two days before *Eagle*, though that was not to be until 1924. Thus *Eagle* was, by a margin of four years, the first starboard island carrier afloat, starting a fashion which, with very few exceptions, was to be followed by the major naval powers for the next thirty years. Of *Eagle*'s two contemporaries, *Argus* and *Furious*, *Argus* never had an operational island and *Furious*, having waited for the outcome of *Eagle*'s tests, strangely was also completed, like *Argus*, flush-decked, though she too was eventually to acquire a small starboard island during one of her many subsequent refits. The Royal Navy, in the immediate postwar years, was clearly substantially ahead of its late ally, the US Navy, in the design and operation of aircraft carriers. It is worthwhile considering the background to the Americans' apparent lack of progress in carriers.

The US Navy, which had pioneered naval flying with Eugene Ely's historic flights, was seemingly reluctant to pursue the early lead that it had established. After the outbreak of World War I, the Royal Navy had, of course, the impetus to strengthen the hand of its aviation protagonists, and the relatively short distances between the belligerents made the use even of the short-range aircraft then developed a feasible proposition; the ever-present Zeppelins over the British fleet also were a constant challenge. The USA, however, was at peace. World War I was to the Americans a European struggle far from their frontiers; it was not until the unrestricted activities of the German U-boats caused the sinking of American ships, with the loss of American lives, together with the interception and decoding in 'Room 40' at the British Admiralty of the celebrated 'Zimmerman Telegram', that the USA became directly involved. Before the USA declared war on the Central Powers, on 6 April 1917, the vast reaches of the Atlantic and Pacific Oceans had seemed to offer little scope for the use of limited-range naval aircraft.

The total number of aircraft on US naval charge on 1 July 1914 was twelve: six 'hydroaeroplanes' (the term 'seaplane' did not come into use in the US Navy until 1916) and six flying boats. However, part of this small force was to be included in the first military operation to use naval aircraft, an event that occurred just before World War I and is now known as 'The Action off Veracruz'.

The United States and Mexico were officially at peace but Mexico was in a state of political unrest with a revolution in prospect. General Victoriano Huerta had declared himself 'the provisional president of Mexico'; a US naval party had been arrested by police whilst ashore and a decision had been made in Washington to mount a punitive expedition to restore the 'dignity and rights' of the United States. A naval force, including the cruisers USS *Birmingham* and *Mississippi*, was ordered to deal with the crisis and sailed with Marines and practically the entire Naval Air Arm aboard, from Pensacola. USS *Birmingham*, the same ship on which Eugene Ely had landed four years earlier, embarked a Curtiss C5 flying boat (AB-5) and a Curtiss A4 hydroaeroplane (AH-2). *Birmingham* sailed to Tampico and rather anticlimactically saw no action. The second flight aboard USS *Mississippi* consisted of a Curtiss C3 flying boat (AB-3) and a Curtiss A3 hydroaeroplane (AH-3); they arrived at Veracruz and began reconnaissance flights from the sea over the city and the harbour, spotting for mines and photographing enemy positions.

The first flight, which was made on 25 April with the flying boat AB-3, piloted by Lt (jg) P.N.L. Bellinger (US Naval Aviator No. 8), accompanied by Ensign Stolz as observer, was the first war patrol of an American naval aircraft. The same pilot flew the AH-3, with Ensign W.D. Lamont as observer, on the first ground-support mission on 2 May; US Marines encamped at Tejar had come under attack and then requested the *Mississippi* Aviation Unit to locate the disposition of the attacking troops. Four days later, Bellinger, this time with Lt (jg) R.C. Saufley as observer, was flying reconnaissance over enemy positions when his AH-3 became the first American naval aircraft to be hit by hostile (rifle) fire. The aircraft was able to alight alongside the *Mississippi*, neither member of the crew being injured.

The Tampico detachment aboard the USS *Birmingham* was also ordered to Veracruz, but by the time it arrived the action was over; US Marines had taken the city and General Huerta was on his way into exile. The US Naval Aviation Unit returned to Pensacola. The commander of the Tampico detachment, Lt J.H. Towers (US Naval Pilot No. 3), was sent to London as Aviation Assistant to the Naval Attaché on the outbreak of World War I on 4 August 1914. To show impartiality — or to obtain all-round information — the following month First Lt B.L. Smith became the Aviation Assistant to the Naval Attaché in Paris and Lt V.D. Herbster (US Naval Pilot No. 4) was appointed to the same post in Berlin.

The United States, as the only western industrial power not directly involved in the early years of World War I, was able, through its military attachés and observers, to watch and profit from the fighting powers' developments in military aviation, though of course certain technical information remained classified and withheld from the Americans. The naval attachés' reports on aviation matters were sent back to Washington, and the importance of naval aviation generally was officially recognised, for on 23 November 1914 the post of Director of Naval Aeronautics was established, Captain Mark L. Bristol being the first officer to hold that title.

There was an appreciation in Washington that the pace of advance in both military and technical terms in Europe was leaving the US behind, and this perturbed the Secretary of the Navy, Josephus Daniels, for he wrote in July 1915 to Thomas Edison, the great inventor: "One of the imperative needs of the Navy, in my judgement, is machinery and facilities for utilising the natural inventive genius of Americans to meet new conditions of warfare."[2] This letter led to the establishment of the Naval Consulting Board, which included civilian advisers and a "Committee on Aeronautics, including Aero Motors".

No doubt the committee did sterling work but the melancholy fact is that "the natural inventive genius of Americans" had only been allowed to provide the Navy with a total of fifteen aircraft by 1 July 1915 and seventeen the following year. When, on 6 April 1917, the United States entered the war, the Navy aircraft muster of all types stood at fifty-five — forty-five seaplanes, six flying boats, three landplanes and one airship. There were forty-eight pilots, including students, and one base: Pensacola, on the Gulf of Mexico.[3]

With the possible exception of the flying boats, these aircraft were quite useless for active service in the European war. The US Army was hardly better placed:

> The Aviation Section of the Army Signal Corps was established by an Act of Congress on 18 July 1914 .... the total number of aircraft delivered to the Army prior to [April 1917] was 224; and all of these were training aircraft not suitable for fighting, bombing, or observation service.
> The personnel complement .... was 52 officers, 1,100 enlisted men .... Although 139 men had received flight training, only 26 could be considered qualified pilots, and then only in training craft.[4]

This did not, of course, include the many American volunteers who flew both with the Royal Flying Corps and the French Service d'Aviation: the famous French fighter squadron *Escadrille Lafayette* was composed entirely of American volunteer pilots. The fighter aircraft which they and the pilots of the American Expeditionary Forces flew were either British or French, the latter predominating.

In the nineteen months from the United States entering the war to 11 November 1918, when the war ended, there was a phenomenal growth in American naval aviation: billions of dollars were appropriated; thousands of young men enrolled in the Naval Reserve Flying Corps and trained as pilots, aircrew and mechanics at scores of bases on both sides of the Atlantic. The production of large numbers of aircraft and aircraft engines was put in hand, though in the understandable haste to get the machines off the drawing boards into service, difficulties were encountered and inevitable mistakes made. In the end the only US-built aircraft to see service over the western front was the DH4, a British De Havilland design; this was not a fighter but a two-seater day bomber. The American version of the DH4 was powered by a 400hp Liberty V12, liquid-cooled engine which was to have been a major contribution of the United States to the air war. The Liberty had its critics on both sides of the Atlantic, who alleged that it was badly designed, too heavy, with a high fuel consumption and prone to overheating.

The DH4, powered by the Liberty, became known as 'The Liberty Plane'; it was produced in fairly large numbers (3,431 by the Armistice), mainly for the US Army, though its effectiveness was less than had been hoped. A total of 155 DH4s (Bureau Nos. A3245 to 3324 and A3384 to 3458) were transferred from the Army to the Navy; fifty-one of these arrived in France and some of them flew with Marine Squadrons 9 and 10 on bombing missions based on Dunkirk.

The main role of the US Navy's Flying Corps in the European war was that of anti-U-boat patrols undertaken with Curtiss long-range flying boats from twenty-seven bases in France, England, Ireland, the Azores and Italy. On the American side of the Atlantic there were twelve flying-boat bases on the US seaboard, two in Canada and one in the Panama Canal Zone. From these bases the US Navy and Marine Corps flew more than 3 million nautical miles on anti-submarine patrols; twenty-five U-boats were attacked and a dozen claimed as sunk or heavily damaged. As a matter of historical interest, the first attack by a US Navy pilot on an enemy submarine was made on 25 March 1918 by Ensign John F. McNamara, flying from RNAS Portland. The attack was reported as 'apparently successful'.[5]

Two days after Ensign McNamara's flight, 3,000 miles away an H16 flying boat, No. A-1049, took off from the Delaware river, Philadelphia. It was a brand new machine and was on its test flight — the first aircraft to be produced at the Philadelphia Naval Aircraft Factory. There was nothing particularly unusual in that, one might think, but even the factory had not existed a year previously; the story is a remarkable one.[6]

When the United States entered the war, the aircraft companies, which had for years been existing on a hand-to-mouth basis, were suddenly innundated with orders and, anticipating a scarcity of production facilities, the Navy prudently decided that it would open an aircraft factory under naval ownership to build a thousand or so simple wooden training seaplanes a year.

The Philadelphia Navy Yard on the Delaware River, which had surplus land, was selected as a suitable site; the man placed in charge of the project was Commander F.G. Coburn of the USN Construction Corps. The first sod was cut in the summer of 1917; 110 days later, in November, the plant was completed and 400 skilled tradesmen and engineers were hired and ready, although only ten of the staff had any previous aeronautical experience. Coburn was now appointed general manager and his orders were not to build the anticipated simple trainers, but what were by the standard of the day large and complex aircraft: Curtiss H16 flying boats. The US Navy, in the light of the success that the Royal Navy had had, decided these were ideal for convoy protection.

It was a daunting undertaking. To make things a little easier for the inexperienced workers, all the working blue-prints for the aircraft — many hundreds — were redrawn, which took two months. Then production got under way, with the result that just 228 days after the work on the factory began, the H16 No. A-1049 took off on its maiden flight; seven days later the flying boat was on its way to England and the war. By November 1918, 150 H16s (Bureau Nos. A-1049— 1098 and A-3459— 3558) and 134 F-5L flying boats (Bureau Nos. 3559— 4035, of which 343 were cancelled) followed the first one from the Philadelphia Naval Aircraft Factory (which was also to produce aircraft and components for World War II). Not all the H16s were produced by the NAF; Curtiss also manufactured them, but all were powered by two 400hp Liberty engines which, despite the critics, gave long and reliable service.

The F-5L is an interesting aircraft because of its convoluted ancestry. It will be remembered that the original H16 supplied to the RNAS had been extensively reworked by Porte at Felixstowe (see p. 103) to become the Felixstowe F2A, incorporating an improved hull design using Curtiss wings and tail units. The US Navy decided to adapt the British variant and built 137 F-5Ls; thus an American original design was improved abroad to come home and be redesigned yet again, this time to incorporate US Liberty 12A engines (in place of Rolls-Royce Eagles) and equipment.

The large Curtiss flying boats were by far the most important contribution made by American naval aviation to the war, though had the conflict continued, the US Navy would also have taken part in the planned strategic bombing of Germany. The growth of naval aviation in the nineteen months of US participation in World War I was impressive; by the armistice it was a force of 6,716 officers and 30,693 enlisted men in naval uniform, together with an additional 282 officers and 2,180 men of the Marine Corps. The number of aircraft had risen from 54 to 2,107 —

695 seaplanes, 1,170 flying boats, 242 landplanes, in addition to 15 airships; 18,000 officers and men and 570 aircraft had served abroad.

Not all the aircraft flown by the US Navy and Marine Corps were built in America; four Sopwith Baby seaplanes (A869—872) came to the United States for evaluation and others were used in Europe (with British naval serials) for training. At least two Sopwith Pups (A5655—5656) were also shipped to the States. The 1½ Strutters were widely used in both Britain and France, flown by US Navy pilots, as were Sopwith Camels. Lt (jg) David Ingalls who flew Camels with 213 Squadron RAF (previously 13 Squadron RNAS), shot down five enemy aircraft in six weeks to be awarded the DFC and the American DSM.

US Navy pilots, many of whom trained in Britain, and those who took part in experimental flying off lighters, no doubt visited British carriers and RNAS bases, where in wardrooms and messes they talked with British naval pilots and RNAS officers of the exciting prospects which aircraft carriers offered for the future of naval aviation. Many of these naval aircrew, like their Royal Navy equivalents, were Volunteer Reservists, signed up for the duration of the war; thus when peace came, most of them, thankful to be still alive, returned to civilian life. Some stayed, as did the career Navy men. These, mostly young officers, who had seen the aircraft carriers and who had flown naval aircraft in action, naturally wanted the US Navy to continue with the developing role which aircraft had shown they could play during the — for the British — four years of war.

For Americans, the task of furthering the claims of naval aviation was in one respect slightly simpler than in Britain: the US Navy was in complete control of its air arm, there being at that time (and for a long time to come) no 'Independent Air Force', no Trenchard to expound the 'Unity of the Air'. If the Navy wanted aircraft and carriers, all it had to do was to pay for them, though to get the money voted by the politicians was admittedly an initial difficulty.

The number of committed naval aviation men who had been in the European war was small when considered as a percentage of the huge two-ocean US Navy; a Navy, moreover, like all navies at that time, dominated by gunnery officers who ideally saw themselves pacing the decks of a battleship. Fortunately — very fortunately, as it was eventually to turn out — there were in America, as in Britain, a few senior officers who were prepared to risk their careers and speak out for naval aviation and the imperative need for the US Navy to acquire at least one experimental carrier.

Commander Kenneth Whiting — a pilot — who had been the CO of the US naval flying-boat base at Killingholme, Ireland, during the war, wrote a memorandum in 1919 to the Committee on Naval Affairs which neatly summed up the situation:

> When the war ended those who had chosen the Navy as a life work, and especially those of the Navy who had taken up Naval Aviation, revived the question of 'carriers' and 'fleet aviation'. They found the sledding not quite so hard as formerly, but the going was still a bit rough.
>
> The naval officers who had not actually seen Naval Aviation working retained their ultra conservatism; some of those who had seen it working were still conservative, but not ultra; they .... wished to be 'shown'. Others, among the ranking officers who had seen, had conquered their conservatism and were convinced.[7]

The 'convinced' group included two admirals and other senior officers on the General Board of the Navy, several members of which had served on the staff of the C-in-C during the war and could thus speak from first-hand experience. The initial problem was to acquire a suitable hull, which meant, in those immediate post-war days, the cheapest possible. There could be no question of laying down a purpose-built ship; for one thing it would delay the project by four years or so, even had the money for such an enterprise been forthcoming, which would have been most unlikely. It is an interesting paradox that, in 1919, the US Navy, with no competing independent air force, had aircraft to burn (literally) and no carrier, while the Royal Navy had three carriers and no aircraft or pilots to call its own.

The search for a ship suitable for conversion was aided by the fact that the US navy, like the Royal Navy, was rapidly changing over from coal- to oil-firing; thus a number of fleet colliers were redundant. One of these worthy vessels, launched in April 1913 — *Fleet Collier No. 3*, later named USS *Jupiter* — was due for scrapping; a bid was made and the ship was acquired by the Aviation Branch for conversion. At first, or indeed at second, glance the 11,500 ton *Jupiter* was far from ideal; however, after closer consideration, the vessel's advantages seemed to outweigh the disadvantages. It is true that as a collier she was slow, with a maximum speed of around 14 knots, though her overall length of 542ft would be adequate for a flight deck and the ship had very large holds, "with high headroom in them, a difficult thing to find in any ship. She had larger hatches leading to these holds than most ships, a factor permitting the stowing of the largest number of planes",[8] wrote Commander Whiting, who was to be her future captain. Another important factor in selecting *Jupiter* was the main machinery: this was turbo-electric — that is her steam turbines drove electric generators, the ship's propulsion being by electric motors (indeed she was one of the first ships so powered, the US Navy favouring the system since it avoided vulnerable long propeller shafts). The arrangement was fortunate, since all *Jupiter*'s machinery spaces, including the boiler rooms, were aft, and this facilitated the conversion. She was envisaged as a flush-deck carrier, similar to *Argus*, and her after stokeholds would make the provision of uptakes near the stern — to clear her flying deck of smoke — a relatively simple operation.

The most compelling reason for *Jupiter*'s selection, apart from the fact that she was available, was the low cost. Commander Whiting wrote: "We thought she could be converted cheaply . . . . that was a mistake, however. In any event, she would have cost less when completely converted than any other ship we might have selected. We thought she could be converted quickly — that was another mistake. The war [was] over and labor, contractors and material [were scarce]."[9]

In spite of a less than ideal hull and other difficulties, the Aviation Branch finally succeeded in persuading the US Navy to complete the conversion. The Naval Appropriations Act for 1920 duly authorised the conversion of the collier, USS *Jupiter* (AC3), into the US Navy's first carrier, USS *Langley* (CV-1). The fiscal year 1920 also provided for the conversion of USS *Wright* (AV1) from a former airship tender into a seaplane tender. It also permitted the construction of the ill-fated airship *Shenandoah (2R-1)* and the purchase, from Britain, of *R38 (US Navy 2R-2)*. In fact *R38* was never delivered, since she broke up during her trials in England without ever being handed over to the US Navy. The same 1920 Act, having granted the above, then limited the total number of naval aircraft bases around the coasts of the United States to six.

In March 1920 *Jupiter* made her last voyage as a collier to the Navy Yard at Norfolk, Virginia, where her conversion was to take place; she was promised to be ready for commissioning as a carrier in January 1921. She was not; the work taking until 1922. In the Navy Yard all the complex coal conveyors were removed, together with all deckhouses and the funnel. A 534 feet long, 64 feet wide, wooden flight deck was built with a hangar below. A lift was installed to convey aircraft from the hangar to the flight deck, together with cranes to hoist seaplanes in and out of the water. The ship had spaces for machine shops, airframe maintenance and spares, including engines. Her accommodation was described as "A bit crowded, but sufficient for the work to be undertaken".[10] Two funnels were fitted to the port side and hinged, so that when flying was in progress they were horizontal.

USS *Langley* bore a superficial resemblance to HMS *Argus* and that ship may well have influenced her design; she had the same unobstructed deck, but the hangar beneath the wooden flight deck was not, as on *Argus*, enclosed. Good ventilation was considered essential, in view of the petrol vapour inseparable from aircraft operations. Because of her slow speed and comparitively short flight deck, it was obvious from the outset that some form of arrester gear would be required if the ship was to operate her aircraft in light airs.

While the major conversion work was proceeding at the Navy Yard in Norfolk, across the channel at the Naval Air Station, Hampton Roads, a strange structure was to be seen; it looked, according to a contemporary report:

> . . . . like the business end of an enormous banjo. One hundred feet in diameter, it was set into a deep circular pit so that it could be rotated flush with the surrounding terrain. The strings were really three-quarter inch wires rigged in parallel . . . . across the banjo. A series of fiddle bridges — wooden boards standing on end — held the wires above the surface. A few cross-deck lines were stretched across the longitudinal ones. [11]

This extraordinary structure was the US Navy's first dummy carrier deck (if one excepts Eugene Ely's tests at Tanforan), built specifically for trials of arrester gear to be installed aboard *Langley*. At 1500 hours on 11 August 1921 the large turntable was turned into wind by a tractor and a US Navy Aeromarine biplane (No. 584), piloted by Lt A.M. Pride, taxied over the dummy deck to begin the trials.

Five hooks were fitted to the axle beam of the Aeromarine to engage the longitudinal wires, to prevent aircraft from being blown over the side of the carrier; a tail hook, which could be raised or lowered by the pilot, engaged the cross wires, dragging sandbags as in Ely's landing on *Pennsylvania* ten years earlier. Admiral Pride recalled in 1976: "I felt Ely's concept was essentially correct . . . . we had to modernize the concept . . . . This indicated the early weight system". [12]

The sandbags were discarded and on the dummy deck at Hampton Roads two 30 feet high towers were built; weights in the form of 11 inch shells were connected by ropes and pulleys to the horizontal athwartship wires, which were held 15 inches above the 'deck'. As the aircraft crossed the wires, the hook beneath the machine engaged a wire and caused the weights to rise in the tower, bringing the aircraft to a progressive halt.

The position of the hook was subject to a good deal of experiment; tail hooks were tried and eventually a position roughly two-thirds down the fuselage was initially chosen. The fore and aft wires were retained for some time (possibly a legacy of Dunning's death on HMS *Furious*), though test pilot Pride had reservations about them:

The British had employed such wires in their experiments and our authorities thought it was a good idea also, to ensure that the plane would be restrained to the deck . . . .

Unfortunately, if the plane bounced at all, it would [lift] one or two wires up and come down with them crossed in front of the axle hooks. This caused a heavy drag on the axle [and] imposed the noseover moment which broke many a propeller.[13]

Lt Pride took the view that the fore and aft wires were not required and that, with the carrier steaming directly into wind, there would be no tendency for aircraft to go over the side. However, those in authority were, in Pride's words, "almost obsessed" with the need for longitudinal wires, and they were duly fitted to *Langley* as a supplement to the transverse arrester wires which, as initially installed, were connected to sheaves leading below deck. The ends of the arrester cables were joined to weights made from boiler plates with holes drilled through them. As the landing aircraft's hook pulled on the wires, the lowest plate rose, collecting additional plates and increasing the total weight until the aircraft came to a standstill.

USS *Langley* was commissioned at Norfolk, Virginia, on 20 March 1922: the US Navy's first aircraft carrier, CV-1. The designation 'CV-1' — 'Cruiser, Heavier-than-air No. 1' — indicated that *Langley* was regarded as a combatant unit, not a fleet auxiliary.

USS *Langley* worked up as a ship, and on 17 October 1922 it fell to Lt V.C. Griffin to make the first take-off, which he did flying a Vought VE-7SF, single-seater fighter, whilst the ship was at anchor in the York river. Nine days later *Langley* was under way off Cape Henry when Lt Commander Geoffrey Chevalier (Naval Pilot No. 7) landed an Aeromarine 39B on the flight deck: the first landing by an American on an American ship since Ely touched down on the deck of USS *Pennsylvania* eleven years previously.

Langley, nicknamed 'Covered Wagon', began three years of intensive experimentation into naval flying and the role of the carrier in the fleet. For the US Navy the early twenties were traumatic years. There were discussions as to whether the United States needed any navy at all; the question of an independent air force was mooted, and a series of tests began in July 1921 to prove both points, when US Army Martin NBS-1 bombers 'attacked' several surrendered German battleships and U-boats. In these tests, conducted by the Army with the active support of the Navy, the battleship *Ostfriesland* and a submarine, *U-117*, were sunk while anchored in Chesapeake Bay. Some days later, the German destroyer *G-102* and the cruiser *Frankfurt* sank under seventy-four bombs from Army and Navy aircraft. The results of the tests were to be widely misunderstood, as one official Navy account testifies:

> The Navy had originally planned the tests to provide detailed technical and tactical data on the effectiveness of aerial bombing against ships and the value of compartmentation in enabling ships to survive bomb damage: the Army participated for the purpose of portraying the superiority of airpower over seapower. The divergence in purposes and resulting differences in operational plans were not reconciled and, in consequence, the Navy's purpose not realised. The significance of the tests was hotly debated and became a bone of contention between a generation of Army and Navy air officers. The one firm conclusion that could be drawn was that aircraft, in unopposed attack, could sink capital ships. [14]

Further tests were carried out in 1923 when two obsolete battleships, the USS *New Jersey* and *Virginia*, were sunk by Army bombers. The result of the bombings was used by the main proponent of airpower as a means of national defence, General 'Billy' Mitchell, as conclusive proof that conventional naval forces were now at the mercy of aircraft and thus redundant and that airpower, and airpower alone, was all that mattered. Mitchell was indefatigable in using the press, or any suitable medium, to get his extreme views to a large public. In 1925, for example, he wrote in his book *Winged Defence*: "In the future, campaigns across the seas will be carried on from land base to land base under the protection of aircraft."

Such views might well have been sympathetically received in the Air Ministry, London; they most emphatically were rejected by the US Navy, which mustered support in Congress; and the 'Mitchell' faction, which was in reality a bid for an American Independent Air Force and a 'Unity of the Air' policy, failed. Disappointed, Mitchell continued a vigorous and increasingly reckless public campaign, overreaching himself eventually when the US Navy airship *Shenandoah* crashed with heavy loss of life in 1925; he then made a statement to the national press in which he accused the Army and Naval High Command of "incompetency, criminal negligence, and almost treasonable administration of the National Defence". It was too much: Mitchell was court-martialled and left the Army Air Service. (He died in early 1936.)

Mitchell's tragedy was, of course, that he was right in the sense that modern aircraft could sink battleships; his prophecies (and indeed those of Glenn Curtiss made fifteen years earlier) were to become a reality during World War II; but his contention that airpower should be in the hands only of an independent air force is less secure. It is of interest, however, that of the four major maritime fighting powers in World War I,

Britain and Germany both had an independent air force; Germany had no operational Fleet Air Arm whatsoever and Britain's was, throughout the war, constantly undermanned, particularly in the skilled aircraft ratings department, and short of specialised home-produced aircraft. Both the United States and Japan, on the other hand, neither of which had an independent air force, had developed highly effective naval wings with aircraft designed and built specifically for carrier operations.

It cannot be said with any certainty that the Mitchell affair had any direct influence on US naval thinking; battleships continued to be commissioned, but from roughly the same time as the controversial 1921 bombing tests, the US Navy began increasingly to take naval aviation seriously. The Bureau of Aeronautics was formed in September 1921 under its first head, Rear Admiral W.A. Moffett, solely to look after naval air interests — an office roughly equivalent to the British Fifth Sea Lord, which post had been abolished on the formation of the RAF in 1918.

The bombing trails and the subsequent statements by Mitchell apart, another event took place in early 1922 which was profoundly to affect international maritime aviation: this was the Washington Treaty, signed in that city on 6 February.

The treaty was an attempt to limit the post-war naval armament of the signatories: Britain, France, Italy, Japan and the United States (Germany being excluded under the provisions of the Treaty of Versailles). The three major maritime powers — Britain, United States and Japan — agreed on a tonnage ratio of 5 — 5 — 3 for capital ships, with a maximum displacement of 23,000 tons allowed for newly constructed aircraft carriers. The actual figures for total aircraft carrier tonnages were: Britain and United States 135,000 tons; Japan 81,000 tons; France and Italy 54,000 tons. There was, however, a clause in the treaty that battleships that were under construction at the time of signature, and which would otherwise have been scrapped under the provisions, could be converted into aircraft carriers of not more than 33,000 tons displacement each.

When the treaty was signed, the US Navy had on the stocks two heavy battle-cruisers, USS *Saratoga* and *Lexington*, 33,500 tons. To have completed these ships — roughly equivalent to the Royal Navy's HMS *Hood* — would have contravened the battleship tonnage provisions of the agreement; however, the clause permitting incomplete hulls to be converted to 33,000 ton aircraft carriers was invoked by the US Navy and on 1 July 1922 Congress duly authorised the completion of the two unfinished cruisers as carriers. The published tonnage of the two carriers was 33,000 tons; the true figure, however, was nearer to 36,000 tons. There was a codicil to the treaty which permitted an additional 3,000 tons of armour against air bombing which could have justified the actual displacement. The US Navy may have been uncertain of the strict legality of the excess, judging by the fact that the official tonnage of the *Saratoga* class was always quoted as the lower treaty maximum figure of 30,000 tons. USS *Langley*, the only existing US Navy carrier, was exempted from the provisions of the Washington Treaty as she was considered to be experimental.

The treaty limits were more or less stuck to by the signatories, though as the British naval historian David Brown has pointed out: "These [Washington Treaty] restrictions led to some remarkable designs which were intended to beat the Treaties, particularly on the part of the Japanese, who contrived to produce an 8,000 ton 590ft carrier capable of stowing 48 aircraft!"[15] If the Americans bent the rules slightly, the Japanese were to abrogate the treaty by the end of 1936. Britain stuck to it, and a later Anglo-German naval agreement, until 1939, with the result that she began World War II with carriers constructed under treaty limits.

In point of fact, numbers apart, other considerations forced naval architects to limit the size of carriers and other warships. The limits imposed by slipways and docking, particularly drydocks, governed the maximum size of British warships to a length of approximately 800 feet, and a beam of around 95 feet, though the Royal Navy had the use of drydocks in Gibraltar, Singapore and Sydney, Australia, permitting world-wide naval operations without undue restriction.

Slipways and docks were not the only limiting factor; the United States, being a 'Two Ocean Navy', had to consider the possibility of transferring warships from the Pacific to the Atlantic and vice versa; thus the locks of the Panama Canal limited her ships to a maximum beam of 108 feet, which in turn limited the practical maximum waterline length to around 850 feet.

The Japanese Navy was mainly interested in operations in the Pacific and, as the United States emerged as their potential enemy, clearly considerations of passing through the Panama Canal did not arise. Japanese docks in the the Home Islands were in fact quite large, but the first aircraft carrier commissioned into the Imperial Navy was modest, with a standard displacement of only 7,470 tons. The carrier was the *Hosho*, a name variously translated as 'Flying Phoenix' or 'Flying Dragon'; 550 feet long with a 60 foot beam, she embarked (in 1923) twenty-one aircraft and had a maximum speed of 25 knots. Though some authorities maintain that *Hosho* was the first carrier built up from the keel as such, this is not so; she was originally laid down in 1919 as a fleet oiler. Work on her was suspended in 1920 while the Japanese Navy, which was then expanding its naval air arm, considered the question of completing the ship as a carrier.

To assist the Navy, the Japanese turned to their late ally, Britain, which in 1920 was the only naval power in the world with operational aircraft carrier experience. A formal request was made to HM Government for the sending to Japan of an Aviation Mission. The British regretted that serving officers of the Royal Air Force could not be spared; however, some ex-RNAS officers, led by a fox-hunting man and former naval pilot, Colonel, The Master of Semphill (later Lord Semphill), formed a semi-official mission and arrived in Japan in 1921 to assist in "The re-organisation, equipment and training of the Imperial Japanese Naval Air Service". This turning to foreigners as air advisers was not without precedent; as early as 1912 Japanese naval officers had been trained in Britain, the US and France. These men had returned to Japan with at least four aircraft: two French Farmans and two Curtiss seaplanes.

Training began at Oppama on the west coast of Tokyo Bay, and among the first six trainees was a young officer, Chikuhe Nakajima, who was to found the aircraft and aero-engine company of that name. Additional British aircraft were bought prior to World War I, including Sopwith and Short seaplanes, some of which, operating from the seaplane tender *Wakamiya*, took part in the capture of the German colony at Tsingtau in September 1914.

In the immediate post-war years, the Japanese Navy bought a number of Sopwith Pups, one of which flew from a temporary ramp on the forecastle of the *Wakamiya*; another took off from a platform mounted on a gun turret of the battleship *Yamashiro*.

The Semphill mission was presumably instrumental in persuading the Japanese Navy to order from Britain some 200 aircraft, together with spares and ancillary equipment. The types delivered were mainly contemporary British naval machines: Sopwith Cuckoo torpedo planes, Parnell Panther spotters and Gloster Sparrowhawk fighters. These last aircraft were unique to the Japanese Navy; the type never flew with the RAF.

The British and other foreign aircraft were only interim; Japanese aircraft constructors, notably Mitsubishi, recruited European designers — men like Herbert Smith of the Sopwith Company, who helped to produce the Mitsubishi Type 10, one of the first indigenous Japanese naval fighters.

Semphill, during his stay in Japan, became an open admirer of the young naval pilots he was helping to train and on his return to England he reported: "The general ability in [Japanese] pilots is distinctly high, possibly higher than we are accustomed to find in this country .... The men are splendid, keen and hard working .... and will often forego leave without being asked in order to finish some work on hand."[16] That prescient observation seems to have been ignored in both Britain and the United States during the years prior to Pearl Harbor ....

The carrier *Hosho* was completed at the Yokosuka Navy Yard on 27 December 1922. At that time she had a small starboard island containing little more than the charthouse and navigation bridge, surmounted by a tripod signal mast; aft of the island there were three short funnels which could be swung horizontal during flying operations. The first flight trials aboard *Hosho* were conducted on 22 February 1922, when a Mitsubishi Type 10, flown by that firm's test pilot — an Englishman named Jordan — landed on. He was followed by another Type 10, piloted by Lt Shun-ichi Kara. Commander Blackley, of the Semphill mission, landed a four-seater Vickers Viking amphibian safely aboard, which must have been a considerable feat, since the aircraft's span of 46 feet and the width of *Hosho*'s flight deck of 60 feet, could not have left much margin to pass the island. The narrowing of the deck abeam of the island proved most unpopular with the pilots; so much so that the island was removed later in 1923, leaving *Hosho* flush-decked like her contemporaries HMS *Argus* and USS *Langley*. The original (1923) armament of *Hosho* was four 14cm guns and two 7.6cm AA guns, plus 7.7mm machine guns.

The first programme of Japanese naval expansion, which began in 1920, was more or less complete with the commissioning of the first carrier; however, Japan in common with other signatories of the Washington Treaty had in her navy yards two incomplete battle-cruiser hulls that would, if proceeded with, contravene the conditions of the Treaty: these ships were the *Amagi* and the *Akagi*. It was decided, as was to be the case with the *Lexington* and *Saratoga*, to convert them as aircraft carriers, as permitted under the international naval agreement. In the event, the *Amagi* was badly damaged while on the slipway at Yokosuka in the disastrous 1923 earthquake and a third redundant battleship hull, the *Kaga*, which was to have been scrapped, was substituted.

Akagi — the name means Red Castle and is actually a Japanese mountain — was launched on 22 April 1925 from the slipway at Kure

Navy Yard and completed by March 1927. *Akagi* was at that time the seventh aircraft carrier in the world and the largest then in service, her declared tonnage being 26,900, just inside the legal limit; in fact the actual displacement was over 30,000 tons. In her 1927 form *Akagi* was a flush-deck carrier with no island, the flight deck extending 632 feet over two large hangars that could accommodate some sixty contemporary aircraft. The uptakes from her nineteen boilers led to two curious funnels arranged horizontally on the starboard side. *Akagi* was a fast ship with a maximum speed of 31 knots, and carried the heaviest armament allowed to an aircraft carrier under the treaty: six 8in (20.32cm) guns, plus twelve 4.72in (120mm) anti-aircraft guns.

Kaga — the name of an old Japanese province — was completed as a carrier on 31 March 1928, though she did not enter operational service until 1930, by which time the original signatories of the Washington Naval Treaty had reconvened in London and a new treaty was signed on 22 April 1930. It retained the general provisions of the original and provided for further reductions in international naval power. The specific proposals affecting naval aviation were the broadening of the strict definition of what constituted an aircraft carrier, and now included ships of any tonnage, the primary role of which was the operation of aircraft, including seaplanes and flying boats. The fitting of 'flying off' or 'landing on' platforms on warships' gun turrets was banned from 1 April 1930.

The Japanese Government of that time was liberal and generally opposed to any large-scale rearmament. However, the new militant faction, eventually to become the Government, was hostile to the provisions of the new London Naval Treaty, naval officers being particularly bitter about the position of implied inferiority, vis-à-vis Britain and the United States, which the 5—5—3 tonnage agreements, signed in Washington and reaffirmed in London, imposed.

As a matter of record, when the London Treaty was signed in April 1930, there were thirteen aircraft carriers in commission with four navies, world-wide. This table lists them in order of commissioning.

Of these thirteen carriers, only one — HMS *Hermes* — had been designed and built from keel to truck as a carrier. Of the others, ten were built on battle-cruiser hulls; one (*Hosho*) was converted from an oiler, another (*Langley*) from a collier, and a third (*Argus*) from a passenger liner hull. All thirteen were to survive to serve (some briefly) during World War II. In the late twenties and early thirties it was from the decks of these carriers that a new concept of naval power emerged which was, during World War II, to enable the aircraft carrier to supplant the

battleship as the capital ship of the major navies, which really came to
mean those of Britain, United States and Japan.

Name	Country	Standard displacement (tons)	Speed (knots)	Flight deck (feet)	Date of commissioning
Argus	Britain	14,000	21	470 x 85	16 Sept. 1918
Langley	USA	11,500	14	534 x 64	20 March 1922
Hosho	Japan	7,470	25	519 x 75	27 Dec. 1922
Hermes	Britain	10,850	26	570 x 90	18 Feb. 1924
Eagle	Britain	22,600	24	652 x 96	24 Feb. 1924
Furious	Britain	27,000	30	570 x 91	Aug. 1925
Akagi	Japan	36,500*	31.5	817 x 100	25 March 1927
Bearn	France	22,146	21	590 x 88	May 1927
Saratoga	USA	36,000*	33	880 x 90	16 Nov. 1927
Lexington	USA	36,000*	33	880 x 90	14 Dec. 1927
Kaga	Japan	29,660	27.5	815 x 100†	31 March 1928††
Courageous	Britain	22,500	30	576 x 91	5 May 1928
Glorious	Britain	22,500	30	576 x 91	10 March 1930

* Actual tonnage — declared tonnage in each case was within Treaty limits, i.e. under 30,000 tons.
† After 1936 rebuild.
†† Date of completion: was not commissioned until 1930.

Although Britain, in 1930, had the largest number of aircraft carriers
in service, with five (*Argus* was in reserve and de-commissioned), she
had in reality fallen behind the United States and Japan in the number of
aircraft that could be operated with the fleet. British carriers, for several
reasons, though mainly because of the design of the small hangars, could
only operate a limited number of aircraft when compared with United
States and Japanese carriers of a similar size. For example, HMS *Eagle*,
displacing 22,600 tons, had a complement of eighteen machines; *Hosho*,
with a displacement of 7,470 tons — a third of *Eagle*'s — could nevertheless operate twenty-one aircraft, the same number as HMS *Hermes*,
10,850 tons. The later British fleet carriers, *Courageous* and *Glorious*,
were supposed to be able to operate fifty-two aircraft each, but an Air
Ministry note which was written opposing any increase in the number of
Fleet Air Arm aircraft, stated: "It has not yet been shown that the larger
carriers [*Courageous* and *Glorious*] can under any conditions effectively
operate their full nominal complement of 52 aircraft each. The maximum
number which has ever been . . . . operated up to the present time [1930]
is 36."[17]

The total number of front-line combat aircraft on charge to the Fleet
Air Arm in 1930 was only 141 machines, all of which could easily have
been embarked on the two US Navy carriers *Lexington* and *Saratoga*,
which operated ninety aircraft each. However, other considerations

apart, since the Royal Navy had a world-wide role in protecting the Empire trade routes and British possessions overseas, it was necessary, in order to provide a carrier for each major fleet, that the tonnage permitted under the treaties be divided among a large number of small ships. The disposition of RN carriers in 1930 was as follows:

HMS *Courageous*: Home Fleet.   9 Flights:*   51 aircraft (nominal figure)

HMS *Glorious*:   Mediterranean Fleet   7 Flights: 40 aircraft (nominal figure)

HMS *Eagle*:   Mediterranean Fleet   3 Flights: 18 aircraft (nominal figure)

HMS *Hermes*:   Far East (China Station)   2½ Flights: 15 aircraft (nominal figure)

* 6 aircraft constituted a Flight.

HMS *Furious* was refitting and *Argus*, as noted, was in reserve. A further fifteen aircraft were embarked in warships equipped with catapults. The total aircraft embarked in June 1930 were 139.

The Admiralty, contemplating additional carriers, complained that "the present strength of the Fleet Air Arm is insufficient for adequate training",[18] and wanted its strength to be increased to a total of 213 by 1935. The Air Ministry, which 'owned' the aircraft, seem to have been incredibly complacent, arguing that the 230 aircraft on RAF bases overseas could meet any threat. They maintained "No country would hazard its Fleet within range of any coast from which it can be easily attacked by .... shore based air forces."[19] The writer seems to have overlooked the Tondern raid, made soon after the RNAS became the RAF, as well as the earlier Cuxhaven operation. The Air Ministry appears to have concluded that in any naval engagement out of range of shore-based aircraft, the Royal Navy was well enough served by the existing strength of the Fleet Air Arm, the 1930 writer arguing:

Shore based aircraft are a potentiality as a reserve for the shipborne aircraft in the sense that pilots, and to some extent aircraft, of shore based units can be drawn on.

[It is] difficult to understand seriously the First Sea Lord's asseverations as "To the shortage of aircraft that exists at the present time".

Japan has two large aircraft carriers each with a complement of 50 aircraft (*Akagi* and *Kaga*) and one small carrier *Hosho* with a maximum complement of 14 aircraft. These considerations suggest that the British Fleet Air Arm at its present strength would have a very substantial superiority over the Japanese Fleet Air Arm.[20]

The above appreciation puts the number of aircraft embarked aboard the two large Japanese carriers at fifty; in fact the number was sixty and *Hosho* embarked twenty-one aircraft, not fourteen. Elsewhere the 1930 paper gives the number of US naval aircraft as 227 front-line machines: the actual figures (published in 1970)[21] put the number on hand, as of 1 July 1930, at 734 combat aircraft, out of a total of 1,081 for the US Navy and Marine Corps. (The figures include reserves.)

When reading the archive material of 1930, one comes to the inescapable conclusion that the British Air Ministry, in pursuing the godhead of 'Unity of the Air', did not really believe that there was any need for a Fleet Air Arm at all and tended to view its formation in April 1924 as unfortunate and part of the 'bride price' to be paid for an independent air force. The Air Marshals took the view that RAF Coastal Command could effectively deal with any maritime air incursion around the British Isles. Elsewhere in the Empire, distance alone precluded any possibility of attack from shore-based aircraft of a major power. Since an attack on British possessions by America was, to say the least, unlikely, that left Japan as the sole possible overseas enemy, and the attitude to the Japanese Navy and its air arm was clearly emerging, both in Britain and the US, as one of amused contempt.

The bitter struggle between the Royal Navy and the Air Ministry for control of the Fleet Air Arm was to dog development during the crucial formative years. The RAF fought against the Navy gaining control of the Fleet Air Arm because, as the Air Ministry put it, it would be "the thin end of the wedge"[22] against the continued existence of the independent air force. In addition to the uncompromising hostility of the RAF, it is an unfortunate fact that the Royal Navy itself was to some extent a divided house on the question of naval aviation: few officers of flag rank had any experience or interest in aircraft, which were not considered 'Navy'. The battleship with 15 inch guns was still the major piece on the board; all else was subservient. That is not to say that the Royal Navy did not have in its ranks dedicated men who were slowly to develop the Fleet Air Arm, often at the cost of their personal careers.

The role of aircraft carriers in the inter-war period was ill defined; the RAF considered them, at most, to be useful as offshore airfields that could extend the range of coastal airspace. Overseas, on the China Station, carrier-based aircraft were used to control the activities of pirates, who still flourished in the Far East. Carriers were also a convenient method of transporting troops and aircraft to distant commands. The carrier was seen by the Royal Navy as part of a Battle Fleet; the aircraft embarked would have three primary tasks. The fighters would fly combat air patrols (CAP) within visual range of the carrier to protect it from air attack; two-seater reconnaissance aircraft would spot for the battleships; and torpedo planes would attack enemy units with the object of slowing them down, when they could be engaged by the guns of the battleships.

The attitude of the US Navy at that time was little different, as Rear Admiral Gallery, an ex-pilot of USS *Langley*, recalled:

> The battleship was still the backbone of the fleet. Most naval officers visualized the final battle of the coming war with Japan as a classic fleet action between main bodies, like Jutland. The two battle lines on parallel courses would slug it out at about 20,000 yards with light forces of the van and rear harassing the main battle lines. Nobody was quite sure what part aircraft carriers would play in this fleet action. Most strategists agreed that they would be kept 50 to 100 miles on the disengaged flank and might be useful in finishing off cripples with bombs![23]

The basic difference between the two English-speaking navies was that the Americans were at least in direct control of their Air Arm; they had opposition only from within their own ranks, though such opposition was real enough. The US Navy was fortunate in having as Head of the Bureau of Aeronautics Rear Admiral W.A. Moffett, a man of insight who did much to further the claims of naval aviation. When the bureau was formed in 1921, it was decreed that at least 70 per cent of its officers should be qualified pilots or observers; Rear Admiral Moffett duly posted himself to the first Naval Aviation Observers' course, which began on 17 June 1922 at the Naval Air Station, Pensacola. Later many captains were required to complete an Air Observer course: that was only a beginning. By April 1926 the Secretary of the Navy had directed that, starting with the class of '26, *all* graduates of the Naval Academy be given a course of at least twenty-five hours of flight instruction during their first year of sea duty.

Considering the late start in carrier operations, the US Navy made extremely rapid progress, and was to introduce not only several technical innovations but new tactical concepts as well. USS *Langley*, after two years as an experimental carrier, joined the Battle Fleet with VF-2 (Fighting Two), the first squadron trained to operate from a carrier, embarked. VF-2 was equipped with Vought VE-7SF single-seater biplane fighters, fitted with flotation gear, and began to work up with the carrier off San Diego during January 1925. In March of that year *Langley* took part in Fleet Problem V (War Games): the carrier's aircraft participation was limited to scouting for the Black Fleet's advance to Guadeloupe Island. The performance of the squadron impressed the Commander-in-Chief, Admiral Coontz, sufficiently for him to recommend the speeding up of the completion of *Saratoga* and *Lexington*. Before those ships were in commission, however, two major innovations were developed by the pilots of the *Langley*, which were eventually to become standard practice in the navies of the US and Britain: the first was the LSO — Landing Signals Officer (Batsman in the Royal Navy); the second development was the safety barrier.

From the outset the *Langley* had both longitudinal and transverse arrester wires (British carriers had discarded the former and would not have the latter until the mid-thirties, relying on their high speed to enable the aircraft to stop). The story of the introduction of the LSOs is interesting because it was accidental. (There have been many accounts as to how Landing Signals Officers evolved: the following version has a better provenance than most.) When the *Langley* began operating her aircraft, the only communication between ship and 'plane was a large flag flown from the after end of the flight deck; a red flag meant 'do not land'; a white, 'land'. One of *Langley*'s original pilots, who retired as a Rear Admiral, J.R. Tate, wrote of the landing techniques of the day:

> We developed no standard approach, but each pilot tried to accomplish a three point landing just short of No. 1 wire in the center of the deck. The major operations were still with Aeromarine 39B training planes but we were starting [in 1923] operations with service types; UO-1, the VE-7 and the TS-2. Most of the landings were still made at anchor with the planes based ashore at NAS [Pensacola].
>
> *Langley*'s regular anchorage lay in the center of a circle of buoyed anchors. Standard practice, on the morning watch, was to hook the stern winch to the appropriate anchor and haul the ship into wind. During the landing, Commander Kenneth Whiting, the exec. and the senior aviator, always watched from the stern of the flight deck, [standing] on the port side. He developed a type of back seat driving and would talk out loud and make appropriate motions. "He's too low, now he's too high, now he's OK." Lt. Pride [the pilot who did the flight tests of the arrester gear] who was just about the hottest pilot on board and whose flying was precision plus, had developed a dragging nose high approach with power on, which was eventually to become the standard approach.[24]

After one such landing, Pride walked across the flight deck and said to Commander Whiting: "'You were wrong on my being too low on the approach. I was over the stern by ten feet and caught the No. 2 wire.' Whiting looked at him in surprise and asked 'Did you see me?' Pride laughed, 'We all watch you waving, too high, too low, OK'."[25]

A meeting of all the *Langley* pilots was called and it was decided to station an experienced pilot on the port side of the after end of the flight deck with a pair of semaphore flags to signal each landing. At first Whiting himself gave the instructions to the flag man: "Too high, too low, OK". The flag-waver signalled the appropriate message: arms up meant "You are too high", arms down "too low", horizontal "OK". Later the "Cut engine" signal was added. The signals were advisory; later when the *Saratoga* and *Lexington* came into service other signals were added and the 'cut' signal became a mandatory order. By that time it was suggested that the signals be reversed: arms up to mean "You are too low — go higher", but by then too many pilots had been trained to react to the original system so it was retained until mirror and light signalling apparatus superseded the LSO after World War II. When, about 1936, the Royal Navy adopted the 'batsman', they did reverse the signals, which became orders rather than advice. This caused considerable difficulties during the war with exchange postings and on other occasions when pilots from one Allied Navy landed aboard the carriers of the other. The Japanese incidently never used LSOs, but started a light signalling system with the *Hosho*.

The second innovation, which was also to influence carrier operations for decades, was, like the LSO, an accidental discovery aboard *Langley*; in this case it was an actual accident. One day in 1926 a Lt Gurley managed to miss all the arrester wires and collided with aircraft which had already landed and had been parked on the forward end of the flight deck. Most of the twelve aircraft of VF-2 were damaged. Captain E.S. Jackson, then *Langley*'s skipper, ordered the ship's bosun to rig some 10 inch hawsers across the deck on sawhorses to protect the parked aircraft should any further machines miss the arresters or pull out a hook. Thus the safety barrier was born.

The importance of these two discoveries is impossible to overstate; the LSO and the safety barrier enabled carriers to operate far more efficiently. For example, before the barrier became standard, it was usual, certainly on British carriers, which for many years had no arrester system whatever, for the aircraft to be struck down into the hangar as soon as it landed. The machine had to be manhandled on to the lift, wings folded (if it had folding wings — by no means all those embarked aboard carriers had this facility); then with warning bells ringing, the lift would convey the aircraft below, where it was pushed by sailors off the platform and the lift would ascend to the flight deck. Then, and only then, would the next machine get the single signal "Land".

In practice, this limited the landing rate, even on the tightest of ships, to approximately one aircraft every two minutes, which was considered satisfactory. On 21 February 1923 tests were conducted aboard *Langley* to establish the minimum recovery interval; the best time for 'landing on' three Aeromarines was seven minutes. (Under ideal conditions, HMS *Courageous* set up something of a record in 1929 when six Fairey Flycatchers were recovered and stowed in 4 minutes 20 seconds.)

Given skilled pilots — and carrier pilots had to be that, for there is no more difficult art in flying than landing a high performance aircraft on a flight deck — the limiting factor in recovery was the time taken to strike the aircraft from the deck down to the hangar below. This was essential, for if a machine missed the arrester wires or broke one, the pilot had little alternative other than opening the throttle and 'going round again'. If, as presumably happened on *Langley*, aircraft were parked ahead of the arrester wires, they were a hostage to fortune, for any aircraft which failed to arrest could not take off again, but simply ploughed into the deck park, as indeed happened to Lt Gurley. The safety barrier prevented that. The temporary hawsers stretched across the deck, supported by sawhorses, were soon replaced by strong double steel cables which could be raised and lowered hydraulically. The standard technique aboard *Saratoga* and *Lexington* and subsequent carriers was that the barrier would be raised when an aircraft was approaching; as soon as it was safely arrested, the barrier would lower and the aircraft, freed from the arrester wire, would taxi under its own power to the deck park. The barrier would then be raised again to protect the park as the next machine 'landed on'. The barrier allowed a far faster rate of recovery than was previously possible, since aircraft did not need to be immediately struck down to the hangar and, with the advent of all-metal aircraft later in the thirties, a permanent deck park was possible, enabling a carrier to increase the number of aircraft on board.

The LSO's contribution was no less revolutionary in the sense that the two developments were complementary: the barrier (until the provision of the angled deck during the 1950s) precluded the possibility of 'going round again'; once a pilot was committed to his landing he had to land. Any serious misjudgement would end with the aircraft hitting the barrier. The role of the LSO became increasingly important with the adoption of the 'nose high, power on' approach; it was difficult — impossible in some aircraft — for the pilot to judge his height above the deck, particularly in a heavy seaway with the carrier heaving up and down. Only the LSO was best able to judge the exact moment to signal the pilot to 'cut', which became a mandatory order. Pilots now relied entirely on the LSO's signals and a high degree of trust had to exist between the carrier pilots and their LSOs; any pilot who was critical of the way he had been signalled down was usually invited to try the job himself: few complained.

Initially experienced pilots would be chosen for the Landing Signals Officers; they would train others on the master—apprentice system. Later, with the expansion due to the war, the US Navy began special training for LSOs and the job became a recognised designation within the US Navy Air Department. (Even today, when carriers use automatic, high precision radar methods (SPN-10) to recover aircraft, each US carrier still has an experienced 'flag-waver' aboard — just in case.)[26]

On 16 November 1927 the USS *Saratoga*, CV-3, the fifth ship of the US Navy to bear the name, was commissioned at Camden, New Jersey, with Captain H.E. Yarnell in command. Four weeks later, on 14 December, USS *Lexington*, CV-2, the fourth ship to bear the name, was commissioned at Quincy, Mass., commanded by Captain A.W. Marshall.

The two new carriers were then the largest in service; indeed their 880 feet flight decks were not only the longest in the world, they would remain so for the next eighteen years. These huge carriers could embark and operate ninety aircraft of the day, in a single range if required. Everything about them was big: fully loaded, they displaced 47,700 tons; they were 901 feet long overall, with 111 feet beam and a draught of 32 feet; sixteen boilers developed 18,000shp, making the two sister ships the most powerful warships in existence prior to 1939. The propulsion was turbo-electric and gave a maximum speed of 33½ knots. The armament (in 1927) was the maximum allowed under the provisions of the Naval Treaties: eight 8in (20.3cm) guns in four twin turrets and twelve 6in (127mm) anti-aircraft guns.

Some idea of the power generated by the machinery of the new carriers can be judged from the fact that, in January 1930, *Lexington* supplied the city of Tacoma, Washington, with electricity for thirty consecutive days when that town's power supply failed, owing to a prolonged drought affecting the hydro-electric generators, a total of 4,251,140 kilowatt hours being consumed by the town.

The cost of the two carriers was about $45 million each. They were at the time considered beautiful ships, but in today's eyes the single enormous starboard funnel gives a somewhat 'out of scale' appearance, making the ships seem top-heavy and smaller than they actually were.

USS *Saratoga* and *Lexington* soon worked up. On 5 January 1928 the first take-off and landing was made on '*Lady Lex*' by Lt Pride, flying a UO-1 while the ship was in the Fore River; four days later another UO-1 took off and landed back on '*Sister Sara*', piloted by that carrier's Air Officer, Commander Marc A. Mitscher, who during World War II, as a Vice Admiral, was to fly his flag in the second *Lexington* in the Battles of the Philippine Sea and Leyte Gulf. Those actions were the culmination of the application of carrier Task Force tactics and will be discussed later. Interestingly the germ of the idea — the use of aircraft carriers as a primary naval force and not the handmaiden of a battleship fleet — had its genesis in the 1929 Fleet Exercises, when *Saratoga* and *Lexington* were with opposing forces. *Saratoga* was with the 'Black Fleet', *Lexington* with the 'Blue'. The Blue Fleet was charged with the defence of the Panama Canal; the Black Fleet was to attack it. The operation as a whole was called Fleet Problem IX.

The exercise was to be held on the Pacific side of the Canal. What followed made US Naval history. The Black carrier, *Saratoga*, was part of a Battle Fleet (BatDiv Five) with the battleships *Colorado*, *West Virginia* and *California*, all armed with 16 inch guns. The Commander Air Battle Force, Rear Admiral J.M. Reeves, was flying his flag in *Saratoga*, having joined the ship from the US Navy War College, where he lectured and was considered an authority on the Battle of Jutland. Possibly because of his intimate knowledge of that fiasco, and probably because his Executive Officer in *Saratoga* was Commander Whiting, who had transferred from *Langley* when *Saratoga* was commissioned, Reeves, though not an aviator, but a brilliant tactician, had become fascinated by the possibilities offered by naval aviation. These included the concept that a large fast carrier like *Saratoga*, with an Air Group embarked, need not be tied to a battleship fleet simply as an ancillary, but could operate against an enemy fleet independently.

In the plan of Naval Exercise IX, these ideas were put to the C-in-C Black Fleet, Admiral W.V. Pratt, for approval; he agreed after some discussion that *Saratoga* could be detached from the Black Fleet BatDiv Five to operate independently, escorted only by the cruiser *Omaha* and two plane guard destroyers whose business was to rescue any ditched airmen. The exercise was to begin on 23 January and end on 26 January, 1929. As soon as manoeuvres began, *Saratoga* sailed away in a wide arc to the Galapagos Islands, far to the south of Problem IX's area and 1,000 miles from the Panama Canal. Owing to the fact that, at the time of the exercise, *Langley* was refitting, her eighteen fighters and six scouts were embarked on *Saratoga*, bringing the total number of operational aircraft on board to ninety-six, the crews of which, together with the rest of the ship's company, enjoyed swimming parties in the quiet anchorages off the Galapagos.

Far to the north, Problem IX was well under way; the Blue Fleet's scouting forces quickly discovered and engaged all the Black fleet units except *Saratoga* and her escorts. The Blue Fleet searched the whole exercise area, signalling to each other "Where is *Saratoga*?" In taking part in the intensive searching, *Lexington* was intercepted by heavy units of the Black Fleet; she came under 16 inch gunfire and was declared sunk by the umpires. Later, no doubt under pressure from above, since she was, in the inexplicable absence of *Saratoga*, the only carrier participating, the umpires amended their decision to "heavy damage" and reduced her speed to 18 knots.

Problem IX was due to end at 0800 on 26 January. At daybreak on 25 January Reeves ordered his small group to weigh and steam north at 30 knots, so as to arrive at a launching point 150 miles from Balboa at 0430 hours on 26 January, but before long, with a freshening wind and a steep sea, *Omaha* made: "Cannot maintain pace." The ship was brusquely told to drop out of formation and to proceed to Panama independently; this left *Saratoga* with just the two plane guard destroyers, which, with the carrier now worked up to 31 knots, were having a very rough passage indeed, though they kept station. Late in the afternoon, a Blue Fleet battleship, *Detroit*, was sighted to starboard; she challenged *Saratoga* with a signal lamp. Reeves ignored the challenge and signalled *Detroit* to position herself 8,000 yards astern, at 31 knots, adding "if you can maintain position, establish yourself as plane guard when we launch".

Reeves' gamble came off; *Detroit* had mistaken *Saratoga* for the 'friendly' *Lexington* and struggled to maintain position astern, as directed. After an hour and as darkness was falling, *Saratoga* signalled *Detroit*: "You have been under fire by both my aft 8 inch turrets for 60 minutes and are adjudged as sunk. *Saratoga*."

The umpires conceded, but judged several of the carrier's planes damaged due to gun blast. *Detroit* reacted, rather unsportingly, by opening up a high-powered radio transmitter, operating on the Blue Fleet net, and broadcast *Saratoga*'s position, course, speed and details of her escorts. Reeves at once protested to the umpires that *Detroit* was sunk and thus could hardly transmit, which in point of fact she continued to do all night, though hard pressed to keep up with the speeding carrier, which continued ploughing north at 31 knots; but now, thanks to *Detroit*, all the Blue Fleet knew where the carrier was, which by 0300 hours was near her designated point off Balboa, when *Detroit* made "*Saratoga* ranging aircraft on deck", and an hour later, "To Commander Blue Fleet: *Saratoga* launching planes".

It was still very dark and there was a note of incredulity about the message, but *Saratoga* turned into wind and launched eighty aircraft from a single range. They formed up and set course across the Gulf of Panama for the targets: the vital canal locks at Pedro Miguel and Miraflores; the Army Air Corps base at Albrook Field, which was charged with the defence of the Canal; and "any enemy battleships found on their way home".

The target area was reached at 0600 hours on 26 January, when Problem IX had only two hours to run. Reeves had told the Air Commanders he would maintain the carrier's course and speed so as to close the distance for the returning flight, adding "by the time we turn into wind to take you aboard the Problem will be over". *Saratoga*'s pilots, in spite of *Detroit*'s transmission, achieved complete and total surprise; as *Sara*'s aircraft roared over the Albrook Field, the Army's machines were lined up below with commendable precision, wingtip to wingtip (as they would be twelve years later at Pearl Harbor). The Army, belatedly realising it was being attacked by planes from *Saratoga*, hastily sent up planes, located a carrier and bombed it. It was adjudged by the umpires as undoubtedly sunk: unfortunately it turned out to be *Lexington*, which thus achieved the unique distinction of being sunk for the second time in the same exercise. *Saratoga*'s air group, which had been assigned the Canal's locks, bombed them without opposition; all the defensive AA guns had their covers on, with the gunners at breakfast. Several battleships, thankfully dropping anchor and thinking that Problem IX was over, were also attacked and 'sunk'.

Saratoga duly appeared over the horizon at 0700 hours and turned into wind, but as she was recovering her triumphant aircraft, she was attacked by the 14 inch guns of Blue Fleet's *Oklahoma* and *Arizona* and judged to be sunk. It did not matter: Reeves had proved his point and the concept that would develop into the Fast Carrier Task Force, which was to dominate the sea war against the Japanese Navy in the Pacific during World War II, was a reality.

The official history of US Naval Aviation sums up the consequences of Fleet Problem IX in a single sentence: "This demonstration made a profound impression on naval tacticians and in the 1930 manoeuvres a tactical unit, built around the aircraft carrier, appeared . . . . for the first time."[27] In fact the row that followed Reeves' unconventional participation in the war games was of considerable heat and proportion. Reeves was accused of resorting to trickery; he defended himself with a masterful exposition of the future role of carriers and thereby made a permanent enemy of every battleship admiral in the US Navy. His reward for his farsighted thinking was to be detached from *Saratoga*, beached, and appointed Inspector of Naval Material: just about the lowest post in the service for an admiral. Two years later, however, he became C-in-C US Fleet.

As for *Saratoga*, she was ordered, the very same day the exercise ended, to have a broad black vertical stripe painted on both sides of her funnel. Even the Army could not mistake her for *Lexington* after that.

The thirteen aircraft carriers in world-wide existence in 1930 represented the span of development of this newest form of warship; but the carriers were only a means to an end. What of the aircraft that were embarked?

As far as Britain was concerned, it must be said that, although the Royal Navy had 'invented' the aircraft carrier, the aircraft operated from British carriers in the decade after World War I were not exactly revolutionary, though it would be unfair to imply, as some writers have done, that all the Fleet Air Arm's aircraft were obsolete RAF types, dumped on the Royal Navy. The late wartime designs — the Pups, 1½ Strutters, Camels and Cuckoos — soon faded out, to be replaced by newer, though not radically different, types. The standard FAA fighter from 1923 to 1934 was a small, stubby biplane, the Fairey Flycatcher. Few aircraft have been more popular with their pilots than this diminutive, agile fighter.

The Flycatcher was designed to Specification 6/22, issued by the Air Ministry in 1922 and which called for a single-seater fighter, designed to be capable of operating from the decks of carriers as a landplane, or from water as a seaplane or amphibian. It was to be powered either by a 425hp Bristol Jupiter or a 400hp Armstrong Siddeley Jaguar radial engine, making the Flycatcher, like its RAF contemporaries, the Gloster Grebe and the Armstrong Whitworth Siskin, among the first post-war generation of front-line aircraft to be powered by radial and not rotary engines. Two designs were built to Spec. 6/22: the Parnell Plover and the Fairey Flycatcher. In competitive trials, the Plover and Flycatcher were to prove of similar performance, but, whereas the Plover was of wooden structure, the Fairey aircraft was of mixed wood and metal construction and immensely strong; indeed its strength was one of the two attributes which became legendary, the other being the noise it produced. Flycatchers could be dived vertically under full power to terminal velocity with impunity; as they were pulled out of their high speed dives the wooden propeller blades fluttered, producing the famed 'blue note' beloved of their pilots, which much impressed the crowds in the late twenties at the annual RAF Hendon Air Shows, where naval Flycatchers dive-bombed the field simultaneously from all points of the compass. Production Flycatchers were powered by one uncowled, fourteen-cylinder, 400hp Armstrong Siddeley Jaguar IV engine, whose multiple open stub exhaust stacks contributed their full quota to the aircraft's stirring orchestration.

Flycatchers were a delight to fly and well suited to their carrier role. An official report of the handling of the prototype after test flights aboard HMS *Argus* in 1923 stated: "The type appears to be highly satisfactory from a deck-landing point of view. Stability is good and is neither too light or too heavy on the controls. No dangerous peculiarities [are evident] and the majority of pilots will have no difficulty in flying the type . . . . The view from the pilot's seat is the best ever experienced for deck landing".[28]

Between 1923 and 1930, 196 Flycatchers (including three prototypes) were constructed; they first entered service with No. 402 Flight in September 1923 and went on to serve on every Royal Navy carrier until finally superseded in 1934 by the Hawker Nimrod. Although primarily fighters, armed with twin .303 Vickers guns, synchronised to fire through the propeller arc, Flycatchers could also operate as dive-bombers, carrying four 20lb bombs suspended from hardpoints beneath the lower mainplanes. For carrier operations, early Flycatchers had steel jaws fitted to the undercarriage spreader-bar to engage the longitudinal wires then in favour; when these were abandoned in 1926, Flycatchers, aided by a low approach speed (70mph), simply landed on without any form of arrest. By 1933, when British carriers had transverse arresters fitted, the Flycatcher had largely been superseded. Possibly because it never had a true arrester hook, the Flycatcher became the first Fleet Air Arm aircraft to be fitted with wheel brakes. Though the type was designed from the outset as a fleet fighter, it did not have folding wings — difficult to arrange with wirebraced biplanes. However, its short (29 feet) span could easily be accommodated on standard lifts and the Flycatcher could also be dismantled, so that no part exceeded 13 feet 16 inches in length, thus aiding the stowage of spare aircraft aboard ship.

No account of Flycatchers would be complete without reference to the famous 'slip flights'. The carriers *Furious*, *Courageous* and *Glorious* had a small hangar under the forward end of the flight deck, with a 60 foot take-off run over the ship's bows. Six Flycatchers could be stowed in the hangar, which opened directly on to the short fore deck. The 'slip flight', as it was called, would run up its engines to full power inside the hangar — the noise must have been memorable — then pop out of the hangar like corks from a bottle to drop over the bows of the carrier, when some of the bolder pilots, out of sight of the bridge, would perform a slow roll . . . .

Flycatchers were only ever flown by the Fleet Air Arm; their performance compared favourably with the contemporary RAF fighters, the Grebes and Siskins, and, although slightly slower (133mph maximum), they were far more rugged, enabling them to withstand the rigours of shipboard operations. The type was finally declared obsolete in April 1935, and sadly none is known to have survived.

There is an old adage among aircraft designers that 'what looks right is right': the Flycatcher, though perhaps quaint to modern eyes, nevertheless had a jaunty air about it and was somehow appropriate to its purpose. In contrast, the Fleet Air Arm's carrier-borne spotter-reconnaissance and torpedo aircraft, which were contemporary with the Flycatcher, could lay claim to being among the most unwieldy machines ever to take to the air. Chronologically the first of three ugly sisters was the Avro Bison, a three/four-seater biplane, 37 feet long and no less than 14 feet in height. The 46 foot, equal span, unstaggered wings could be folded manually to facilitate stowage in the carrier's hangars. The inordinate depth of the fuselage was to accommodate a navigator; this was considered essential in the days before the development of efficient radio-homing aids. The navigator, or observer as he was then called, was ensconced in a roomy enclosed cabin with two large windows, fitted with a chart table, sextant, compass and the other tools of the trade. Often a telegraphist with his heavy wireless set was also accommodated within the Bison's interior; abaft the cabin there was a gunner's position, roughly midway between the trailing edge of the wings and the tailplane. A single .303 Lewis gun was carried on a Scarff ring. Because of the depth of the fuselage, the pilot, who sat in an open cockpit just ahead of the top wing, had an eye-line of 45° to the propeller boss; indeed the Bison had a sight fitted on the fore and aft line of flight to assist the pilot in landing and taking off.

The fabric-covered wooden structure of the Avro Bison tipped the scales at 6,336lb, fully loaded. To heave this monstrosity around, one 450hp Napier Lion II, liquid-cooled engine was provided; that it managed to do so, at up to 110mph, remains an aerodynamic mystery.

Between 1921 and 1927 the Avro Company built about forty Bisons — some for the RAF; the carrier-borne version of the type, the Mark II, entered service with the FAA in 1923 and, until declared obsolete in 1929, they served at home and abroad on FAA shore stations and on most of the Royal Navy carriers of the day.

The Blackburn Aeroplane and Motor Company Limited of Leeds, Yorkshire, made the second of the trio, oddly named the Blackburn Blackburn: it was, if possible, even more unattractive than the Bison.

The general arrangement was similar to the earlier machine, including the large office for the observer/navigator and the telegraphist; the only noticeable difference was that the picture windows were replaced with rather appropriate portholes.

The type entered service with the Fleet Air Arm in 1923, with No. 422 Fleet Spotter Flight aboard HMS *Eagle*; subsequently Blackburns embarked on *Furious, Argus* and *Courageous.* Unlike the Bison, which briefly equipped No. 3 Squadron, Coastal Command, RAF, the Blackburn was never issued to an RAF unit; it served solely with the Fleet Air Arm, which makes its lack of wing-folding strange. Stranger was its very limited range — only 210 miles, hardly adequate for a reconnaissance aircraft, but an inevitable consequence of the very poor aerodynamics of the ungainly design. The performance, too, was limited to a speed of 100 mph, which was the maximum its 450hp Napier Lion V could confer. Blackburn Blackburns served with the Home and Mediterranean Fleets and the China Station; about fifty are thought to have been built, the type finally disappearing in March 1933. The last of those flying was probably the side-by-side, dual control trainer variant — the Blackburn Bull — which it is alleged took 10 minutes to reach 1,000 feet.

The final aircraft of the trio was not a reconnaissance machine but a torpedo carrier — the Blackburn Dart — and, although as unlovely as the others, was in fact a fairly satisfactory design. It replaced the Sopwith Cuckoo and served with the fleet from 1923 until 1933, but continued in a training role until 1935. The Dart had no internal cabin and carried only a pilot; it was powered by the same long-suffering Napier Lion and could carry an 18 inch naval torpedo at 110mph. It was considered easy to fly, and one Dart (N9804) made the first night landing on a British carrier — HMS *Furious* — on 1 July 1926.

It is, of course, easy now to scoff at these early aircraft; but they were produced at a time when money for armaments was very grudgingly granted by tight-fisted governments, imbued with pacifism and facing up to the economic reality of the years of depression. Though short in numbers and only able to support a tiny industry, the Navy, with each new aircraft acquired, added a little to the sum of its knowledge; new techniques were worked out, particularly in torpedo and dive-bombing attacks, and a new generation of carrier pilots and aircrews trained.

The US Navy was in much the same position, as far as finance went, as Britain's; though the number of naval aircraft on hand was larger and the quality of the machines was high. One manufacturer was to dominate the first decade of US Navy carrier flying: Boeing. Apart from a very small batch of Curtiss F6C-1 Hawks, the classic ship-borne fighters were the

(*Above*) USS *Saratoga*, CV-3, 47,700 tons. *Saratoga* and her sister ship, *Lexington*, could operate 90 aircraft each and were, for eighteen years after they had been commissioned (in 1927), the largest carriers in the world (*Boeing*); (*below*) the two large US Fleet carriers could, and did, remain at sea for long periods; thus all maintenance of the aircraft embarked had to be done on board. This picture, taken in a workshop in *Saratoga* around 1929, shows a Pratt and Whitney engine being prepared for installation in an F2B-1 (*US Navy*)

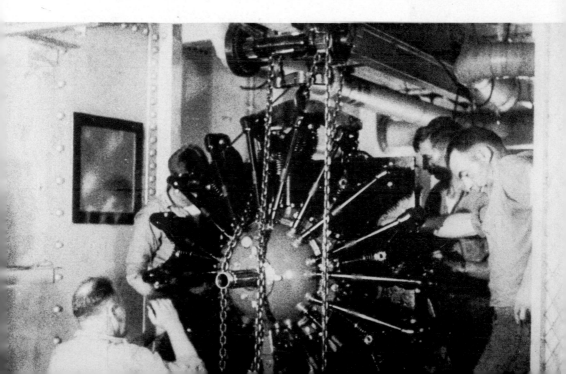

(*Above*) A Fairey Flycatcher on forward lift of HMS *Glorious*. Though Flycatchers did not have folding wings, the airframe could readily be dismantled so that no part exceeded 13ft 6in, enabling spare machines to be stowed, some slung from the hangar-roof girders (*Flight International*); (*below*) the best that can be said of the Avro Bison is that its intrepid pilot had an excellent view of the flight deck when landing his 46ft charge. N9848 was a Bison II and is shown here sometime in 1928, when serving as '21' of 423 Flight, embarked in HMS *Eagle* (*RAF Museum*)

(*Above*) In any short-list of all-time great military aircraft designs, the Fokker DVII must secure a high place. Not especially fast, the Fokker was nevertheless very manoeuvrable at high altitudes and was immensely strong. It was powered by a 180hp water-cooled Mercedes, giving a maximum speed of 124mph. After the Armistice twelve of the type were allocated to the US Navy and these influenced the design of the Boeing FB-1 (*Author*); (*below*) the first true Boeing carrier fighter, the FB-1. This photograph, taken 9 September 1926, is probably the first production aircraft, A7101 (*Boeing*)

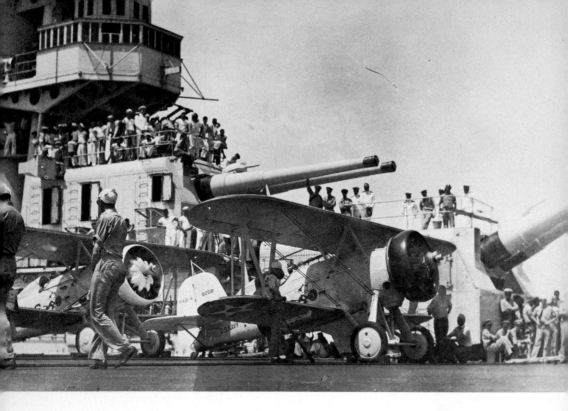

(*Above*) Boeing F4B-4s of VF-6B (Fighter-Bomber), with the famous Felix the Cat badge, prepare to take off from *Saratoga* on 17 May 1934. The squadron is flying off over the carrier's stern — a most unusual occurrence (*US Navy*); (*below*) a Glenn-Martin T4M-1 torpedo-scout-bomber, with its arrester-hook extended prior to landing, circles over USS *Saratoga* (*US Navy*)

(*Above*) The island and flight deck of USS *Ranger*, 14,500 tons, which was commissioned in 1934. The aircraft on *Ranger's* elevator are two F3F-1s of Fighting Four (*US Navy*); (*below*) K-1085, a Bulldog II of No 17 Squadron (Upavon and Kenley), photographed in June 1930 (*Aeroplane*)

USS *Enterprise* during the United States occupation of the Gilbert Islands. A TBF Avenger circles over the ship, awaiting the clearance of the flight deck (*US Navy*)

(*Above*) A Curtiss F11C-2 photographed on 14 April 1936. The role of the Curtiss Hawks had changed from that of fighter to bomber, hence the designation 'BFC' on the rudder of 9273 (*US Navy*); (*below*) A Vought Vindicator (SB2U), of Bombing Three, takes off from the teak flight deck of USS *Saratoga* in 1939, just prior to the outbreak of the war in Europe (*US Navy*)

(*Above*) The Martlet (the RN name for the Grumman F4F) has the distinction of being the first American fighter to shoot down an enemy aircraft during World War II. An early Martlet 1 — G36A — in 1940 FAA markings (*Grumman*); (*below*) the cockpit of an F4F-3A. Unlike British aircraft, which had a standard flying panel common to all service aircraft, US aircraft had individual panels (*Grumman*)

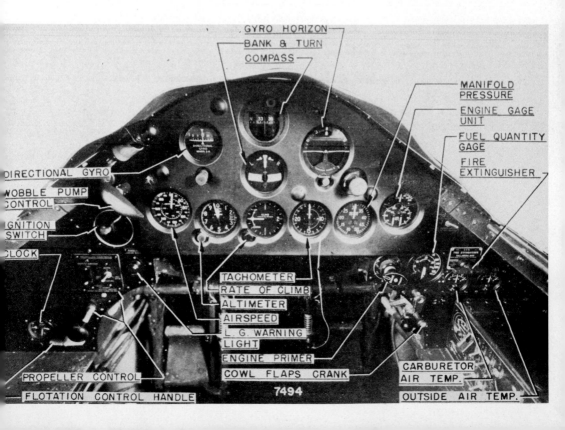

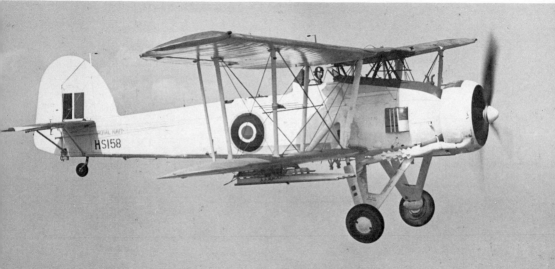

(*Above*) A Fairey Swordfish I of 820 Squadron about to take off from HMS *Ark Royal* in 1939 (*RAF Museum*); (*below*) the German light cruiser *Königsberg*, 5,600 tons, built at Wilhelmshaven in 1937 and, in 1940, flagship of the Senior Officer Scouting Forces, Norway (*BfZ Stuttgart*)

(*Opposite top*) L2728, a Fairey-built Swordfish photographed in 1937 (*Flight International*); (*opposite centre*) HS158, a MkII, built by Blackburn; this aircraft served with 816 Squadron aboard an Escort carrier at the height of the Battle of the Atlantic (*British Aerospace*); (*opposite below*) the Blackburn Skua, the first British aircraft designed from the outset as a dive-bomber. In 1936, 190 Skuas were ordered by the Air Ministry for the Fleet Air Arm (*British Aerospace*)

(*Above*) An FAA Martlet II with manually folding wings being manhandled off the lift into the hangar of HMS *Illustrious* in 1941 (*IWM film still*); (*below*) HMS *Illustrious*, 23,000 tons. It was from this carrier that the Swordfish strike against Taranto was flown off. The photograph was taken in 1947, when *Illustrious* still had a conventional (non-angled) flight deck (*British Aerospace*)

(*Above*) Fairey Swordfish being armed with 1,610lb naval torpedoes. For the first strike of the Taranto attack, six aircraft were bombed-up with torpedoes; the remaining six carried 6 x 250lb armour-piercing bombs or flares (*IWM film still*); (*below*) a Swordfish is ranged on a British carrier at the time of the *Bismarck* action. The 1½ metre ASV Yagi radar antenna is visible on the foremost interplane strut (*IWM film still*)

(*Above*) The first shots in the Pacific War were fired by a US Navy PBY Catalina flying boat of Patrol Squadron 14. The time was 0650, 7 December 1941 (*US Navy*); (*below*) 'Battleship Row', minutes after the first wave of the Japanese Kates had released their torpedoes at pointblank range. The *Arizona* was hit by an AP bomb which detonated her magazines, killing over 1,000 of her sleeping crew (*IWM film still*)

(*Above*) CV-10, USS *Yorktown*, recovering an F6F over her bow whilst steaming astern, a most unusual manoeuvre (*US Navy*); (*below*) HMS *Indomitable* in 1943; a British carrier which saw a great deal of action in World War II. The wartime censor has deleted the radar antennae on both the ship and the Swordfish (*MOD* (*RN*) *Crown Copyright*)

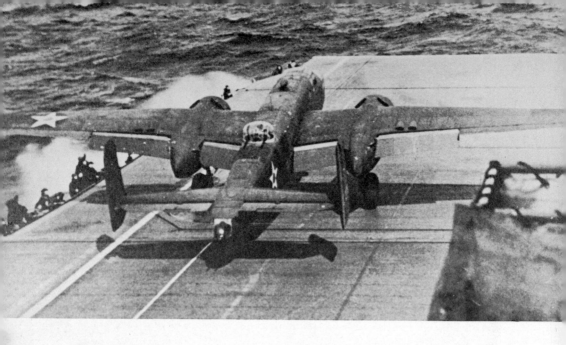

(*Above*) The lead B25 on the Tokyo raid, piloted by Lt-Col Doolittle. All the B25s took off successfully, but all were lost in forced landings, mostly in China (*US Navy*); (*below*) Japanese Staff during the Battle of the Coral Sea; this was the first major naval action in which the opposing ships never sighted each other, all the fighting being done by carrier aircraft (*IWM film still*)

Boeing biplanes of the FB series. The origin of these excellent machines was in 1923 when both Boeing and Curtiss decided to build fighters for the Army and Navy. However, with large stocks of unused wartime aircraft available in store, the US Government made it clear that it would not finance the design or development costs of any new ventures.

Since only a vigorous military market offered the hope of continued existence, the two companies decided to finance their projects from their own slender resources. Boeing in those days was a small company building wooden Thomas-Morse MB-3A fighters for the US Army, which at least occupied the production lines and kept together a skilled work force. Boeing's engineers felt they could do better; they had made a detailed study of what was arguably the best fighter design of the war, the German Fokker D VII, a number of which had been captured and brought to the United States. The German fighter, which had been designed by the Dutchman Antony Fokker and his chief draughtsman, Rheinhold Platz, had an enormously strong yet light structure made up of welded steel tubing, fabric-covered.

Boeing's engineers were impressed with the simplicity and ease of manufacture of metal tubing when compared with the laborious carpentry of a wooden aircraft with its many intricate, small metal fittings; they decided to produce a single-seater, high-performance fighter, using welded steel tubing for the primary structure. They were confident that they could improve on the Fokker, partly because they had the advantage of the then newly perfected electric arc welding techniques, which were much superior to the gas welding the Germans had used. The result was the Boeing Model 15, which was to serve as a prototype for a production line of a whole series of excellent modern fighters for the Army and Navy. Three prototypes were to be built, without official support, though using engines and instruments 'borrowed' from undisclosed sources.

The prototype Model 15, known to the US Army as the XPW-9 (*Ex*perimental-*P*ursuit *W*ater-Cooled Type *9*), was first flown on 2 June 1923 and tested by Army pilots at the McCook Field. The Model 15 was a neat biplane with unequal span wings and generally clean lines, spoilt perhaps by the rather bulbous cowling of the in-line, liquid-cooled, 435hp Curtiss D-12 engine. It offered an excellent performance, however: maximum speed 159mph and a rate of climb of 1,630 feet per minute, with a ceiling of nearly 19,000 feet.

The Army, impressed, immediately ordered thirty machines, including the three prototypes, of which only one was then completed. Shortly after the Army orders, the Navy ordered fourteen Model 15s (Bureau

Numbers A6884—6897), the first ten being identical to the Army XPW-9, though they were designated FB-1 in compliance with the current naval nomenclature (F-Fighter, B-Boeing, Type 1). The batch was delivered during December 1925. These aircraft were to be shore-based and assigned to a Marine Corps Squadron.

The first true carrier variant of the Boeing fighter was the FB-5, Boeing Model 65, essentially similar to the original but with a redesigned undercarriage and an arrester hook, and a more powerful engine (520hp Packard 2A-1500), which increased the maximum speed slightly and improved the ceiling. Twenty-eight FB-5s (A7101—7128) were delivered to the US Navy in January 1927, being floated on lighters across the river which fronted Boeing's Seattle plant, then winched aboard USS *Langley*, to equip VF-1B and VF-6B aboard that carrier.

The Boeings immediately proved very popular with their pilots; they were extremely manoeuvrable, fast fighters with a landing speed of only 60mph, which in view of *Langley*'s low speed was fortunate. Their armament was to provide a standard for US naval biplanes, consisting of either twin .30 or one .30 and one .50in machine gun, synchronised to fire through the airscrew arc.

The FB-5s had a short carrier life, for the US Navy had decided to standardise on air-cooled radial engines for all future shipboard aircraft. Careful studies had shown that a high percentage of failures with liquid-cooled engines was due to the vulnerable plumbing and radiators, not to mechanical breakdown of the engine components. One of the air-cooled radials selected for fighters was the 425hp Pratt and Whitney R-1340B Wasp, an engine which owed its existence to some extent to the great success that was being enjoyed in Europe with the similar Bristol Jupiter, which was then powering RAF fighters. (European fighters, however, would increasingly swing back to the liquid-cooled engines, e.g. Rolls-Royce, Napier, Daimler Benz and Junkers, while the Americans would continue with air-cooled radials up to the jet age.)

With the availability of the new engine, the Boeing Company radically redesigned the FB series, the new aircraft being designated Model 69. The prototype was built as a private venture and tested as the XF2B-1 (A7385); it handled well and proved to have a speed of 154mph. The US Navy at once ordered thirty-two of the type, designating them F2B-1 (A7424—7455), deliveries commencing in January 1928. It was a most attractive biplane with constant chord wings of unequal span, the upper planes having a pronounced sweep-back. The 425hp Wasp R-1340B was uncowled. The airframe was a metal structure, though now of bolted dural in place of welded steel; the control surfaces, including the fin, rudder and horizontal stabiliser were made of semi-monocoque corrugated aluminium.

The initial batch of F2B-1 fighters was assigned to VF-1B and VB-2B aboard *Saratoga*, which is confusing since the second of those squadrons was a bombing unit. The designation of the US Navy's carrier aircraft during the late twenties and early thirties was most complex. Because of financial stringency, it was decided that multi-role squadrons would serve aboard US carriers. Thus the F2B-1s and a later development, the F3B-1s, could be assigned to either a dive-bombing, fighter or even a scouting unit. The pilots, being rotated and trained in each role, often flew the same aircraft, which as a bomber could carry up to five 25lb bombs under wing.

Unit marking consisted of tail and (when fitted) engine cowling colour; an heraldic shield or cartoon figure was painted as a squadron emblem on the fuselage, plus the squadron number, function (B-Bomber, F-Fighter) and the aircraft's individual number within the squadron. Because of the constantly changing roles, the system became, to say the least, confusing; for example, in 1929 VB-2B, one of the first squadrons to equip with the F2B-1 aboard *Saratoga*, had as its unit marking Felix the Cat — a silent movie cartoon character of the time — holding a bomb; though entirely appropriate for a light dive-bomber unit, the emblem was retained when a year later the squadron became a fighter squadron, VF-6B. Thus 'Bombing Two' became 'Fighting Six', but .... it had also been known as 'Fighting Two' before 1929, with a completely different emblem. Confusion notwithstanding, it made for the most colourful aircraft flying anywhere at that time.

Other emblems in use were the High Hats of VF-1, 'Fighting One', and the Chief Petty Officer badge of 'Fighting Two'. That squadron was also known as 'The Fighting Chiefs' because the majority of its pilots were Chief Petty Officers. The US Navy, by law, had then to have enlisted men form 30 per cent of the pilot strength (later reduced to 20 per cent). Many of these CPOs, because of the time it took to rotate through the different squadrons, flew operationally for as long as fifteen years and logged thousands of hours, to achieve a very high standard of flying. One pilot who joined 'The Fighting Chiefs' — one of the top squadrons — in 1929, G.F. Ocskay, wrote: "Most pilots had 1,000 or more hours when they transferred to 'Fighting Two'. I only had about 800, so I was transferred to them with the understanding that if it did not work out I would not remain with them."[29] When one considers that, during World War II, most RAF pilots first flew operations with about 200 hours behind them, one can gauge the standard aboard the US carriers. (Incidentally, G.F. Ocskay did remain with 'Fighting Two' — for nine years.)

Apart from the carriers' fighters, the other squadrons embarked were Torpedo and Scouting (reconnaissance) units. The aircraft complement in the late twenties aboard each of the two large carriers, *Saratoga* and *Lexington*, was four squadrons: one fighter, one dive-bomber and one each for scouting and torpedo-dropping. Each squadron had a nominal strength of eighteen aircraft. Boeing F3Bs would fulfil the dive-bomber role, and their pilots had developed a tactic known as the 'split dive', in which all bombers of a squadron would dive vertically on to their target simultaneously from three different directions. Although fighters would operate as dive-bombers for some time, by 1930 specialised aircraft were beginning to supplant the F3B fighter/bomber.

The first true US Navy dive-bomber was the Glenn Martin BM-1, designed to Bureau of Aeronautics specification No 77, issued in 1928, which called for a two-seater dive-bomber to carry a single 1,000lb bomb (or a torpedo — the Navy was still economy-conscious) and strong enough to pull out of a terminal velocity dive with the bomb still attached, should it fail to separate (which was not unknown). The BM-1 prototype flew in March 1930.

The Glenn Martin Company also built the US Navy's carrier torpedo bombers, typical of which were the T3M-2s, the last US naval aircraft powered with in-line engines: liquid-cooled 770hp Packards. T3M-2s were embarked aboard *Lexington* and *Langley* in mid-1927. When compared with its British contemporary, the Blackburn Dart, there was not a lot to choose between them; the maximum speed of the T3M-2 — 109mph — was within 1mph of the Dart, which possessed a much better service ceiling of 15,000 feet, against 7,900 feet for the Martin. The US design, however, had a far greater range: 634 statute miles, more than twice the Dart's.

With the vast distances of the Pacific Ocean before them, most US Navy aircraft possessed a long-range capability. Even the fighters, which normally did not fly from the sight of the ship, had provision to carry long-range tanks, which greatly extended their range.

When Pratt and Whitney radial engines became freely available, Martin re-engined the T3M-2 and revised the airframe so extensively that it became a new sub-type, T4M-1, which was ordered by the Navy in substantial numbers (102), to serve in the carriers *Saratoga* and *Lexington*, replacing the earlier aircraft, which were then relegated to shore-based squadrons, including those of the Marine Corps. The T4M-1, powered by an R-1690-24 radial engine of 525hp, was a much improved machine; the speed went up to 114mph and the service ceiling to 10,150 feet.

Torpedo bombers were expected to operate up to 100 miles from their carriers and were equipped with radio, though only telegraphy. The use by US naval aircraft of airborne radio was increasing, and in June 1929 all aircraft delivered to the Navy had efficient shielded ignition, suppressed sparkplugs, magnetos and bonding to improve the use of radio aids.

Radio-equipped aircraft could operate far from the ship and navigate back to it with some degree of confidence. To achieve this the Navy used a technique called 'Point Option', which was an imaginary point moving at a fixed speed and course, given to the pilots when briefed before take-off; the carrier kept her aircraft informed of her position relative to this point by radio. The message, had it been intercepted by an enemy, would have been quite useless as the Point Option, its course and speed, would have been unknown to them. It seems to have worked well enough, but on one occasion during an exercise a navigational error on the part of a carrier almost cost an entire strike of T4Ms and scouting aircraft and forced the squadron to make night carrier landings, which few if any of the pilots had been trained for. The story, which is told by one of *Lexington*'s pilots, now a Rear Admiral, is interesting because it gives an insight into the difficulties of operating carrier aircraft far from land in those pre-radar days:

Captain E.J. King who was the skipper of *Lex* and who took great pride in being a great disciplinarian .... sent *Lex*'s air group out one afternoon on a strike. He called all the squadron C.O.s on the bridge to brief them. It was to be a four hour operation with the landing about one hour before sunset. He finished his briefing with what [position] the ship would maintain in reference to Point Option.

At the end of the four hours, the combat air patrol, which had stayed within sight of the ship, landed. The fighter commander went to the bridge just as Lt. Cdr. Haddon, the (TM-4) torpedo squadron C.O., sent a message saying he was at the rendezvous and no ship. The fighter commander made a remark that *Lex* wasn't within 40 miles of the rendezvous. King told him to get off the bridge. There was much excitement: radio bearings, *Lex* made smoke, full power, more radio bearings — an hour passed. Haddon called the group into tight formation; *Lex* turned all searchlights into the dark sky; Haddon prepared to ditch the air group while they still had some light and gas.

Suddenly one of the scouts radioed he had *Lex*'s searchlights in sight. Though few pilots had qualified for night landings, the air group got back aboard. But as one of the T4M pilots remarked after he had landed on: "I can drink all the gas left in that airplane."[30]

Thus, by trial and error, a great naval air arm was being forged.

Across the Pacific Ocean the future enemy, Japan, was also perfecting a naval air service. It is difficult today to obtain any accurate information about the formative years of the Japanese Fleet Air Arm; the Japanese naval archives were badly damaged in the great fire raids on Japan towards the end of World War II, and at the end of the conflict the surviving documents vanished, either — as is usually claimed — deliberately burned, or hidden by the defeated Japanese. If the latter case is true, then they have yet to reappear.

In September 1931, when the Japanese began fighting in Manchuria, contact with the west virtually ceased; there is therefore a long period of obscurity. Prior to 1931 information was available, though by then the western powers had tended to dismiss the Japanese as mere copyists. It should be realised, by a generation that has come to regard 'Made in Japan' as synonymous with high-quality optical and photographic goods, cars, television, transistor radios and other electronics, that this was not always so. Far from it; pre-war 'Made in Japan' usually indicated designs copied (often without licence) from other countries.

The Semphill mission, which returned to England at the end of 1922, began the process of denigration. Though Colonel Semphill himself had a high regard for individual Japanese pilots whom he had trained, even he did not think that the Japanese Naval Air Arm had a particularly bright future. The British Naval Attaché in Tokyo, Captain R.M. Colvin, wrote in a report, following a visit to the Japanese Naval Air Arm training base at Kasumigaura, where he had talked to Col. Semphill: "[Semphill] is making good progress though he forecasts that when he leaves the naval air services will collapse owing to lack of policy."[31] This very nearly became true, for after the departure of Colonel Semphill, who incidentally held the honorary rank of Captain in the Imperial Japanese Navy, the British Vice Consul, M.E. Dening, noted after a visit to the base, that there was at Kasumigaura "A distinct deterioration in discipline and efficiency".[32] Semphill, in conversation with Colvin, had been critical of the Japanese-built (though designed by an Englishman, H. Smith) Mitsubishi Type 10, which he considered "very poor", though it should be pointed out that Semphill was hoping to sell more British-built aircraft to the Japanese Navy.

The PRO File which contains the Foreign Office papers on the Semphill mission has some revealing marginal comments by FO officials, one of whom wrote at the foot of the Colvin report: "Interesting. The Japanese do not seem to be promising airmen."[33] Another read: "The Japanese are not really high flyers and never will be."[34]

It is of significance that the Admiralty, in a secret letter to the Under

Secretary of State, expressed concern that the Semphill mission had divulged classified information that had been of material help to the Japanese Navy: "It happens that the Japanese carrier *Hosho* has had various structural details such as the configuration of the flight deck modified by the members of [the Semphill mission]." The Admiralty letter went on to ask for an assurance that "The Secretary of State will not lend his support to the continued retention of the mission in Japan, and will consider what measures are practical to prevent the future disclosures . . . . of British ideas and technical details in Naval aviation".[35]

In fact, since the Semphill mission was a purely civilian and commercial affair and in Japan at the request of what was then a friendly Government and late ally, it was decided that the Foreign Office could do little; to withdraw the mission would cause offence to the Japanese and, more pertinently, open commercial opportunities to the French and Americans. The interesting point is that when Semphill did return to England later that year (1922), no one at the Air Ministry or the Admiralty interviewed him; presumably those august authorities did not think they could learn anything of value from the Japanese Naval Air Service. The dangerous slide into complacency about the ability of the Japanese to create an efficient naval air service which could in any way threaten the Royal Navy, or British spheres of influence, had begun. The Air Ministry, in a 1930 appreciation,[36] put the front-line strength of Japanese carrier aircraft at seventy-two machines (the Admiralty put the figure at 118); the Air Ministry figures must be treated with reserve, since they were part of a paper to deny the Admiralty any immediate expansion of the Fleet Air Arm. The true figures may well have been higher even than those in the Navy's estimates.

Most of the Japanese Navy's aircraft up to the late twenties were either direct imports, mainly from Britain, such as the Sparrowhawk and Nightjar, or licence-built versions of foreign designs; for example, the Nakajima A1N1 was a Gloster Gambet, constructed in Japan, with a duralumin airframe and powered by a Nakajima Bristol Jupiter VI radial. Designated Navy Type 3, it was a single-seater fighter and offered a better performance than either the contemporary Fairey Flycatcher of the Fleet Air Arm or the RAF's Gloster Gamecock, which it closely resembled (though the Gamecock was of wooden construction).

About 1928 the Japanese stopped openly buying foreign aircraft and licences; instead they began progressively to obtain the latest western military designs by devious methods. By way of example, the case of the Bristol Bulldog is typical. The facts are as follows.

In 1929 the Bristol Bulldog, a single-seater fighter, had just entered front line service with the Royal Air Force. It was, for its day, an excellent machine, easy to fly with good manoeuvrability; it was powered by a 490hp Bristol Jupiter VIIF radial and could climb to 20,000 feet in 14½ minutes, having a top speed of 174mph. That same year, 1929, the company had a request from its agents in Japan, Mitsui, for a Bulldog II airframe; a machine from the RAF production run, c/n 7341, was crated up and delivered. The airframe was erected in Japan, fitted with a Nakajima licence-built Bristol Jupiter and successfully flown. The Japanese Government then requested a working party to be sent from the Bristol factory to Japan to supervise a licence-built version of the Bulldog; 7341 was to be the pattern aircraft, but the Japanese version was to have its airframe stressed to be constructed from locally produced grades of steel. Hoping for substantial orders for either additional Bulldog airframes or a least licences for production, two Bristol engineers, L.G. Frise and H.W. Dunn, arrived in Tokyo in March 1930 to supervise the design and construction of a similar, but revised, Bulldog, which became the Japanese Naval Type 701. Powered by a Nakajima Jupiter VII, it proved capable of 196mph, over 20mph faster than the RAF's squadron Bulldogs. A second similar machine was built and tested, and proved as good as the first. Although these two machines were in every way successful, no orders or licences were placed and Frise and Dunn returned home. Later, according to the Bristol Company, several features of the Bulldog design appeared in subsequent Japanese aircraft, without any acknowledgement or payment.

Thus, for the cost, in foreign exchange, of one engineless airframe, the Japanese had acquired the services of a highly experienced design team and had been able to produce an aircraft, built with indigenous materials, in advance of the type currently flying with the front-line fighter squadrons of the RAF.

One cannot say how much of the Bristol engineers' know-how went into the next Japanese naval fighter, the Nakajima A2N (Type 90), which appeared in 1931, but it too was powered by a Nakajima Jupiter and could achieve 200mph on its 450hp, which was good going for a biplane in 1931 and would not be bettered by the British Fleet Air Arm aircraft until the Gloster Gladiator fighter entered service in February 1939.

The mainstay of the Japanese Navy torpedo squadrons from 1924 was the Mitsubishi B1M, of which some 450 were built over a span of nine years. The type, which was extensively used in the war against China, was a biplane, similar in appearance to the US Navy T3M-1; it had a 48 foot 6 inch span, with a length of 32 feet. The power plant was either a licence-built, liquid-cooled British Napier Lion or a French Hispano-

Suiza, both of 450hp, which gave the B1M a maximum speed of 130mph when armed with a naval torpedo. The aircraft was also used as a bomber and for reconnaissance and was destined to have a long service life. It was to have been supplanted in 1932 by a later Mitsubishi design, the B2M, but that machine proved a disappointment and, in the event, the B1M outlived its intended successor.

Taking 1930 as the end of the formative decade, the three principal naval air arms had each developed carriers which suited their projected needs, embarked with the best aircraft that financial stringency and 'state of the art' permitted. The twenties had been, above all, a period of experimentation; of defining the parameters of naval air power and in working out new tactics — Battle Problem IX, for example — and the techniques of operating aircraft, still largely fabric-covered, in the hostile environment of salt water and the cramped confines of carriers.

The Fleet Air Arms of the three major navies, were by 1930, very roughly equal, though in the next and final decade before World War II some would become decidedly more equal than others.

Early development of carriers in the following decade from 1930 was still governed by treaty: under the provisions of the London Naval Treaty, the United States, after the commissioning of *Saratoga* and *Lexington*, still had 69,000 tons of new construction available for aircraft carriers. After long consideration, the US Navy decided to obtain congressional approval to order a carrier with a displacement of 13,800 tons each year for five years, the carriers to have a design capacity of seventy-six contemporary aircraft. Although there were dissenting voices as to the usefulness of such comparatively light ships, Congress, on 13 February 1929, authorised the construction of the first 13,800 ton carrier (CV-4).

The ship, to be named USS *Ranger* — the name had originally been assigned to one of the cruisers scrapped under the provisions of the Washington Treaty — was laid down on 26 September 1931, on the stocks of the Newport News Shipbuilding and Drydock Company. *Ranger* was the first US carrier to be built as such from the keel up. The original design was for a flush-decked carrier, similar to *Langley*, but during the construction it was decided to incorporate a starboard island, since experience with *Saratoga* and *Lexington* had pointed to the advantages of that configuration. However, unlike the *Saratoga* class, which incorporated the funnel as part of the island structure, the uptakes from *Ranger*'s six boilers connected with six stacks, arranged three by three, on both sides of the after flight deck. When she was steaming, the hinged funnels were upright, standing about 15 feet above the flight deck; when she was operating her aircraft, the six funnels could be swung horizontally outboard.

USS *Ranger* was launched in February 1933 and commissioned on 4 June 1934. The carrier's standard displacement proved to be 14,500 tons (700 tons overweight); her wooden steel sheathed flight deck was 750 x 80 feet, with two lifts to a 500 x 70 foot hangar, which had two large doors to permit aircraft to be hoisted directly inboard, either from the quay or from lighters alongside, two cranes being provided for this duty. Seventy-six aircraft of the day could be accommodated in the hangar, which, considering the fact that none of the fighters had folding wings, must have made the stowage something of a problem. Six transverse arrester wires, one set near the bows, were installed; these, following experiments aboard *Langley*, were operated hydraulically (as by this time were *Saratoga*'s and *Lexington*'s).

In service, *Ranger* was to prove really too small and lightly built; even before launching, the General Board of the Navy had decided that the minimum effective size of fleet carriers was 20,000 tons. Nevertheless, *Ranger* could operate a full air group of four squadrons, and she was to see a great deal of naval aircraft development in the last years of peace. The first deck landing on *Ranger* was made by Lt Commander A.C. Davis on 21 June 1934, flying a Vought O3U-3 observation plane. After several take-offs and normal landings the ship went full astern and aircraft were recovered by the bow arrester wire alone.

The US Navy quickly came to the conclusion that *Ranger*, though a welcome addition to the fleet, was too slow, insufficiently armoured and deficient in defensive fire power; the necessity to save weight had only allowed for the provision of eight 5 inch AA guns, disposed in pairs on either side of the gallery below the flight deck. In view of the decision made by the General Board of the Navy that the minimum standard displacement for an effective carrier was around 20,000 tons, the four projected 18,000 ton sisters of *Ranger* were cancelled; instead, a request was made that, in the Building Programme for 1934, which was issued in September 1932, two 20,000 ton carriers be built. To finance the ships, the Secretary of the Navy asked President Roosevelt for funds to be made available under the National Recovery Act. The appeal was successful and $238 million was allocated to the Navy for new ships, including two carriers. In less than two months contracts were awarded for CV-5 and CV-6.

The designs of the new carriers, which were to be known as the *Yorktown* class, were the subject of considerable thought, in view of the shortcomings of the earlier *Ranger*, as the files of the Bureau of Aeronautics reveal:

> The Department has approved a new building program with two aircraft carriers similar to *Ranger*, but before embarking on this new construction, it is suggested that a careful examination may show many design changes are desirable.
>
> The particular improvements in the *Ranger* design that should be considered are: speed increase to 32.5 knots; addition of underwater sub-division to resist torpedo and bomb explosions; horizontal protective deck over machinery, magazines, and [aviation] fuel tanks; improvement in operational facility (this includes hangar deck devoted exclusively to plane stowage, four fast elevators, complete bomb handling facilities, possible use of two flying-off decks, and improved .... anti-aircraft defense).[37]

In 1934 the two carriers, CV-5 and CV-6, which were to be named *Yorktown* and *Enterprise* respectively, were laid down in the same yard that built *Ranger*: Newport News, Virginia. *Yorktown* was launched, appropriately by Mrs Franklin D. Roosevelt, on 4 April 1936, followed by *Enterprise* on 30 October, in the same year. They were the fifth and sixth US Navy carriers, though that September, USS *Langley*, CV-1, was detached from the Battle Force for conversion to a seaplane tender, for which purpose the forward third of her flight deck was removed and a normal ship's bridge substituted.

While the *Yorktown* and *Enterprise* were still being fitted out, the Navy's General Board tried to replace *Langley* with an additional 15,200 ton carrier. But since *Langley* had been classed as 'Experimental', to have done so would have violated the provisions of the London and Washington Treaties.

To return to the *Yorktown* class, *Yorktown* herself was commissioned on 30 September 1936. Her standard displacement was 19,800 tons, and she was a true starboard island carrier; her unarmoured wooden flight deck was 781 x 80 feet above a 560 x 75 foot hangar, capable of housing some seventy aircraft. The hangar had, like earlier and subsequent US carriers, large apertures in the sides, partly to provide adequate ventilation against the inevitable aircraft fuel spillage. Two additional openings to port and starboard permitted the launching of aircraft directly from the hangar by catapult — the American equivalent of the British 'slip flight' — though this facility was little used.

Yorktown and her two sisters (a third, USS *Hornet*, CV-8, was built out of sequence after the termination of the Naval Treaty, in 1939, as an answer to Japanese carrier expansion) were fast ships, capable of 33 knots and a radius of 8,220 nautical miles at 20 knots. USS *Enterprise*, essentially identical, was commissioned on 12 May 1938. CV-7, USS *Wasp*, was the next carrier chronologically; she was ordered on 27 March 1934, to utilise the remaining 15,000 tons available under the original treaty limitations, and commissioned on 25 April 1940. Her displacement was 14,700 tons, though by then the treaties which had limited her size had ceased to be valid, owing to Japan's abrogation.

In 1931, when Japan had begun military operations in Manchuria, she became increasingly withdrawn; large areas of Japan were forbidden to foreigners and there was a growing distaste for western influence. The restraints imposed by the Naval Treaties were becoming unacceptable to Japanese naval officers, who particularly resented the inferiority the 5-5-3 ratio imposed on them. There was to have been a further naval conference held in London in 1935, at which it was hoped to persuade the major powers to agree to a measure of disarmament. The Japanese Navy was not in the least interested in disarmament, however; rather it demanded immediate naval parity with the United States and Great Britain, citing among other reasons the existence of the British naval base at Singapore and the extending of US bases in Hawaii. In a growing atmosphere of concern at Japan's by now overt expansionist attitude, neither of the western powers could agree to Japan's demand for naval parity. As a consequence, Japan, on 29 December 1934, gave the correct formal two years' notice that she would no longer consider herself bound by the Washington and London Treaties after 31 December 1936.

The last Japanese carrier to be constructed with any pretensions of remaining within the 1930 Treaty limits was *Ryujo* (Prancing Dragon), which was built to the most stringent weight limits possible, in order to obtain the maximum size within the tonnage remaining under the treaty. By sacrificing armour and even stability to that end, the Japanese naval architects managed to produce a carrier displacing only 8,000 tons, yet capable of operating forty-eight aircraft. *Ryujo*, the first Japanese carrier to be built from the keel up as such, was laid down in 1929 at Yokohama; she was launched on 2 April 1931 and completed in May 1933 as a flush-deck carrier. During construction the naval staff required an increased aircraft complement; the only solution to the demand was to incorporate an additional hangar on the top of the original flight deck, which reduced the stability and compromised the ship's seaworthiness. After being in service for about a year, *Ryujo* was considerably modified, though her subsequent career was to be limited both by her lack of armour and her inability to operate most of the later front-line aircraft.

Ryujo demonstrated to the Japanese the unsatisfactory result of trying to build a carrier large enough to operate forty or more aircraft, yet with a

displacement of under 15,000 tons; therefore the Japanese declared their
next construction at 15,900 tons, which left them with only 4,200 tons
for further carriers under the treaty obligations. This was an additional
factor in Japan deciding to abandon the 1930 constraints.

The 15,900 ton carrier was one of two approved for the Imperial Navy
in 1932. *Soryu* — Blue Dragon, the Japanese name for the constellation
Ursa Major — was laid down in 1934 with a starboard island and was
designed to operate up to seventy-one aircraft as a fast (34.5 knots)
carrier. To save weight, she was not heavily armoured; her two hangars,
one above the other, were lightly constructed boxes, supported by pillars
fore and aft. The flight deck was 711 x 85 feet, with nine transverse
arrester wires and two safety barriers aft of the wires. Three lifts were
fitted to enable aircraft to be ranged as quickly as possible, though, as
David Brown, the naval historian, has pointed out, at the cost of a
considerable loss of usable hangar space — 3,000 square feet — and a
compromising of the structural integrity of the flight deck caused by
stress concentrations, particularly around her large midships lift. [38]
There was also a lack of headroom in the two hangars. The upper hangar
was only 15 feet high and the lower a foot less, which was to make the
servicing of the wartime aircraft difficult, since the wings of most types
could not be either folded or spread when the aircraft was in the hangars,
a deficiency which made even routine maintenance difficult. Though
designed to carry seventy-one aircraft, *Soryu* operated only sixty-three.

So as not to reduce the available width of the flight deck, *Soryu*'s
starboard island was built on an overhanging extension; two funnels
exhausted the oil-fired boilers horizontally abaft the island and just
below the level of the flight deck. The armament of this latest Japanese
carrier consisted of three twin 5in (127mm) dual-purpose guns which
could either all be directed from a fire-control position atop the island, or
independently. There were also twenty-eight 25mm guns in twin
mountings, for close range anti-aircraft defence.

Soryu was completed at the Navy Yard, Kure, on 29 December 1937,
and represented to the Imperial Japanese Navy the fruits of the years of
experiment and trial, as did the *Yorktown* class, which had been
commissioned the previous September, to the US Navy. The Royal
Navy's *Ark Royal*, still under construction, was to be the British
equivalent; all three classes would serve as the basic pattern for subsequent
developments, and all represented the considered foreseeable requirements
of the navies that built them.

Freed from the intolerable — as they saw it — restrictions of the 1930
Treaty, the Japanese, before the two year's notice to abrogate had

expired, laid down in 1936 the second carrier allocated under the 1932 expansion programme; this was *Hiryu* — Flying or Heavenly Dragon — which was a stepsister to *Soryu*. Although the second carrier was originally conceived as a true sister ship, the withdrawal from the 1930 Treaty enabled the Japanese to increase her displacement to 17,300 tons, exceeding by some 13,000 tons the collective limits still technically in force for Japanese carriers.

The main difference between *Soryu* and *Hiryu* (completed in July 1939) was that the latter had, most unusually, a port island and a 3 foot increase in beam: the actual figures were *Soryu*, 69 feet 11 inches; *Hiryu*, 73 feet 3 inches, which does not seem very great, but it permitted *Hiryu* to carry 20 per cent more fuel oil in her bunkers, thereby increasing her operational radius to 10,330 nautical miles at 18 knots, 4,580 nautical miles greater than *Soryu*.

Hiryu's port island is interesting for it gives the clue to the strategic thinking of the Japanese Navy in the late thirties. By 1939 two carriers had port islands: *Hiryu* and *Akagi* (the latter had been extensively modernised and had emerged from the Sasebo Navy Yard with a small port island in 1936). The idea was that a port island vessel would operate in formation with a starboard island carrier in a strike force; the alternate islands would enable two ships, when steaming in close company, to operate their aircraft in compact formations, with opposite traffic patterns, the port island carrier flying right-hand circuits, the starboard island ship left-hand, without mutual interference. (During the war RAF airfields in Britain that were in close proximity often adopted opposite handed circuits to minimise the risk of collisions.) *Hiryu* was completed two months before the outbreak of the European war, and did in fact operate as planned in company with *Soryu*, the two ships forming Carrier Division Two.

From mid-1936 the Japanese Cabinet had been dominated increasingly by military men who ruthlessly pushed ahead with an ambitious policy of rearmament, developing the heavy industrial base that such a programme required. As a comparatively small island power with aspirations far from her own coasts, Japan had early realised that carriers were to be of vital importance, and the naval air service was to be expanded dramatically in order to restore the inbalance vis-à-vis the western powers that the treaties had imposed.

During 1937, the year that the Sino-Japanese 'Incident' became a shooting war, two further carriers were laid down: the *Shokaku* class, comprising *Shokaku* (Flying Crane) and *Zuikaku* (Lucky Crane), each with a standard displacement of 25,675 tons. These ships, unlike any previous Japanese carriers, were heavily armoured (though oddly the flight decks were not) and were built to be a match for any existing US carrier. They were designed to operate up to eighty-four aircraft, though the usual number embarked was less, at seventy-three. Strangely, though they were planned to operate as a single unit, and did, they both had starboard islands. They were fast ships, capable of over 34 knots, with a large radius of action of 9,700 nautical miles.

Shokaku was to be completed at Kokosuka Navy Yard on 8 August 1941 and *Zuikaku* at Kobe on 25 September in the same year. Although some fifteen further aircraft carriers would be commisioned into the Japanese Navy the *Shokaku* and *Zuikaku* were the last to enter service before the Pearl Harbor attack.

With their open aggression against China in 1937, the Japanese, like the German Condor legions in Spain, had a unique opportunity to test and perfect in action the weapons and aircraft for which the carriers existed. The experience of aircraft manufacture gained prior to the Sino-Japanese conflict — one way or another — during the years when Japan openly depended on western ideas and designs, was being put into practice. Japan now had an industrial and design capability to construct her own aircraft, which, far from being mere copies, were to prove to be in advance of anything available to the English-speaking navies.

The Nakajima Type 97, which was to be code-named 'Kate' by the Allies, appeared in 1937; it was the world's first stressed skin, all-metal, carrier-borne monoplane and immediately outclassed, by a very large margin, any other torpedo bomber then in service. A clean design, when powered by a single 1,000hp Nakajima Sakae radial, it had a maximum speed of 235mph, with the unbelievable range of 1,238 miles. The Type 97 could be armed with a torpedo or 1,760lb of bombs; it was to become the standard Japanese attack aircraft of World War II, 1,149 being built between 1937 and 1943.

The Type 97 was the first carrier-borne monoplane torpedo aircraft, but it was not the first Japanese Navy monoplane; that distinction must go to a fighter, the Mitsubishi A5M, Type 96. This aircraft was designed by a brilliant aeronautical engineer, the 31-year-old (in 1934) Jiro Horikoshi — a name then unknown outside Japan — who had realised that the biplane fighter was obsolete. That it continued in service with the major navies was due to a belief that the extreme manoeuvrability and low landing speed, both the product of a light wing-loading, were advantages of the biplane's configuration that outweighed the dis-advantages, particularly the limited maximum speed attainable — roughly 250mph — owing to the drag of two wings and the associated bracing wires and interplane struts.

Horikoshi's A5M (code-named Claude by the Allies during the war) first flew on 4 February 1935 at Kaganigahara airfield, north of Nagoya. The prototype was a small, gull-winged, radial-engined monoplane, with a neatly faired fixed undercarriage. The original engine was a Nakajima Kotobuki of 550hp, on which it was later claimed the A5M managed 280mph in level flight; however, subsequent flight testing revealed certain deficiencies due to the gull wings. A second prototype, with straight wings, crashed following a spin, but the promise the type held was sufficient for the Navy to order it into production as the A5M-1 with a 585hp Nakajima engine.

The A5M-1 was first flown operationally from the carrier *Kaga* on 18 September 1937, escorting bombers attacking Nanking. The opposing Chinese fighters were British-built Gloster Gladiators, American Boeing P.26s, Curtiss Hawks, Italian Fiats and Russian Polikarpov I-16s — all, with the exception of the Boeing P.26s, biplanes (the P.26s were monoplanes but externally wire-braced). The Chinese fighters were mainly flown by western mercenary pilots, who found that the agility and speed of the A5Ms made them exceedingly difficult to best, with the result that the new monoplanes quickly established air superiority, enabling the bombers to attack with impunity. The only limiting factor the A5M-1s possessed was a relatively short range of around 750 miles; thus, when the bombers attacked targets deep inside China, they had to fly unescorted, with heavy losses. However, Jiro Horikoshi had a successor to the A5M-1 — the A6M Mitsubishi Zero Sen — which had flown for the first time on 1 April 1939. The story of that astonishing fighter belongs to the next chapter; let it suffice for the moment to say that this and other Japanese types gave their pilots priceless combat experience over China, the result of which was, to put it at its lowest, to surprise the Western Allies when the Pacific War started.

Though one suspects that the reports filtering from the Sino-Japanese conflict were available to western naval intelligence, there is no doubt that these were discounted as exaggerated. There was a reluctance to concede that the Japanese could possibly produce such advanced aircraft; it was even widely believed that the Japanese had difficulty in training sufficient pilots, owing to the high incidence of ethnic myopia.

Whether or not the reports from China were under estimated, there was no doubting the obvious expansionist aims of Japan. These alarmed the United States and in May 1938 Congress authorised an additional 40,000 tons for US aircraft carriers; this was to lead to the *Essex* class, which eventually embraced no fewer than thirty-two fleet carriers, none of which, however, would be commissioned before the United States entered the war.

The Japanese Navy was certainly the first to use carrier-borne monoplanes in service. The Americans, however, had also come to the conclusion, like the Japanese, that the biplane fighter was nearing the end of its development; interestingly, though, the first monoplanes which operated with the US Navy were not, as might be supposed, fighters but dive-bombers. The dive-bomber function was, until 1936, performed in the main by the multi-role small biplane fighter/bombers, of which the Boeing F4B-4 — the last of that line — and the Curtiss F11-C Hawk are examples. In 1934 it was decided that specialised dive-bombers would be employed, partly because a heavier weight of bomb was deemed necessary and also the envisaged radius of action of around 1,000 miles required the services of a navigator. Several manufacturers were invited to tender for contracts, including Curtiss, Martin, Brewster, Vought and Northrop. It was the latter Californian concern which was the successful contender, its prototype, designed by John K Northrop, the XB7-1 (Bureau Number 9745), being chosen for limited production for both the Navy, as the BT-1, and the US Army Air Corps, as the A-17

The BT-1 was powered by a single 825hp Pratt and Whitney R-1535-94 engine, which gave it a maximum speed of 222mph, a service ceiling of over 25,000 feet and a range of 1,150 statute miles. It was normally armed with a 1,000lb bomb. The Northrop was a cleanly designed, low wing, all metal monoplane with a retractable undercarriage and variable pitch propeller. Though only fifty-three were ordered (in two batches, Nos. 0590-0626, 0628-0643), thirty-nine of these were embarked on two carriers between late 1937 and 1940, serving with Bombing Six on *Enterprise* and with Bombing Five on *Yorktown*. By the time of Pearl Harbor all had been withdrawn.

The BT-1, apart from being the first operational monoplane to fly from US carriers, was to have a far more significant role, since it was to lead, by development, to the Douglas SBD Dauntless, when that firm took over Northrop. The Dauntless was to prove one of the most successful dive-bombers in the world and virtually enable the US Navy to win the Battle of Midway. The Dauntless apart, many people believe, with some truth, that the Northrop Gamma, a fast mail plane from which the BT-1 evolved, was used by the Japanese as a basis from which to design the B5N 'Kate'.

A second monoplane scout/dive-bomber, which, with its low wing, radial engine and large greenhouse cockpit canopy, looked superficially like the BT-1 (or for that matter the Japanese Kate), was the Vought

SB2U Vindicator. This bomber, ordered in 1936, was delivered to the US Navy in December 1937; succeeding production batches served with Bombing Two, VB-2, on *Lexington*; VB-3 on *Saratoga*; VB-4 with Scouting Squadrons VS-41 and 42 aboard *Ranger*; and VS-71 and 72 on *Wasp*. The Vindicator, powered by an 825hp Pratt and Whitney R-1535-02 engine, had a maximum speed of 243mph at 9,500 feet and a range of 1,120 statute miles. Unlike the BT-1, the wings of which were fixed, the SB2U had upward folding wings, operated manually.

The Vindicator was to serve with the US Navy until 1942; it was also supplied to the US Marine Corps and was exported: in 1938 a large order was received from the French Government for the V-156 (SB2U-3). Most of those delivered were either captured or destroyed by the Germans, but the last fifty aircraft of the contract were diverted to Britain in 1941 and served with the Royal Navy as the Chesapeke 1 (AL908— AL957), though, owing to a long take-off run, the type was not used aboard the small British Escort carriers as had been envisaged, and the Chesapeke was confined to shore-based training duties and Fleet Requirement Units.

The third monoplane to see service pre-war with the US Navy was the Douglas TBD Devastator. The Devastator was the result of trials of comparable biplane and monoplane types to determine which was the most suitable as a carrier-based torpedo bomber. After tests against the Great Lakes biplane, XTGD-1, the prototype from the Douglas Company, the XTBD-1 (9720) monoplane was selected for production; 114 TBD-1s (0268— 0381) were ordered in February 1936 and a further fifteen (1505-1519) in August 1938. It was a three-place, rather ungainly looking, low wing monoplane, with a partly retracting undercarriage. Powered by an invariable radial, in this case a 900hp Pratt and Whitney R-1830-64, it was endowed with a maximum speed of only around 200 mph. It carried a 21 inch Bliss-Leavitt torpedo externally, which reduced its range to 435 miles.

The Devastator was a big improvement on previous biplane torpedo bombers and was to serve in five squadrons on all pre-war carriers and several wartime ships, though, following heavy losses, it was to be withdrawn from front-line service after the Battle of Midway.

Boeing, the name most closely associated with the biplane fighters of the late twenties and early thirties, ceased to supply aircraft to the US Navy when the final version of the FB4-4 was withdrawn from front-line service in 1937. Another name, dating from the earliest days of United States naval flying — Curtiss — supplied the BFC-1s which were assigned to Fighting Five aboard USS *Ranger* in October 1934, but they served only a few months before being withdrawn and were the last Curtiss fighters to serve with the US Navy. A new name had appeared in 1931: Grumman, a company which had secured naval contracts for a series of characteristically tubby biplane fighters which culminated in the F3F-3s, delivered in December 1938 to VF-5 aboard USS *Yorktown*. They were to be the last of the US Navy's biplane fighters.

The twenty-seven F3F-3s represented the final development of the US biplane fighter. Its 950hp Wright R-1820-22 engine gave it a maximum speed of 264mph; to help to mitigate the inevitable 'built-in head wind' of the type, it had an enclosed cockpit, a variable pitch propeller and an undercarriage that was fully retractable. The Grumman F3F would continue in front-line service with the US Navy until 1940, at which time in Europe the Battle of Britain was being fought out with monoplanes at 350mph at 30,000 feet. However, American aircraft designers were as well aware as any others that the days of the biplane fighter were fast running out. In 1936 the Grumman Company had come to the conclusion that any significant improvement over their existing biplane designs would be impossible and they therefore produced a monoplane prototype, the XF4F-2 (0383). The XF4F-2 was flight-tested in 1937 and achieved only 290mph on a 1,050hp radial, and it was not selected at that time for the US Navy; instead — and very mistakenly — the Navy picked a rival design, the Brewster F2A Buffalo, which was chosen for production in 1938 and was later to be totally outclassed in action.

Fortunately the Grumman design had not been totally rejected; the Navy issued development contracts for the prototype, which was re-engined with a two-stage supercharged twin Wasp. The airframe was modified and detailed design changes made, all of which boosted its peformance to 333mph at 21,300 feet when test-flown in February 1939; these figures were near those of the contemporary European fighters. Fifty-four F4F-3s were ordered in August 1939 for the US Navy. The first production aircraft was rolled out from the Bethpage, Long Island, plant only six months later, in February 1940.

Though the F4F-3s were planned to serve initially aboard USS *Ranger* and *Wasp*, the first to fly in naval service were a French order of an export version, the G36A. The entire contract of eighty-one machines was taken over by Britain and the aircraft, known to the Royal Navy as the Martlet 1, were delivered to the Fleet Air Arm in June 1940, before the US Navy began to accept their F4F-3s to equip VF-4 aboard *Ranger*. British Martlets had the distinction of being the first US aircraft to shoot down an enemy plane during World War II, when a Ju88 fell to two

Martlets of 804 Squadron on Christmas Day 1940.

The F4F, or Wildcat, to use its American name, was to serve throughout the war, and over 7,900 were built; its story properly belongs to the next chapter; let it suffice for the moment to say that it was one of the timely naval fighters of World War II. Typical performance figures for an F4F-3 were — engine, one 1,200hp Pratt and Whitney twin Wasp radial; speed, 328mph at 21,000 feet; service ceiling, 37,500 feet; range, 1,150 miles. It was usually armed with four .50in machine guns, and could carry two 100lb bombs suspended from hardpoints beneath the wings.

Between the wars Britain pursued a policy known as the 'One Power Standard', which meant that the Royal Navy must at least equal in strength any other navy in the world. The standard was maintained in so far as battleships were concerned, but it was obvious by the early thirties that in aircraft carriers Britain was falling behind the United States. (She was also behind Japan, but that was not, it seems, officially admitted.)

Germany, having repudiated the Treaty of Versailles, concluded a Naval Treaty with Britain in 1935 which allowed the German Navy a total of 42,250 tons of aircraft carrier construction. The treaty, however, forbade the formation of a separate Naval Air Arm: the 'Unity of the Air' was forced on the Germans. One wonders if this was a conscious attempt thereby to inhibit her carriers' ultimate effectiveness. Two German carriers were laid down — Flugzeugträger A and B, each with a declared tonnage of 19,250 standard; in fact the true displacement projected was 23,430 tons standard. Only one of the two carriers — Flugzeugträger A, named *Graf Zeppelin* — was ever launched (at Kiel on 8 December 1938), but she was destined never to be completed.

The small amount of money voted for the British Naval Estimates had precluded the contemplation of any new construction before 1934, when an additional carrier was authorised. Though Britain was within the still valid treaty limits, the new carrier ordered in 1935 was to be of only 27,000 tons standard displacement, because of financial considerations.

Ark Royal, the first carrier to be laid down in Britain for seventeen years, was launched from Cammell Laird's yard at Birkenhead on 13 April 1937. During her construction extensive use had been made of arc welding, in place of the traditional rivets, which resulted in a considerable saving in weight; even so, her two hangars could only accommodate sixty operational aircraft. Though not, when compared with later British designs, heavily armoured, *Ark Royal* did have some measure of protection afforded to the aviation fuel stowage, magazines and machinery spaces.

Three lifts pierced the flight deck. Two of them, however, were only 22 feet wide, which precluded the ranging of any non-wing-folding aircraft, and all three lifts served only the upper hangar directly; aircraft had to be manhandled on to adjacent lifts to gain access to the lower hangar — a time-consuming operation even in a calm sea.

Ark Royal was 800 feet long overall, with a waterline beam of approximately 95 feet; her steel flight deck was 720 x 95ft. Eight arrester wires were installed and a crash barrier, the first to be fitted to a British carrier. '*Ark*'s' maximum speed from her three geared turbines was 31.75 knots, with a radius operation of 7,620 nautical miles at 20 knots. Her armament consisted of eight 4.5 inch, dual purpose guns in twin mountings, supplemented by six 8-barrelled pompoms and a similar number of .50 inch quadruple machine guns.

By 1935, the year in which *Ark Royal* was laid down, it was obvious that the international situation was fast deteriorating; once Hitler had assumed power in Germany in 1933, he quickly repudiated the Versailles Treaty and the country began openly to rearm. The possibility of a new European war became — to the realists — a distinct probability.

As a result, in 1935, even before *Ark Royal*'s keel had been laid, the Admiralty argued that it was imperative that the four existing carriers — *Argus*, *Eagle*, *Hermes* and *Furious*, all of which dated more or less from World War I — be replaced. Overt German rearmament speeded up the programme, loosening the tightly drawn government purse strings to the extent that two new carriers a year, commencing in 1937, could be built. The carriers were to be of the *Illustrious* class: heavily armoured ships capable of sustaining considerable damage and yet still remaining operational. To that end, out of a standard displacement of 23,000 tons, the class had no less than 5,000 tons of armour, 1,500 tons of which constituted the flight deck. The armoured construction of the hangar — only one was possible — reduced the number of aircraft embarked to thirty-three, but the decision was to be proved correct, for *Illustrious*, commissioned in 1940 — the first of her class — was herself to absorb more punishment than any other warship during World War II, while still remaining afloat and capable of steaming.

Although the Admiralty was able to order new carriers out of the Naval Estimates, paradoxically it could not order the aircraft to equip them; these were still in the gift of the RAF and, from 1936, the principal preoccupation of that service was to make good the years of neglect that had run it down to a dangerously low level of operational efficiency. As a consequence, the Fleet Air Arm came a poor second to the RAF in the matter of modern aircraft.

Between 1930 and 1939 the biplane was the Fleet Air Arm standard. The Flycatcher fighters were replaced in 1932 by Hawker Nimrods, which were a naval variant of the RAF Fury 1, that most elegant biplane, which had first appeared in squadron service in May 1931. Powered by an inline Rolls-Royce Kestrel of 525hp, it was the first RAF fighter to exceed 200mph in level flight (maximum 207mph at 14,000 feet). The Nimrod was not, as is often assumed, a navalised version of the Fury, but a distinct sub-type, built to Specification 16/30. It had a slightly longer span than the Fury, which increased its wing area from 252 square feet (for the Fury) to 298.5 square feet. The addition of naval equipment, arrester hook and flotation gear, increased the all up weight to 4,258lb, fully loaded — an addition of 768lb compared with the RAF fighter. Consequently, though both aircraft had virtually the same 525hp Kestrel engine, the Nimrod's maximum speed was 181mph at 13,000 feet, compared with the Fury's 207mph. Later a MkII Nimrod, with an uprated 640hp Kestrel V engine, managed 195mph, though again it compared unfavourably with its RAF counterpart, the Fury II, which, also powered by a 640hp Kestrel, had a maximum speed of 223mph. Thus the Navy's versions of what was an almost identical design remained 20mph slower.

The Fleet Air Arm Nimrods — fifty-six were produced, including prototypes — like so many of the biplane fighters of that era, were a delight to fly. They operated from HMS *Courageous*, *Glorious* and *Furious* and had a long service life; the type was still embarked in *Glorious* as late as May 1939.

The Fury and Nimrod came from the drawing board of the chief designer of the Hawker Company, Sydney Camm; he had earlier produced the two-seater Hawker Hart, a light bomber which entered squadron service with the RAF in 1930. The Hart airframe proved most adaptable and resulted in a number of 'Hart variants' which included the Hawker Audax, an army co-operation machine; the Demon, a two-seater fighter; the Hind, an improved Hart light bomber, later much used for training; and the Osprey, which entered service with the Fleet Air Arm in November 1932 as a two-seater carrier-borne fighter/reconnaissance aircraft.

The Osprey was also constructed as a seaplane, and extensively operated from catapults aboard HM cruisers. In landplane form it was embarked in most pre-war British carriers. The Osprey's maximum speed of 174mph was low for its fighter role, and the endurance of 2½ hours must have limited its reconnaissance operations. However, although it was a modification of a basic RAF type, the naval version had folding wings, built-in buoyancy and, after the initial batch, a stainless steel airframe to combat salt-water corrosion. There were four main production batches and in all some 109 Ospreys were produced, which continued in front-line service with the Fleet Air Arm until 1939.

One of the aircraft displaced by the Osprey was the Fairey IIIF, a large three-seater spotter which was powered by a Napier Lion engine. In the late twenties Charles Lindbergh, the famous American aviator, flew a IIIF from the RAF airfield at Northolt, just west of London: he was not, it seems, impressed: "A cavernous cockpit filled with nothing but smell and noise and me, supported by great shuddering wings strung together with random struts and string. The engine sounded somehow as if it had been running since the beginning of time, but that it would go on until the end."[39]

The IIIF was a viceless aircraft and popular with the FAA pilots. It was easy to land on board, though its low approach speed of 44 knots made, in a 20 knot wind, for long finals and some apprehension on the part of its pilots as to whether the carrier would ever be caught up at all.

Another Fairey aircraft, which first entered RN service with 825 Squadron, embarked in HMS *Glorious* in July 1936, was destined to become legendary. It was the Swordfish. It would serve with distinction throughout World War II to become Britain's last operational biplane.

The Swordfish was a torpedo/spotter/reconnaissance aircraft. The first prototype, TSR-1, was lost in a flying accident; the second, revised, aircraft TSR-2 (K4190) flew on 17 April 1934. Eighty-six Swordfish were ordered for the Fleet Air Arm in April 1935; by the time the aircraft was withdrawn from service in 1945, 2,391 machines had been built.

Affectionately nicknamed 'Stringbag', the Swordfish, though obsolete for much of its service life, was one of the great fleet aircraft, taking part in such actions as Taranto, the sinking of the *Bismarck*, and the Battle of the Atlantic, which will be recounted later. The flying qualities of this big, friendly biplane, with its single 690hp Bristol Pegasus radial engine, is best summed up by a man who flew them, Terence Horsley:

You could pull a Swordfish off the deck and put her in a climbing turn at 55 knots. It would manoeuvre in a vertical plane as easily as it would straight and level and even when diving from 10,000 feet her A.S.I. never rose much above 200 knots. The controls were not frozen rigid by the force of the slipstream and it was possible to hold the dive within 200 feet of the water .... The approach to the carrier deck could be made at a staggeringly low speed, yet response to the controls remained firm and insistent. Consider what such qualities meant on a dark night when the carrier deck was pitching the height of a house.[40]

There was really less excuse for the retention of the biplane naval fighter in Britain than in the United States. The Americans had no obvious enemy — Britain had Germany. The Luftwaffe was quickly built up and by 1936, when the new German Air Force was rehearsing over Spain, most of Britain was within modern bomber range. The prototype of the Messerschmitt Bf 109 fighter — powered incidently by a Rolls-Royce Kestrel V — was flown in September 1935 and quickly went into production. Two months later, the Hawker Hurricane, one of the Me 109's future opponents, had also flown, powered by a new 1,025 hp Rolls-Royce engine, the PV12, later to be named 'Merlin'. Using the same type of engine, another prototype, the Supermarine 300, first flew on 5 March 1936 from the factory airfield at Eastleigh, near Southampton; the beautiful monoplane achieved 349.5mph and went into production — for the RAF — as the Spitfire.

The Schneider Trophy had been won outright for Britain by the Supermarine S6B monoplane, which then went on to set up a world speed record of 406.99mph *in 1931*. It would seem reasonable to assume therefore that, given some government backing, the British aircraft industry could have produced something as good as, say, the Japanese Claude for the Fleet Air Arm long before the first monoplane actually appeared on a Royal Navy carrier in November 1938.

The monoplane in question was the Blackburn Skua. It was not, as might be supposed, a single-seater fighter, but a two-seater dive-bomber/fighter — an unusual and, indeed, near impossible classification, for the Skua, a neat enough, low wing, all-metal monoplane was far too slow to be an efficient fighter, even with four fixed machine guns in the wings. Its maximum speed was only 225mph at 6,100 feet, 20mph slower than the Gloster Sea Gladiator biplane fighter, which went into service with the FAA *after* the Skua in May 1939. The Skua also had the very modest ceiling for a fighter of 20,200 feet. As a dive-bomber, however, the Skua was a not unreasonable proposition for its day, though the bomb load was only 500lb and its range was limited to 760 miles. Skuas were embarked in *Ark Royal* to replace Nimrods and Ospreys at the end of 1938, part of an Air Ministry order of 190 machines (L2867 to L3056).

The Blackburn Company developed the Skua airframe into the Roc to meet another Air Ministry specification which had called for a two-seater carrier-borne fighter with a Bolton and Paul power-operated turret, armed with four .303 inch Browning machine guns. A total of 136 Rocs were ordered, though none served in their intended role for the very good reason that their performance as fighters was lamentable: the drag of the gun turret cut the maximum speed down to 194mph — less than that of the discarded Nimrod II, designed nine years earlier. Most Rocs spent their short service existence with their turrets removed, towing gunnery drogues.

Thus the biplane Sea Gladiator, already mentioned, was the fighter with which the Fleet Air Arm was to go to war against Germany in September 1939. It was not the fault of the Admiralty, which had unremittingly endeavoured to wrest control of its Air Arm from the Air Ministry since 1919. On 1 April 1939, twenty-one years to that not inappropriate day, the total control of the Fleet Air Arm became, at long last, the Admiralty's concern. It could not have occurred at a worse time: with World War II only five months away, the Fleet Air Arm overnight lost its RAF carrier pilots, including a number of squadron leaders and also, in the long run, more seriously, most of its highly skilled technicians — the engine and airframe fitters, radio men, instrument specialists. As the RAF flight sergeants and tradesmen trooped off the carriers to inland airfields, the Fleet Air Arm was left with a desperate shortage of trained ratings, which was to bedevil the service throughout the war. Five months was not long enough to reconstruct Britain's Naval Air Arm and make good all the long years of compromise and neglect. The total number of front-line aircraft on charge to the Navy on 3 September 1939, when Britain declared war on Germany, was 232 machines, of which only eighteen were modern Skua monoplanes.

4

World War II

At 1115 hours on the morning of 3 September 1939, practically the entire adult population of Great Britain was listening to the BBC, as the Prime Minister, Neville Chamberlain, announced the result of an ultimatum which had been presented to Germany that day:

> This morning the British Ambassador in Berlin handed the German Government an official note stating that unless we heard from them by 11 o'clock that they were prepared at once to withdraw their troops from Poland a state of war would exist between us. I have to tell you now that no such undertaking has been received and that consequently this country is at war with Germany ....

History has not treated Chamberlain kindly; his name is forever associated with the craven policies of appeasement; of the selling out of Czechoslovakia and Poland; of the false hopes of 'Peace in Our Time'. The Munich crisis a year earlier had been to many Britons a time of deep shame, but it gained for the country a priceless year in which to face up to the realities of Hitler's ambitions in Europe and to enable the Government of Britain to complete a series of modernisation plans, begun in 1936, to put the RAF on a footing of at least paper equality with the German Luftwaffe.

At the outbreak of war the RAF strength was 9,343 aircraft of all types, in 135 squadrons at home and overseas, manned by about 174,000 officers and men. The eight-gun monoplane fighters, the Spitfires and Hurricanes, were at last in squadron service — the latter in significant numbers.

No one today would deny that the expansion of the RAF was a vital necessity: what can be questioned is the eleventh-hour panic which caused the RAF growth to be at the expense of the Fleet Air Arm. Had the Admiralty been in charge of its Air Arm earlier, there is reason to think that the desperate lack of modern aircraft embarked in British carriers could have been alleviated. It is, of course, easy now to place the blame four-square on the shoulders of the Air Ministry; in point of fact there were still many admirals in the late thirties who did not think that the carrier was anything but an ancillary to the guns of the battle fleet. The subsidiary role of carriers had, in spite of all the advances, trials and war games in the years of peace, changed little in many minds from 1918.

Britain had nearly lost the war in 1917 owing to the ruthless unrestricted attacks by enemy submarines on her mercantile shipping, on which, as an island, she depended, not only for the effective prosecution of the war, but also for very survival. Thus, right from the beginning of World War II, the U-boat was regarded as a deadly opponent. To meet the very real threat of the German submarines, the Royal Navy had, between the wars, perfected an underwater detection system, ASDIC (later known by its American name of SONAR). The system, which worked with high-frequency underwater sound pulses, was eventually to prove successful, but the Royal Navy overestimated its effectiveness, particularly under operational conditions. Confident — overconfident — that the highly secret ASDIC would enable the escorts easily to detect any U-boats, three of Britain's six operational fleet carriers set sail on 3 September 1939 on anti-submarine patrols, covering the vital mercantile shipping lanes.

Ark Royal was to search in the North West Approaches; *Courageous* and *Hermes* in the South West. Before the ships were on station, however, SS *Athenia* was torpedoed and sunk without warning by *U-30* off Rockall, with the loss of 128 passengers and crew. The convoy system, which had stemmed the runaway victories of German U-boats in 1917, was at once reintroduced, but while it was being organised, aircraft from the three carriers flew anti-submarine searches in the hope that they could locate and attack any U-boats discovered; at the very least, the appearance far out to sea of aircraft would cause enemy submarines to dive for safety.

With the hindsight that bitter experience soon taught, it was a supreme folly to hazard half the Royal Navy's carrier force, ill-protected, in waters where German U-boats were known to be active. The insufficiency of ASDIC-equipped escorts, which were distracted away from the carriers they were supposed to be screening, while chasing false echoes and SOS messages from merchant ships, resulted in *Ark Royal* narrowly missing being torpedoed by a U-boat on 14 September 1939. Alerted by the near miss, the destroyers HMS *Faulkner*, *Foxhound* and *Firedrake*, of the *Ark*'s escort, located and sank *U-39*, the type IXA, 1,153 ton, ocean-going submarine which had attacked *Ark Royal*.

It was true that the escorts screening *Ark Royal* sank the U-boat, but the fact remained that *U-39* manoeuvred into an attacking position and was able to fire torpedoes at the carrier — which it very nearly hit —

despite the screening escorts and air cover. This should have been warning enough; it was not to prove so.

The same day as the attack by *U-39*, two Skuas of 803 Squadron embarked in *Ark Royal* caught another U-boat on the surface; this was *U-30*, a type VIIA, which had earlier sunk SS *Athenia*. The two aircraft dived to the attack and released anti-submarine bombs from a low altitude — too low, it would appear, since both the Skuas were brought down by their own bomb blast, the crews suffering the mortification of being rescued by the submarine they had ineffectively attacked, which, undamaged, returned triumphant to Germany. (*U-30* seems to have enjoyed a charmed life; it was one of the few pre-war U-boats to survive the war, being scuttled at Flensburg in May 1945.)

Three days later, while on patrol in the South West Approaches, HMS *Courageous* with two squadrons (811 and 822) of Swordfish embarked, intercepted an SOS from the passenger steamship SS *Kaliristan*, which had been stopped — strictly according to the Geneva Convention, it must be said — by a surfaced U-boat whose captain had ordered the passengers and crew to take to the boats, prior to the ship being sunk by the U-boat's gunfire. The ship's radio officer had only time to send a brief message which indicated a somewhat vague position on the Atlantic edge of the Bay of Biscay, roughly 100 miles to the south of *Courageous*. Eight Swordfish, armed with anti-submarine bombs, took off from the carrier; they fanned out to search for SS *Kaliristan*. One of the Swordfish sighted the sinking ship and attacked the still surfaced U-boat, causing it to dive; the aircraft then sent a W/T contact report to *Courageous*. Immediately two of the four destroyers that formed the carrier screen left *Courageous* and raced to rescue the passengers and crew from the *Kaliristan*'s boats.

As the final Swordfish of the search, down to its last drop of fuel, landed on, *Courageous* was protected by only two ASDIC-equipped destroyers. The pilot of that Swordfish, Charles Lamb, had just seen his aircraft struck below and was stepping into the wardroom with his observer, Lt Wall:

> I said to him: "What are you going to have?" but he had not time to answer because at that moment there were two explosions, a split second apart, the like of which I had never imagined possible. If the core of the earth exploded, and the universe split from pole to pole, it could have been no worse. Every light went out immediately and the deck reared upwards .... In the sudden deathly silence which followed I knew the ship had died.[1]

Twenty minutes later on that 17 September 1939, HMS *Courageous*, her hull ripped apart by two torpedoes from *U-29*, raised her stern high in the air and slipped out of sight into the cold waters of the North Atlantic, taking 500 of her crew with her. One of her escorts, HMS *Impulsive*, in spite of the known presence of the U-boat, hove to and picked up the survivors. HMS *Courageous* had the melancholy distinction to be the first aircraft carrier to be sunk.

The fact that *Courageous* was torpedoed in daylight by a U-boat, which incidently escaped unharmed in spite of an ASDIC escort and with anti-submarine aircraft aloft while the U-boat was manoeuvring into an attacking position, pointed to the dangers of using precious fleet carriers in submarine-infested waters. The Admiralty had no alternative but to call off the anti-submarine sweeps.

The loss of HMS *Courageous* was sustained in circumstances that could only be put down to inexperience and were not likely to be repeated. That aircraft carriers could be used effectively in an anti-submarine role was correct; that irreplaceable fleet carriers be so employed was not. Small expendable escort carriers built on merchant hulls would be developed for convoy protection and would play a vital role in the Battle of the Atlantic later in the war.

The next action to involve British aircraft carriers was the ill-fated Norwegian campaign, when, for the first time since 1918, the carriers and their aircraft would be in direct contact with the enemy.

Up to April 1940 the Scandinavian countries — Norway, Sweden and Denmark — were still neutral. The first military action to involve them was the *Altmark* incident. The *Altmark* was a German tanker suspected to have British prisoners on board; the vessel had taken refuge in a Norwegian fjord, pursued by a British destroyer, HMS *Cossack*, which, on 16 February 1940, cornered the ship, boarded her and released 300 prisoners of war. In so doing, *Cossack* had violated Norwegian neutrality.

It was not prison ships which really worried the British Government but iron ore, which Sweden was quite openly supplying to Germany through the port of Narvik and Norwegian territorial waters. There had been diplomatic moves to stem the flow of that vital war commodity but these had failed. Britain decided to adopt the only practical counter and laid minefields in the iron-ore shipping routes, having first warned both the Norwegian and Swedish Governments of her intentions some days prior to the actual mining, which had taken place on 8 April 1940. The timing was unfortunate, for it gave Hitler the excuse he needed: that the Allies were about to occupy Norway to prevent her supplying Germany with Swedish iron ore. The next day German forces struck; Denmark was invaded and in hours surrendered.

The same day as the Danish occupation — 9 April 1940 — a massive,

well-organised sea and airborne German forced attacked Norway,
troops quickly taking Oslo, Bergen, Trondheim and Narvik. The
Norwegians, unlike the Dutch, put up a spirited resistance and managed
to sink the German heavy cruiser *Blücher*, 13,900 tons, by gunfire and
shore-launched torpedoes, in Oslo Fjord. Britain's response was to
launch an immediate naval operation and to make preparations to land
Allied troops on the Norwegian mainland, supported by carrier-borne
aircraft.

At the time of the German invasion of Norway and Denmark the only
carrier in home waters was HMS *Furious*, which was on the Clyde
completing a refit. *Glorious* was with the Mediterranean Fleet, as was
Ark Royal (she was at Alexandria training her Swordfish pilots).
However, the *Ark*'s two Skua squadrons, Nos 800 and 803, had been
disembarked and were based at Hatson in the Orkney Islands, from
where sixteen of them took off to attack the German cruiser *Königsberg*,
5,600 tons, the Flagship of the Senior Officer, Scouting Forces, Norway,
which was moored to the Skoltegrund mole at Bergen, repairing minor
damage caused by a hit from a 6 inch shell. The position of the German
cruiser had been radioed to London by the Norwegian Underground,
and the Admiralty had decided to attack the temporarily immobilised
ship at first light on 10 April.

To attack the *Königsberg* involved a round trip flight of 600 miles,
most of it over water. The sixteen Skuas — seven from 800 Squadron and
nine from 803 — each armed with a single 500lb semi armour-piercing
(SAP) bomb, and fuelled to the brim, staggered from Hatson's short
runway and, in a loose formation, set course for Bergen 300 miles to the
east. In the rear cockpits the observers, with their 'Bigsworth' plotting
boards on their knees, checked and rechecked the dead reckoning track,
in the knowledge that the aircraft were operating at the very limit of their
endurance and that any navigational errors would inevitably result in
ditching in the cold northern waters.

After two hours of flying over the sea, a landfall was made, exactly on
course and within one minute of the planned ETA. As they sighted the
coast, the sun rose above the hills surrounding Bergen harbour; flying
towards the sun, the two squadrons climbed to 8,000 feet and then dived
at 60° towards the unsuspecting *Königsberg*. The leading Skua was down
to 4,000 feet before the cruiser's anti-aircraft guns opened up. (The
German crew's slow reaction was probably due in part to a standing
order that any single-engined aircraft over Bergen must be assumed to
be friendly.)

The long years of training that the naval pilots of the two squadrons had put in now paid off. Three bombs were direct hits; others were very near misses and damaged the cruiser with their mining effects; five bombs hit the stone mole alongside, raking the ship with fragments; two bombs burst between the ship and the mole. *Königsberg* instantly became a blazing hulk, slowly rolling over to one side; then, as the last of the Skuas pulled away, the German warship's magazine exploded, blasting the hull into two. The Skuas, having pulled out of their dives, zigzagged up the fjord, shooting up any shipping they encountered. They reformed over Lyso Island and set course for home; three had been hit and one aircraft, which was last observed pulling out of its dive and flying up the fjord was not to be seen again. Four and a half hours after taking off, fifteen Skuas landed back at Hatson; the loss of only one aircraft and slight damage to three others was remarkable, in view of the heavy curtain of fire put up by the German gunners, who had continued to serve their guns until their ship disintegrated.

The raid had been a classic example of a well co-ordinated dive-bombing attack; the mean bombing error was less than 20 yards. The *Königsberg* was the first enemy warship to be destroyed by aircraft alone, thus vindicating the prophesies of Mitchell and Curtiss. Ironically, the previous *Königsberg* had also made history, being the first ship sunk with the use of aircraft, twenty-five years earlier in the Rufigi Delta. Strangely, in spite of the triumph of the Skuas at Bergen and the success of the screaming German Stukas, which had blasted their way through Poland and would shortly do the same in France, neither the Fleet Air Arm nor the Royal Air Force pursued the concept of dive-bombing, preferring to use high-level bombing, which was far less effective against such small and mobile targets as ships.

The *Königsberg* attack was to be the only major success of British forces during the unfortunate Norwegian campaign. By 12 April 1940, just three days after first attacking, German forces had more or less complete control over Norway. To some extent they had been aided by sympathisers from within, led by Vidkun Quisling, whose name was to become synonymous with 'traitor', who set up a puppet pro-German Government. The Germans now had control of every major port and airfield, many crammed with fighters, bombers and transport aircraft.

Although with hindsight it is obvious that Norway was already a lost cause, at the time it was decided — disastrously as it turned out — to try to retrieve the position, not only with further air strikes, but also by landing British and French troops on the Norwegian mainland. Small forces were landed at Namsos and Åndalsnes, roughly 80 miles to the north and south respectively of Trondheim, to advance on that port. A further force landed at Harstad, 45 miles to the north-west of Narvik. All three operations were to fail: the Germans counter-attacked the forces advancing on Trondheim, forcing them back to their starting points. Further north, the troops at Harstad were impeded by deep snow and irresolute leadership. While the ground forces were fighting, the Navy flew in support; Swordfish from *Furious*, which had hurriedly rejoined the fleet, flew on strike and reconnaissance sorties for the Allied ground forces. However, on 25 April, *Furious* was damaged by near misses from German bombers and had to withdraw to Scapa Flow. She was replaced by HMS *Glorious*, and *Ark Royal*, which had been sent to the area from the Mediterranean Fleet.

The two carriers arrived off Norway on 24 April; *Ark Royal* had Skuas embarked, *Glorious* has six Skuas and also eighteen Gloster Gladiator biplane fighters of No. 263 Squadron RAF. The Gladiators, led by a Fleet Air Arm Skua, took off from *Glorious*, while 180 miles off the Norwegian coast, and flew to the frozen lake Lesjaskog, 45 miles south-east of Åndalsnes, which, with all the available airfields occupied by the enemy, was the only alternative base. It was an impossible location. Within an hour of their arrival, the machines were frozen to the ice; their engines were solid, as were the controls. Ground crews and pilots managed to unfreeze two of the fighters and miraculously start their reluctant engines; the fighters immediately took off to fly in support of the hard-pressed Allied troops in the Åndalsnes area. As soon as the first two Gladiators were airborne, the Luftwaffe arrived over the frozen lake and Ju88s and Heinkel 111s commenced bombing it. In spite of German attacks, the toiling men of 263 Squadron eventually got the remaining fighters into the air, though by now the lake was badly cratered with large holes in the ice caused by the German bombs. In continuous action from 24 until 28 April, 263 Squadron, fighting the élite of the Luftwaffe with obsolete biplanes, had destroyed fourteen enemy aircraft without losing a single fighter in combat; but ten machines had been put out of action on the lake, one had suffered engine failure and, with all fuel exhausted, the squadron was ordered to retire. The remaining three serviceable aircraft were set on fire by the ground crews as they withdrew.

While the RAF was desperately fighting from Lake Lesjaskog, fighter aircraft from *Ark Royal* and *Glorious* also began to fly over Åndalsnes and Namsos in support of the troops, the carriers being protected by combat air patrols of Sea Gladiators and a small number of Rocs embarked in *Ark Royal*. Swordfish and Skuas took part in strikes near Trondheim; Skuas from both ships heavily attacked Vaernes airfield and a seaplane base at Trondheim. By 28 April, though only three Skuas were shot down in action, *Ark Royal* had lost ten machines in accidents and had to withdraw. On 1 May *Glorious* beat off Luftwaffe shadowers with her Sea Gladiators while her Skuas protected a troop convoy; in the evening of that day, *Glorious* withdrew to Scapa. By 2 May the ground forces from both Åndalsnes and Namsos, their position hopeless, were evacuated by sea, the Royal Navy still having control of the approaches.

The evacuation of the forces that were to have converged on Trondheim did not end the ground fighting in Norway. *Ark Royal* sailed from Scapa on 4 May and from 6 to 19 May operated her Skuas and Swordfish, supporting the fighting in the Narvik area by giving the army reconnaissance cover and dive-bombing enemy rail traffic. The midnight sun of the Arctic summer meant that the aircraft were flying practically twenty-four hours a day. Reinforcements arrived off Norway on 18 May when *Glorious* and the repaired *Furious* returned. *Furious*, in addition to her own aircraft, had the irrepressible 263 Squadron embarked, with eighteen replacement Gladiators; *Glorious* had another RAF Squadron, No. 46, on board with sixteen of its Hurricanes.

The RAF fighters, led by a navigational Swordfish, flew off in the early hours of 21 May for the airfield at Bardufoss. The two RAF squadrons, supported by Fleet Air Arm aircraft, claimed thirty-six enemy aircraft shot down, and they managed to establish local air superiority long enough for the Allied troops to capture Narvik on 28 May.

Norway had become, by then, a sideshow: the Germans had begun their offensive in France on 10 May, ending the so-called 'phoney war', and events moved rapidly to the crisis that was to culminate in the Dunkirk evacuation. In the light of the disaster in France, the Allied Expeditionary Forces in Narvik began to withdraw on 2 June, after destroying the harbour installations. By 7 June the last of the Allied forces in Norway had left. Ten surviving RAF Gladiators at Bardufoss, though unhooked, were flown out to *Glorious* and safely landed on. Seven Hurricanes had also survived; their pilots were ordered to destroy them since it was not considered feasible to land these high-performance, unhooked fighters aboard the carrier, but aware of the desperate shortage of modern fighters in Britain and in view of the expected invasion of the country following Dunkirk, the RAF pilots pleaded to be allowed to try to land on the carrier. None of these men had ever made a deck landing, yet all seven got their Hurricanes on board without damage. It would have been a major feat even had the aircraft been equipped with arrester gear; these RAF Hurricanes had none.

The following day, *Glorious*, screened only by two destroyers, with all her aircraft struck below, was intercepted by the 31,000 ton German cruisers *Scharnhorst* and *Gneisenau*. Her 4.7 inch armament was no match for the 28cm (11 inch) guns of the enemy and, although the two destroyers made valiant efforts to cover the carrier's escape with smokescreens, all three ships, outgunned, were sunk. Of the RAF pilots who had fought from the frozen lake and who had tried to save their precious aircraft, only two survived. One of the destroyers, *Acasta*, managed just before she sank to hit *Scharnhorst* with a single torpedo, which caused that ship to make for Trondheim for emergency repairs.

After *Glorious* had been sunk, *Ark Royal*'s aircraft searched for the German battle-cruisers but she was herself spotted by enemy aircraft on 8 June. The next day, German high-level bombers attempted to attack the carrier, but the *Ark*'s Skuas drove them off, shooting one down.

Once the troopships that were carrying the Narvik forces back to Britain were out of enemy waters, *Ark Royal* left the convoy and launched a strike of Skuas to dive-bomb *Scharnhorst*, now known to be at Trondheim; fifteen aircraft took part, flown by picked crews. From air reconnaissance reports, the enemy was aware of *Ark Royal*'s presence; thus a surprise attack, such as was made on the *Königsberg*, was unlikely. To distract the enemy, RAF bombers were to attack simultaneously the German fighter base at Vaernes. Long-range Blenheim fighters from Britain were to provide an escort for the Skuas. Co-ordination was lost when the RAF bombed the airfield too soon, which alerted the defences; since the Skuas had to fly over occupied territory for some 40 miles en route to the target, this gave the defenders at Trondheim a precise ETA. In the face of intense flak and droves of fighters, the Skuas dived on to the *Scharnhorst*. Only one bomb was a direct hit and that failed to explode. Eight of the dive-bombers were shot down; the remaining seven fought running battles with German fighters as they withdrew and, when clear of the Trondheim area, met their RAF escort coming in. Among the pilots shot down were the COs of 821 and 823 Squadrons; all the pilots lost were experienced and irreplaceable men.

The Royal Navy made another attack on *Scharnhorst* on 21 June, as the German ship was returning to Kiel to complete repairs to the damage inflicted by *Acasta*'s torpedo. Six Swordfish from Hatson flew to the limit of their endurance to deliver a torpedo strike — the first by aircraft against a capital ship at sea — but sadly the crews had received little training in that most specialised form of naval warfare and no hits were scored. As they dropped their torpedoes, two of the Swordfish were shot down by the ship's secondary armament.

So ended the Norwegian campaign. It had one unforeseen and ironical outcome: the early failure of the attack on Trondheim caused the downfall, not, as might be expected of Winston Churchill, who as First Lord of the Admiralty had assumed responsibility for the disastrous campaign, but of the Prime Minister, Neville Chamberlain, who was forced to resign.[2] Churchill became Prime Minister and Minister for Defence on 10 May. It was, as A.J.P. Taylor wrote, "the appropriate moment. A real war had begun that morning: German armies had invaded Holland and Belgium".[3] On 13 May Churchill made to the House of Commons the first of his wartime speeches, which began: "I have nothing to offer but blood, toil, tears and sweat . . . .".[4]

The engagement off Norway had shown that, in spite of the opinions lately held by the Air Ministry, that fleet operations would be impossible within range of enemy shore-based air power, the carrier-borne aircraft of the Royal Navy had in fact done relatively well. The Luftwaffe possessed some 600 modern machines based in Norway, yet even the obsolete Gladiators and the heavy Skua dive-bombers had been able to give a good account of themselves.

The Gladiators had fought well, far better than their performance would have indicated: however, the Sea Gladiators flying combat air patrols over the carriers were, in the main, dealing with slow shadowing aircraft, of which the Dornier Do18 was typical. That flying-boat had a maximum speed of only 160mph at 5,800 feet and was easy meat for the agile biplanes. The German bombers were a different proposition: Gladiators could only just catch the Heinkel He111s, provided they had not dropped their bomb load; once they had, the He111s could outfly the biplanes. Against the Ju88, the Gladiator was totally outclassed; even the early Ju88A-1 series could achieve 286mph at 17,000 feet, so unless Gladiators could dive on to them, the bombers could easily outpace the 245mph fighters.

The Skuas, though they performed well as dive-bombers, were at 225 mph too slow, with a poor ceiling, to be effective as fighters, though in the fighting over Norway they, like the Gladiators, performed valuable service. The Norwegian campaign was really the high point of the Skuas' service life; they were progressively withdrawn soon afterwards, few remaining operational after the early months of 1941, though many continued with Fleet Requirement Units as trainers and target tugs.

Of the Fairey Swordfish which also took part in the Norwegian campaign, what can one say? It was by any test obsolete then, yet in fact it was only at the beginning of its active service career; it would outlive the Albacore, its intended successor, and would be flying in greater numbers

at the end of the war than at the outset. It was one of the great naval
aircraft.

Whatever the achievements over Norway, in the summer of 1940 it
was obvious that the existing aircraft, particularly the fighters embarked
in HM carriers, were inadequate against the Luftwaffe. The RAF pilots
of 46 Squadron had demonstrated the feasibility of operating the then
high-performance Hurricane from carriers, but as the Battle of Britain
was at that time being fought, and the RAF was scraping the bottom of
the barrel for any fighters (including Gladiators, which were defending
Plymouth), no Hurricanes, much less Spitfires, could be expected by the
FAA for some time. Thus the Navy had had to turn elsewhere for the
monoplane fighters which were about to be delivered to the Fleet Air
Arm: in mid-June 1940 HMS *Furious* was bringing the first of the
American Grumman Martlet 1s from the USA — French-ordered
G-36s — the beginning of a continuous stream of American naval
aircraft without which the Fleet Air Arm would frankly have been
unable to prosecute the war. *Furious* also ferried a small number of
Brewster Buffalos — B-339s — that were to have been exported to
Belgium. These aircraft, equivalent to the US Navy's F2A-2, were
evaluated at Hatson by 804 Squadron in July 1940; it was then discovered
that with armour and ammunition they could manage only about
270mph at 6,000 feet, against the 'brochure' figures of 313mph at 13,000
feet. The only operational use of the F2A-2 by the FAA was of a small
number in the Mediterranean, when the majority of their pilots declared
a preference for Gladiators.

The Martlet, on the other hand, was completely successful, though it
too had a somewhat reduced performance under active service conditions.
Nevertheless, its figures in operational trim of 305mph at 15,000 feet
made it the fastest carrier fighter available until mid-1941.

The Martlet had a rugged airframe which could be dived safely up to
400mph, and was manoeuvrable enough to turn inside most enemy
aircraft. The standard armament of four fixed .50in calibre machine
guns in the wings gave the Martlet a hitting power that no contemporary
adversary could long withstand.

The MkI fighter was powered by a 1,200hp Wright Cyclone G-205A,
air-cooled radial engine, which, like all radials, could withstand considerable rifle calibre machine-gun fire without stopping. The range on
standard tanks was 1,150 miles, or nearly twice that of the Sea Gladiator.
All in all, the Martlet was, without doubt, the best naval fighter available
in 1940; its only real drawback was its fixed wings, which made stowage
difficult. However, in late 1940, a folding version was ordered by the

Royal Navy; this, the MkII had a different engine — a 1,200hp Pratt and Whitney S3C4-G Twin Wasp, a power plant of monumental reliability. The MkII also had its armament increased to six .50 guns.

Though the Fleet Air Arm had to turn to America for the Martlet, it was not an entirely one-way traffic. The US Navy learnt a great deal from the operational experience of the Royal Navy and was able to incorporate its hard won lessons into subsequent Grumman fighters.

The Martlet first entered FAA service with 804 Squadron at Hatson in September 1940, replacing Sea Gladiators, three months before the US Navy issued the equivalent F4F-3 Wildcat to its fleet. Two of 804's Martlets shot down a Ju88 over Scapa Flow on Christmas Day, 1940, which, as previously mentioned, gained the type the distinction of being the first American fighter to destroy an enemy aircraft during World War II.

The first *British* monoplane fighter to serve with the Fleet Air Arm was the Fairey Fulmar. This eight gun, 1030hp Rolls-Royce Merlin powered aircraft, was a clean, two-seater, low wing machine which had a family resemblance to the ill-fated 'Battle' bomber, also produced by the Fairey Company. Unlike the Battle, which gained the distinction of suffering the highest loss rate, in relation to its numbers, of any bomber on RAF charge, the Fulmar was to prove successful.

The prototype Fulmar (N1854), which, one is pleased to report, still exists, preserved in the Fleet Air Arm Museum at Yeovilton, first flew at Ringway, Manchester, on 4 January 1940. Production orders for 250 Fulmar Is followed. They were true fleet aircraft with folding wings and full naval equipment. The type entered service with the FAA with No. 808 Squadron at Worthy Down in June 1940; later with 808 Squadron embarked in *Ark Royal*.

The Fulmars were to prove formidable naval fighters in spite of a low maximum speed of 280mph; their armament of eight Browning .303in machine guns gave them considerable hitting power. Terence Horsley summed them up well:

> There was never anything wrong with the eight-gun Fulmar. It was a fine aeroplane, manoeuvrable, with a good take off, moderate climb, and plenty of endurance. It satisfied the demands for a navigator's seat and [the] several wireless sets considered essential for Fleet work. It merely lacked the fighter's first essential quality — speed. Unless the pilot's first burst made a kill, he rarely got a second chance.[5]

Nevertheless, during their service life, Fulmars shot down some 112 enemy aircraft.

The appearance of the new fighters was timely: on 10 June 1940,

Mussolini, sensing easy pickings, had taken Italy into the war on the side of Germany. If at that time the Italian army might have been considered a joke, her navy was not; it contained several modern fast battleships and cruisers, though no operational carriers — Italy had an independent air force. Seen as a potentially strong naval power, Italy could effectively disrupt the Royal Navy command of the Mediterranean, closing the Suez Canal route to India and the Far East and preventing supplies to the British army in the Middle East, this at a time when, following the loss of HMS *Courageous* and *Glorious*, the Royal Navy carrier strength was seriously depleted.

HMS *Argus*, which had been in use as a deck-landing training carrier in the Gulf of Lions, of Toulon, departed temporarily from the Mediterranean for aircraft ferry duties and was, for a time non-operational. To compensate for this and the early losses, the first of the four *Illustrious* class armoured carriers under construction, *Illustrious* herself, was commissioned and had worked up off Bermuda to join the Mediterranean Fleet on 2 September 1940, with fifteen of the new Fulmars (806 Squadron) and sixteen Swordfish (815, 819 Squadrons) embarked. HMS *Eagle* was already in the Mediterranean, with two Swordfish squadrons (813, 824) and a fighter flight of six Gladiators on board.

The two carriers, which had a total strike capacity of thirty-six aircraft and an efficient defensive capability, aided by the new radar with which *Illustrious* was the first carrier to be equipped, enabled Admiral Cunningham to attack the Italians almost at will.

The attacks began on 4 September, just two days after *Illustrious* joined the fleet, with a strike on enemy airfields on the island of Rhodes, followed by extensive mining and dive-bombing attacks on Benghazi, which sank two destroyers and four merchant ships. *Illustrious* also took part in providing air cover to the almost continuous supply convoys to Malta. During these operations the Italian Navy showed little inclination to interfere; indeed, it was rather cruelly joked aboard the British ships that, whereas the Royal Navy drank rum, the Italians preferred to stick to port. That being so, in port was where the Italian Navy would be attacked. The result of that decision was a night action against the enemy Battle Fleet, which lay in the harbour of Taranto, in the heel of Italy.

The Battle of Taranto, as it is now known, was originally planned to take place on 21 October, Trafalgar Day, but an unfortunate fire in *Illustrious*'s hangar had damaged several Swordfish and the operation was postponed until the night of 11-12 November 1940. By that time *Eagle* was out of action; the mining effects of several near misses from bombs had shaken her venerable hull to the extent that her aviation fuel system was out of order. Six of her Swordfish were therefore flown on to *Illustrious* to bring the force up to twenty-one aircraft, it being planned to fly the strike in two flights an hour apart, but owing to a recent loss only twenty aircraft were to be available for the attack.

The initial strike of twelve aircraft was to be led by the CO of 815 Squadron, Lt Commander Kenneth Williamson. Six of the Swordfish were armed with torpedoes which had 'Duplex' pistols: that is, they could be detonated either by contact or magnetically, being set to run under the known anti-torpedo nets and explode beneath the Italian ships' hulls. Of the six remaining aircraft of the first strike, five carried 6 x 250lb AP bombs and one carried 4 x 250lb bombs and sixteen parachute flares to illuminate the target.

During the early evening of 11 November, with the lift warning bells clanging, the Swordfish of the first strike were brought up from the hangar and ranged on the flight deck. The wings unfolded, the Swordfish were armed and fuelled by the flight deck crews. Only pilots and observers were to be carried; the telegraphists/air gunners were left behind to save weight. Soon after 2000 hours, struggling ratings swung the handles of the inertia starters, faster and faster; the pilots clutched in the engines and one by one the 690hp Bristol Pegasus radials clattered into life, sounding more life farm implements than aircraft engines. As *Illustrious* turned into wind under an overcast sky, the pilots ran up the motors, testing magnetos, the aircraft behind rocking in the slipstream.

At 2030 hours Lt Commander Williamson, the strike leader, opened the throttle and the Swordfish slowly gathered speed; its tail rose and as the aircraft left the flight deck the next was already rolling. The Battle of Taranto was on.

As the aircraft slowly climbed over the Ionian Sea on their 2½ hour flight to the target, they soon ran into heavy cumulus cloud which persisted up to 7,500 feet, when in bright moonlight above the clouds only nine of the twelve Swordfish that had left the carrier could be seen. The three missing aircraft made their way independently to the target; unfortunately one of them did so at sea level, with the result that when it arrived outside the harbour fifteen minutes early, it was detected by sound locators which alerted the defenders. In spite of this, both the first and second strikes were highly successful and only two Swordfish — one from each flight — were shot down, with the loss of one two-man crew.

Charles Lamb was one of the pilots in the first wave who had been charged with the flare dropping. He thus had a grandstand view of the raid from 5,000 feet and clearly saw the Swordfish attacking below:

. . . . flying into the harbour only a few feet above sea level — so low that one
or two of them actually touched the water with their wheels as they sped
through the harbour entrance.

Nine other spidery biplanes dropped out of the night sky, appearing in a
crescendo of noise in vertical dives from the slow moving glitter of the yellow
parachute flares.[6]

The result of the raid was impressive. The battleship *Conte di Cavour*
sank; the new battleship *Littorio* was struck by at least three torpedoes,
putting her out of action for six months; the battleship *Caio Duilio* had to
be beached to prevent her from sinking; and three cruisers and a
destroyer were hit, in addition to extensive damage to harbour installations.

For the loss of two men and two aircraft, the Royal Navy had inflicted
a crushing blow on the Italian Navy. Had Taranto been fought as a naval
battle between surface vessels, it would have been hailed as one of the
greatest victories since Trafalgar. As it wasy, twenty Swordfish had
achieved more in one hour than the entire Grand Fleet at Jutland, which
had cost the Royal Navy over 6,000 men killed and the loss of fourteen
ships. The significance of the raid was perhaps underrated by the Royal
and US Navy's implacable battleship men, who lost no time in pointing
out that the Italian ships were unable to manoeuvre; the the results of the
Battle of Taranto were studied in the minutest detail in Japan . . . .

Following the Taranto action, the respect the Italians now held for the
Fleet Air Arm was such that the mere appearance of naval aircraft was
sufficient for the Italian fleet to break off action and run for home, as
happened during the inconclusive Battle of Cape Spartivento, when *Ark
Royal*'s Swordfish unsuccessfully attacked two Italian battleships, which
promptly turned and retired. *Ark Royal* was, in turn, attacked by Italian
high-level bombers, which flew in impeccable formation, greatly aiding
the task of the ship's anti-aircraft defences and fighters. Four of the
bombers were shot down by the *Ark*'s Fulmars, at which several
bombers jettisoned their bombs; others, however, continued with the
attack on the carrier, which disappeared in a great upheaval of water
from which she emerged unscathed.

By the end of 1940 the only limitation placed on the activities of the
Fleet Air Arm in the Mediterranean was a shortage of aircraft; though in
point of fact few had been destroyed by enemy action, the extremely
arduous life imposed by carrier operations under active service conditions
had reduced the numbers remaining serviceable. The Navy was always
short of skilled artificers and of replacement aircraft and spares; stocks of
reserve aircraft, particularly Fulmars, were exhausted and Sea Gladiators
were embarked once again in *Eagle* and *Illustrious* to keep the fighter

squadrons up to strength. It was at this time, early 1941, that Admiral Cunningham was offered the Brewster Buffalos, but they were declined as inferior even to the obsolete Gladiator biplanes. Apart from the physical difficulties of supplying aircraft from the United Kingdom, the temporary cessation of British naval aircraft construction, which had been imposed during the desperate summer of 1940, when the factories were fully engaged producing fighters for the Battle of Britain, was now being felt and would continue to restrict the replacement programme for some time to come. But the shortages did not prevent aircraft from *Illustrious* flying in support of the British Eighth Army advancing in Cyrenaica against the Italians and, at the eastern end of the Mediterranean, aircraft from *Eagle* (now repaired) raiding Tripoli, as did Swordfish from *Illustrious* some days later. These aircraft had also torpedoed two Italian supply ships off Sfax.

Following the successful British operations against the Italians, an older and very much more formidable adversary was now to join in the fighting in the Mediterranean — the Luftwaffe.

In January 1941 Hitler had decided to send the Afrika Korps into the North African desert to try to save his Axis partner from humiliating and total defeat in Cyrenaica. To enable the desert army to be supplied, convoys would have to cross the central Mediterranean, which was now very much under the domination of the Royal Navy. The Germans knew that if they were to wrest control of the vital convoy routes, the Royal Navy, and in particular its aircraft carriers, would have to be countered.

The man appointed to achieve control was General der Flieger Geisler, commanding a crack Luftwaffe anti-shipping unit, Fliegerkorps X, which had been serving in Norway. The sinking of the *Königsberg* in Bergen Harbour by Fleet Air Arm Skua dive-bombers may have influenced the general, or it may simply have served as a confirmation that the best form of airborne attack on shipping was not the high-level bombing favoured by most air forces, in spite of contrary evidence, but dive-bombers, and Fliegerkorps X had some of the most effective: 150 Junkers Ju87 Stukas.

The Stukas, under the command of Oberst Harlinghausen, were based at Comiso and Catania, in Sicily, where they began to train on a floating target which represented a British carrier. Oberst Harlinghausen was of the opinion that four direct hits with 500lb bombs could sink a carrier, and that that number of hits could be achieved without too much difficulty. The killing ground was to be the Sicilian Narrows between the island of Sicily and Cape Bon, on the North African coast. Sooner or later a convoy from Gibraltar to Malta or Alexandria would have to pass through the Narrows and such convoys were invariably escorted by a carrier. The Stukas trained and bided their time on their Sicilian bases . . . .

They did not have long to wait. On 6 January 1941 five large modern merchant ships, loaded to capacity with supplies for the desert armies, set sail from Gibraltar for the Egyptian port of Alexandria, initially escorted by elements of Force 'H', comprising the battleships HMS *Renown* and *Malaya*, the carrier *Ark Royal* and several destroyers. The convoy, code named 'Excess', was by dawn on 10 January to be taken under the protection of the Mediterranean Fleet in the Sicilian Channel by HMS *Warspite* and *Valiant*, the cruisers *Gloucester* and *Southampton* and the carrier *Illustrious*. The transfer of escorts duly took place without incident and Force 'H' returned west to Gibraltar. The convoy steamed into the Narrows and the trap that was about to be sprung.

On their airfields in Sicily the Stukas were bombed up ready for the convoy, which had been reported by German reconnaissance aircraft and shadowed since the previous day. At dawn on 10 January two Italian destroyers, presumably unaware of the impending attack, sailed into the area and were immediately engaged by the surface fleet; one, *Vega*, was blown out of the water and the other left the scene at full speed. By now the convoy was off the island of Pantellaria and the first phase of the air attack commenced. At 1223 hours Italian Savoia-Marchetti SM79 Sparviero (Hawk), three engined bombers of the Aerosiluranti (torpedo unit) of the Regia Aeronautica, attacked HMS *Valiant*. The battleship avoided all the torpedoes but the subsidiary purpose of the low-level attack, to lead *Illustrious*'s Fulmars away from the carrier, succeeded. Five of the fighters chased the SM79s: as they were using up their ammunition on the bombers, at 1230 the radar on *Illustrious* reported a large formation of enemy aircraft approaching from the south. At this, Captain Boyd of *Illustrious* asked permission to turn the carrier into wind to recover his airborne fighters for rearming and to launch a fresh range of six Fulmars which were waiting to take off. Before the ship could turn, the Stukas of Fliegerkorps X attacked.

Forty-three Ju87s, in two groups — 11/St.G2, led by Major Enneccerus, and 1/St.G1 under Hauptmann Hozzel — were rolling over into near vertical dives from 12,000 feet. Ten aircraft broke away to attack the battleships, but the main target was the long broad flight deck of *Illustrious*. Fascinated, the men on the ships watched as the Stukas, defying murderous anti-aircraft fire, split into a clover leaf formation. It was a superb display of disciplined flying. Admiral Cunningham, watching in his flagship *Warspite*, later wrote:

One was too interested in this new form of dive-bombing attack to be frightened, and there was no doubt that we were watching complete experts .... We could not but admire the skill and the precision of it all. The attacks were pressed home at point blank range, and as they pulled out of their dives some were seen to fly along the flight deck of the *Illustrious*·below the level of the funnel.[7]

The Stukas released their bombs from between 1,200 and 800 feet. From that first attack, which lasted for only a few minutes, the dive-bombers scored six direct hits and three very near misses. *Illustrious* had a 3 inch armoured deck which was designed to withstand a direct hit from a 500lb bomb, but some of the Stukas were armed with 1,100lb armour-piercing bombs and no practical armour could withstand that.

The firs two 1,100lb bombs did not hit the flight deck directly: one landed on a gun position, a second plunged through decks and set fire to a paint store. A third bomb of 500lb hit the after lift, which was unarmoured; the blast sent the lift platform crashing down into the hangar. Another 500lb bomb crashed down the same lift well, exploding inside the hangar, setting fire to several aircraft; a fifth bomb struck the forward lift directly, and a sixth penetrated the armoured flight deck by the lift. The ship was now blazing from numerous fires, her hangar an inferno; the steering was out of action, owing to 'whip' effect and the flooding of the steerage flat, caused by the numerous hoses of the fire fighters and sprinklers.

The armoured construction of *Illustrious* was such that throughout the action her machinery spaces were undamaged, which left her pumps and fire-fighting mains operational; her hull remained intact and the aviation fuel coffer dams unbreached. Well trained damage control parties toiled effectively in conditions of extreme hazard and the ship was gradually got under control, though for a time steering was only possible by means of the engines. The ship's speed never fell below 18 knots, in spite of the stokehold temperature rising to 140°. Throughout the action the gunners continued to fire the remaining guns and four Stukas were shot down by the combined efforts of the ship's guns and Fulmar fighters.

Later that day high-level bombers ineffectually attacked the ship. But *Illustrious* got to Malta under her own steam to tie up in Valetta's Grand Harbour, to quell the fires, to make emergency repairs and collect her dead.

Even when at Malta, *Illustrious* was far from safe, though RAF Hurricanes based on the island and the carrier's Fulmars — which, unable to land on, had flown the 60 miles from the point of attack to

Malta — intercepted the attacking bombers to such good effect that nine were shot down and the subsequent dive-bombing was not pressed home with quite the same precision of the earlier raids. Nevertheless, in harbour, the carrier sustained two more direct hits and near misses, causing further damage and bringing her total casualties to 126 killed and 91 badly wounded.

By the evening of 23 January 1941 *Illustrious* was made seaworthy and left Malta, working up to 25 knots, and two days later passed through the Suez Canal and steamed for the Norfolk Navy Yard in America, where extensive repairs were put in hand. During the bombing *Illustrious* absorbed punishment that would have sent most contemporary carriers to the bottom; as it was, her damage was so extensive that she was not to return to action until late 1941.

Although *Illustrious* survived the dive-bombing attacks in the sense that she was still afloat and able to steam, she had effectively been put out of action for nearly a year. The ship had therefore not lived up to the expectations of the Naval Staff who had ordered the armoured carriers. The whole point of the original concept was that a carrier could take the heaviest bomb likely to be used against it on its flight deck and yet still be capable of operating its aircraft. To achieve that end the flight deck, which of course formed the hangar roof, was constructed of 3 inch armour plate, as was the floor of the hangar; the sides were also armoured, though not so heavily, but were additionally protected by fire-proof lobbies. Thus the whole hangar was an immensly strong box, capable of withstanding considerable weight of explosives. The price that had to be paid was first and foremost a much reduced aircraft storage capacity. The carriers of the *Illustrious* class could accommodate about thirty-three aircraft in their hangars, yet the earlier *Ark Royal*, which had only an armoured floor to her hangar, could embark 40 per cent more aircraft than the *Illustrious* class, although her displacement was, at 22,000 tons, 1,000 tons *less* than the armoured ships.

Had *Illustrious* been able to continue to operate her aircraft, the decision might have been justified; as it was, it was doubtful if the armour was worth the reduced operational capacity. It could be argued that if *Illustrious* or her sisters had been built with the American type of lightly constructed hangar, which was the full width of the ship, they could have embarked enough fighters to enable them to drive off the dive-bombers. The Stuka was a good less formidable than German propaganda claimed: after the initial shock of their rather theatrical performance in Poland and France, they were found by RAF fighter pilots to be easy targets, and were shot down in such numbers during the

Battle of Britain that they were soon withdrawn from that conflict. To be fair to the Naval Staff, it must be remembered that armoured carriers were designed in the mid-thirties and the protection envisaged was against plunging fire from 6 inch naval shells, or the heaviest bomb that the RAF could drop, which at that time was 500lb — somewhat lighter and with far less penetrating power than the German armour-piercing bombs of 1941.

The enforced departure of *Illustrious* for America left the Royal Navy with only the old and unarmoured *Eagle* in the Central Mediterranean; it was a difficult time, for the vital Sicilian Narrows were now virtually closed to the convoys by the German dive-bombers, certainly during daylight. The situation was rendered the more serious by the Afrika Korps, under General Erwin Rommel, advancing through Cyrenaica, making the supply of reinforcements and war materials to the hard-pressed British forces and urgent priority.

The only possible solution was to withdraw *Eagle*, which was in any case overdue for a refit, and replace her with the second of the new armoured carriers, HMS *Formidable*, which had been commissioned the previous November and which was now (January 1941) in the South Atlantic hunting German surface raiders. The decision having been made, *Eagle* escorted her last convoy from Port Said to the Greek port of Piraeus in February and then turned round to pass through the Suez Canal into the Indian Ocean. While the ship was slowly sailing through the canal, her Swordfish were disembarked and, in the space of a fortnight or so, supported British ground forces with considerable success before rejoining *Eagle* at Suez.

Formidable joined the Mediterranean fleet on 10 March 1941 with four FAA squadrons embarked: 803 with twelve Fulmars, 806 with eight Fulmars, 829 with nine Albacores and Swordfish, and 826 with twelve Albacores. The Fairey Albacore was a new aircraft which had originally been ordered in 1937 as a Swordfish replacement. It was in many ways an improvement on the earlier aircraft: though it was still a biplane, it nevertheless boasted an all-metal monocoque airframe with an enclosed cabin for its crew of three. The engine was a 1,065hp Bristol Taurus II, a smooth running and noticeably quiet sleeve-valve radial, which gave the Albacore a maximum speed of 161mph at 4,000 feet. Its range was 930 miles with a 1,610lb warload, usually an 18 inch torpedo or 4 x 500lb bombs slung under the wings. Despite the marked advantages over the 'Stringbag', the older aircraft was to outlive its intended successor. The first major engagement for the Albacore was to be the Battle of Cape Matapan in March 1941.

Admiral Cunningham, now that he once again had an armoured
carrier at his disposal, lost no time in seeking out the elusive Italian fleet.
Late on 27 March an RAF reconnaissance aircraft based on Malta
sighted elements of the enemy fleet steering for Crete; Cunningham,
flying his flag in the battleship *Warspite*, immediately sailed from
Alexandria in pursuit, accompanied by the battleships *Valiant* and
Barham, the carrier *Formidable* and destroyers. *Formidable* flew off
Albacores, at dawn to search; the enemy was located and shadowed all
that day, 28 March 1941. RAF high-level bombers attacked, with no
hits being scored, but the raid at least distracted the Italian anti-aircraft
fire from a torpedo strike by Albacores from *Formidable*, which resulted
in the battleship *Vittorio Veneto* receiving a torpedo, which slowed her
down to 8 knots. Her crew, however, patched her up and she managed to
escape. Another torpedo strike was launched from the carrier as darkness
was falling; in the absence of the *Vittorio Veneto*, the heavy cruiser *Pola*
was hit and stopped. The Italian commander of the cruiser force,
Admiral Angelo Iachino, being denied air reconnaissance by the Fulmars
from *Formidable*, did not know the position or composition of the British
fleet; he detached two of his 8 inch gun cruisers, *Fiume* and *Zara*,
together with two destroyers, to stand by the crippled *Pola*.

Meanwhile, Admiral Cunningham's fleet raced at full speed to try to
intercept the damaged battleship *Vittorio Veneto*, not knowing that the
ship had made good its escape. At around 2215 hours *Valiant* reported a
radar contact which was identified visually with night glasses as two
heavy cruisers of the *Zara* class, with another vessel ahead. On sighting
the enemy, the Battle Fleet turned into line ahead.

Using their radar, Cunningham's fleet rapidly closed on the Italian
ships. As the 15 inch guns of the British ships trained on the enemy
cruisers, it was obvious that, lacking radar, in the darkness of the
moonless night, they were unaware of the presence of the Royal Navy,
for their guns were seen to be trained fore and aft.

Admiral Cunningham, on his bridge in *Warspite*, never forgot the next
fifteen minutes of what was to be one of the last battleship actions fought
by the Royal Navy:

> In the dead silence, a silence that could almost be felt, one heard only the
> voices of the gun-control personnel putting the guns on to the new target
> . . . . looking forward, one saw the turrets swing and steady when the fifteen
> inch guns pointed at the enemy cruisers. Never in my whole life have I
> experienced a more thrilling moment when I heard a calm voice from the
> director tower — "Director layer sees the target".

The next moment there was a blinding orange flash and numbing concussion as the *Warspite* fired a broadside from six 15 inch guns at the point blank range of 3,800 yards. The destroyer *Greyhound* illuminated the target. The first salvo was a direct hit; the Italian ship disintegrated, with complete turrets blown skywards. The other British ships also pounded the unfortunate enemy and in fifteen minutes both Italian cruisers and a destroyer were sinking. *Pola* was discovered wallowing in the calm sea and HMS *Jervis* went alongside. Conditions on board were indescribable; no resistance was offered since many of her crew were preparing to abandon their ship. *Jervis* took off the Italians and then sank the *Pola* with torpedoes at 0410 hours on 29 March 1941. So ended the Battle of Cape Matapan. Five enemy ships had been sunk for the loss of one Albacore and its crew.

The battle was a demonstration of the classic use of an aircraft carrier as envisaged before the war: the shadowing and torpedoing of enemy battleships to enable the guns of the fleet to engage. On only one further occasion would this happen, the last great gunnery engagement between battleships — the action now known as 'The Sinking of the *Bismarck*'.

Bismarck was one of two heavy battleships (the other being *Tirpitz*) which were the biggest units of the German Navy, *Bismarck* displacing 41,700 tons (though to comply with the Anglo-German Treaty of 1935, the tonnage was announced as only 35,000 tons). The ship had over 12 inches of main armour, carried eight 15 inch guns and was capable of 30 knots. It would not be an exaggeration to state that in 1941 *Bismarck* was the most formidable capital ship afloat.

Although the sinking of *Bismarck* was to be achieved by the guns and torpedoes of the Royal Navy's surface units, without the Fleet Air Arm it is most unlikely she would have been brought to action, for aircraft were to play a decisive role from first to last.

From intelligence reports it was known that *Bismarck* was, in May 1941, anchored in the Polish, German-occupied, Baltic port of Gdynia; what was not known was that on 18 May *Bismarck*, wearing the flag of Admiral Lutjens, had sailed in company with the heavy cruiser *Prinz Eugen* (13,900 tons). The first intimation that the Admiralty had of the battleship's movements came on 20 May when a Norwegian agent radioed London that heavy German naval units were passing through the Skagerrak. Confirmation came when a reconnaissance aircraft of RAF Coastal Command located and identified the two ships anchored in Norwegian waters in Kars Fjord, to the south of Bergen. The Admiralty at once alerted Coastal Command to keep a watch on the two ships. It was a very serious situation, for if, as seemed the case, *Bismarck* were to slip out into the Atlantic, she would be able to attack the vital convoys like a fox in a hen run. The Battle of the Atlantic was going badly enough as it was, with German U-boats sinking shipping at a rate that was fast getting beyond the capacity of the yards to replace; *Bismarck* and *Prinz Eugen* could tip the balance irretrievably in the enemy's favour.

The difficulties facing the Naval Staff were daunting. *Bismarck* could well remain secure in the Norwegian fjord indefinitely, posing a 'Fleet in Being' threat that would tie down heavy British forces; she could, on the other hand, sail into the Atlantic at any moment. If that was accepted as the most likely eventuality and the Royal Navy sent a heavy force to sea to stand by to intercept, they could well find themselves short of fuel at the critical moment. The only possible tactic was to keep the German units under a twenty-four hour surveillance and then, when they had been seen to leave, to put the fleet to sea to intercept and engage.

At Scapa Flow the Home Fleet, comprising the battleships *King George V, Prince of Wales, Hood* and the fleet carrier *Victorious*, with attendant cruisers and destroyers, was ordered to immediate readiness.

The task of maintaining the air reconnaissance was entrusted to RAF Coastal Command. However, on 22 May, the RAF informed the Admiralty that, owing to deteriorating weather, none of their aircraft had been able to fly over Kars Fjord and that further flights were out of the question since the searching aircraft were now grounded.

At the Fleet Air Arm station at Hatson in the Orkneys the CO, Captain Fancourt, had a strike of torpedo-armed Albacores standing by to intercept *Bismarck* the moment she was reported to have left the Fjord. The captain became increasingly impatient with the RAF's apparent lack of concern; since they could or would not fly a reconnaissance sortie, he decided that the Fleet Air Arm would fly one for themselves.

Under his command were several highly experienced men for whom bad-weather flying over the sea was no deterrent; one was Commander Rotherham, an ex-observer whose active flying had been in pre-radio-aid days and whose ability to navigate by dead reckoning was exceptional. A pilot, Lt N.E. Goddard, the CO of 771 Squadron — a target-towing unit — volunteered with a telegraphist named Armstrong to fly an unarmed target tug, an American Martin Maryland, over the Fjord, 300 miles away.

It was an epic flight, mainly on instruments in 10/10ths cloud which was at times down to sea level; Lt Goddard, the pilot, skimming above the sea to enable Commander Rotherham, who was navigating, to obtain the vital drift sights to keep on track. Flying low in cloud at 270mph towards a mountainous coastline required a dedication to duty and a confidence in the navigation that was exceptional. That confidence was justified: the cloud base lifted sufficiently for the primary landfall, Marsten Island, to appear exactly on time and only a few hundred yards to starboard. After 300 miles of blind flying over the sea, that flight, as a feat of dead reckoning navigation, must rank very highly.

The cloud base was still less than 1,000 feet as Goddard turned the twin-engined Maryland into Kars Fjord. It was deserted, with no sign of *Bismarck* or *Prinz Eugen*; after flying up and down other nearby, equally deserted, fjords, Rotherham suggested seeing if the two ships had put into Bergen. As they flew over the harbour, the aircraft shuddered from repeated near misses from heavy and accurate flak, which riddled the fuselage and wings with shrapnel, miraculously without holing the non-self-sealing fuel tanks.

Seeing no sign of the battleships in the harbour or the adjacent Bergen Fjord, Goddard thankfully pulled the Maryland up into the cloud and set course for Hatson. Commander Rotherham was now convinced that *Bismarck* and *Prinz Eugen* had sailed. Realising that the aircraft could have sustained serious damage that might well manifest itself at any moment, or that the German fighters, which would most certainly have been scrambled, would find the unarmed plane in a break in the clouds, Rotherham decided that the extreme importance of the news of the departure of the German ships outweighed any considerations of radio silence; he therefore instructed his telegraphist to signal a report.

Armstrong tried for some time to contact RAF Coastal Command but he could not get any reply; so he then retuned his set to the training frequency of 771 Squadron and called them. In the target-towing aircraft flying around Scapa Flow the bored telegraphists heard the faint morse and found themselves copying one of the most vital operational signals of the entire war. The message was at once passed from 771 Squadron Operations Room to Sir John Tovey, C-in-C Home Fleet, at Scapa Flow. The Fleet was ordered to sea and thus began one of the greatest chases ever staged on the high seas.

The opening moves of the *Bismarck* action were classic. First cruisers scoured the ocean to try to make contact with the enemy and then report and shadow. Such was the respect the Admiralty held for the all-powerful *Bismarck* and her smaller consort that the entire Home Fleet, comprising four capital ships, three battle-cruisers, fourteen light cruisers and no fewer than five destroyer flotillas steamed north to intercept.

At 0722 hours on 23 May 1941 the county-class cruiser *Suffolk* reported a radar contact with *Bismarck* and *Prinz Eugen* in the Arctic waters of the Denmark Strait between Iceland and Greenland. The two

German ships were skirting the pack-ice off Greenland, steaming at high speed. *Suffolk*, accompanied by another cruiser, *Norfolk*, shadowed all that day by radar, through heavy seas in blinding snowstorms. The two German ships tried desperately to throw off the shadowing British cruisers, which grimly maintained contact at full speed, their radar operators intent on the shimmering green screens, constantly reporting changes of course and speed of the enemy, which were signalled to the C-in-C Home Fleet.

This phase of the operation was an early example of electronic war. Smoke, blizzard and darkness made not the slightest difference: their radars enabled the shadowers to remain in contact yet to be out of effective gunnery range of their quarries. This was the new warfare, which the men of the old Dreadnoughts could not have comprehended.

Acting on the reports from the shadowing cruisers, the C-in-C, Admiral Sir John Tovey, dispatched the old battle-cruiser *Hood*, 42,500 tons, and the battleship *Prince of Wales* — two of the most powerful units of the Royal Navy — to engage the enemy.

At 63°12′ North, 32°00′ West, as dawn broke the next day, 24 May, *Hood* and *Prince of Wales* made radar contact with *Bismarck* and *Prinz Eugen*. HMS *Hood*, the ship which epitomised the supremacy of the Royal Navy, opened fire at a range of 15 miles, her salvoes quickly straddling *Bismarck*, which at once returned fire with deadly accuracy, for, to the horror of the men on the British ships, *Hood* simply disintegrated when her magazines exploded under direct hits from *Bismarck*'s 15 inch guns. Of her crew of 1,419 officers and men, only three were picked out of the freezing sea alive by the destroyer *Electra*.

Prince of Wales continued to engage *Bismarck* and scored hits, but was hit in return; one 15 inch shell demolished the bridge, killing everyone on it except Captain Leach and a signaller. *Prince of Wales* was a brand new ship which had been working up when ordered to engage *Bismarck*; she still had many civilian technicians on board, such was the haste in which she had left Scapa. Defects in the gun turrets and mounting damage from repeated hits and near misses compelled *Prince of Wales* to break off the action. *Bismarck* did not attempt to pursue her, for she too was damaged and was leaving a long trail of fuel oil in her wake, though she was still capable of steaming at 28 knots and her fighting efficiency was unimpaired.

When the loss of HMS *Hood* was reported to the C-in-C and to the Admiralty, it was received with total incredulity, for *Hood* was by far the most famous ship then serving with the Royal Navy. After the initial shock and sorrow, however, another motive for destroying *Bismarck* now entered into consideration. Revenge. The fleet was scoured for ships; escorts were withdrawn from convoys; ships working up and training were ordered immediately to join the fleet; Force 'H' sailed from Gibraltar into the Atlantic. In all, thirty-nine warships were now hunting for one — the *Bismarck*.

While all these vessels were converging, the two cruisers *Norfolk* and *Suffolk* continued to maintain radar contact with *Bismarck* (*Prinz Eugen* had separated and took no further part in the action), while *King George V*, *Repulse* and the carrier *Victorious* closed from the south-east. Admiral Sir John Tovey, worried that *Bismarck* could give her numerous pursuers the slip during the night, ordered *Victorious* to fly off a strike of torpedo aircraft to try to slow down the German ship. When ordered into the *Bismarck* action, *Victorious* had been on the point of sailing for Gibraltar with a cargo of crated Hurricane fighters for the RAF in her hangars; she had only nine Swordfish torpedo planes and six Fulmar fighters operational — hardly an adequate force to attack the most heavily armoured battleship in the world.

The nine Swordfish of 825 Squadron were ranged and flew off in very bad weather and poor visibility. Few, if any, of the young pilots had any experience of action, for the squadron had only just been formed and had been working up prior to joining the new *Victorious*. Nevertheless, the attack was pressed home and one, possibly two, torpedoes struck *Bismarck* amidships where her armour was thickest, killing one man — the first German casualty of the action — but causing little damage, though the violent manoeuvres at high speed to avoid the Swordfish attack caused the earlier damage to a fuel tank, inflicted by the *Prince of Wales*, to become more extensive. More fuel oil was being lost and speed was reduced to about 18 knots.

After the first Swordfish attack, two Fulmars from *Victorious*, which were shadowing the *Bismarck*, were lost at sea (one crew being luckily rescued from their dinghy by a merchant ship that was miles off her intended course). Then, in the absence of the reconnaissance aircraft, the worst happened: the two cruisers, *Norfolk* and *Suffolk*, which had for so long maintained radar contact, now, in mountainous seas, lost touch with *Bismarck*. This would not be regained for another thirty hours.

The forces of the Home Fleet vainly searched an area north-west of the last known position of *Bismarck*; from his air reconnaissance Admiral Sir John Tovey knew that she had been hit, slowed down and was losing oil, but was uncertain as to the extent of her damage. He thought it possible, however, that the Atlantic foray would be called off and that the battleship could well be heading for the German-held Atlantic port of Brest for repairs. If this were so, then Force 'H', steaming at full speed from the south, would be in a good position to intercept; to this end, at first light on 26 May, Swordfish from *Ark Royal*, with long-range tanks, flew divergent search patrols ahead of the ships.

Before the Swordfish made contact, by chance, at 1030 hours, a Catalina of RAF Coastal Command on an anti-U-boat patrol sighted *Bismarck* steaming at 20 knots towards the French Atlantic coast, which at that speed was only some eleven hours distant. Within a very short time the damaged battleship would be able to count on the protection of long-range German fighters based in France. The position, radioed by the RAF Catalina, showed that *Bismarck* was some 80 miles to the north-west of Force 'H'.

The Catalina continued to shadow the battleship for forty-five minutes, when heavy flak damaged the RAF plane; however, almost at the same time, the searching Swordfish from *Ark Royal* made contact and maintained it, supplemented by the cruiser *Sheffield*. The fact that *Sheffield* was also shadowing *Bismarck* was unfortunately unknown aboard *Ark Royal*, owing to a delay in deciphering a signal to that effect; the result was that the first strike by torpedo-armed aircraft from the *Ark* attacked the British cruiser, eleven of the fifteen aircraft dropping their torpedoes. It was fortunate that all missed their intended target; several were seen to explode on contact with the sea, the remainder *Sheffield* evaded. The torpedoes that exploded prematurely had been fitted with Duplex pistols which had been set for magnetic actuation and had been detonated by the steep Atlantic swell.

Realising that they had attacked a British ship, the chastened Swordfish crews landed back on *Ark Royal* in weather such as few of them had ever experienced; yet, within an hour, the Swordfish were refuelled and rearmed with their torpedoes set to fire only on contact.

Conditions aboard *Ark Royal* were unbelievable: seas were breaking green over the bows; the wind over the flight deck was measured at 55 knots — a speed at which a Swordfish could become comfortably airborne. Each aircraft had handling parties of up to forty men to hold the machines on to the slippery, rolling deck against the gale. As the Swordfish on the carrier were preparing to fly off on a second strike, the shadowers were grimly keeping in contact with the target. It was a far from easy task: scudding clouds often hid the quarry from sight; if they got too close, the ship opened up with radar-laid flak which followed them into the clouds, yet they could not fly too far away for fear of losing contact. Though the torpedo planes received all the glory, it should be remembered that the men in these shadowing aircraft flew in open cockpits for hours, knowing that if their single engine failed them, there was no possibility of survival in the icy seas.

At 1910 hours *Ark Royal* turned into wind and one by one the fifteen torpedo-carrying Swordfish, led by Lt Commander Esmonde, took off. As they passed over HMS *Sheffield* on their way to the target, she made with an Aldis lamp the rather pointed signal: "The enemy is twelve miles dead ahead."

The biplanes were fitted with an early 1½ metre ASV (Air to Surface Vessel) radar and could therefore attack from cloud cover. The formation climbed in sub-flights and in the cloud became spread out, so the attack was not well co-ordinated. Even so, in very bad visibility and in the face of intense German gunfire, two torpedo hits were obtained; one on the main armoured belt did little if any damage, but the second was to prove fatal, for it exploded aft and jammed the giant battleship's rudder 20° to port. From that moment *Bismarck*, for all her massive armament, was doomed.

Miraculously all the fifteen Swordfish of the strike were able to return to *Ark Royal*, though practically every machine was damaged, some badly, with several members of their crews injured. The landing on was to be difficult: a full gale was still blowing down the carrier's flight deck, which was awash with spray and sleet and was lifting up to 50 feet in the heavy seas. The plane guard destroyers were at times half submerged in the long green waves.

The landings were a tribute to the pilots, the wonderful flying qualities of the Swordfish and, by no means least, the skill of the LSO, the 'Batsman' on the *Ark*, Lt Commander Stringer. Time after time a Swordfish approached, practically hovering in the high wind, one moment 60 feet above the plunging flight deck, the next several feet below it. Stringer coolly judged the precise moment to signal the vital 'Cut', when the pilot chopped the throttle and the aircraft dropped to the deck. Time after time he waved aircraft round again — one made seven circuits — but all fifteen 'Stringbags' were eventually recovered without loss of life, though several aircraft were badly damaged. That any were recovered at all was, in itself, a miracle.

The immediate result of the Swordfish strike against *Bismarck* was that the ship, with rudders jammed, turned away from France and back into the Atlantic towards the ships of the Royal Navy. The jammed rudder was to some extent corrected with the engines, but this slowed the ship down and only made the radius of her circling larger.

During the night of 27 May 1941 Admiral Sir John Tovey, flying his flag in *King George V*, led his force of battleships to close in around the crippled German ship. Also that night *Bismarck* was attacked with torpedoes by the 4th Destroyer Flotilla and further damaged.

At first light, 0843 hours, HMS *Rodney* opened fire with a full salvo from her 16 inch guns at a range of 12 miles. Seconds later, *King George V* fired her forward turrets at the distant target; the flight time of the shells was fifty-five seconds, but before they landed, orange flames rippled along the length of *Bismarck* as she, in turn, returned fire with her 8 x 15 inch guns.

The British ships could manoeuvre: *Bismarck* could not. She was first straddled then hit repeatedly by crashing salvoes of 15 and 16 inch shells. Huge fires broke out; her decks were a shambles but her armoured hull was unbreached and she remained afloat. Her return fire was unco-ordinated, owing to hits on the director, and the British battleships advanced to point-blank range, broadside after broadside slamming into the blazing hulk of the German ship, as fast as the sweating gunners could serve their guns.

Bismarck, her engines wrecked, was now lying stopped; fires below decks were turning her plating red hot. Most of her crew were by now dead or wounded, but still the ship would not sink; some guns continued to fire and her battle ensign still flew. At 1045 Sir John Tovey made the signal to Admiral Somerville, C-in-C Force 'H': "Cannot sink her with gunfire." The cruiser *Dorsetshire* then closed in and fired several torpedoes into *Bismarck* at 2,500 yards range; as she did so, the cruiser signalled to the C-in-C: "I torpedoed *Bismarck* both sides . . . . she had ceased firing but her colours were still flying."

At 2227 hours *Bismarck* slowly rolled over to port, then disappeared into the Atlantic Ocean at 48°10' North, 10°12' West. Of her crew of 2,402, only 110 were picked out of the sea: HMS *Hood* had been avenged. *Bismarck* had fought and died well, a fitting memorial to the long era of supremacy of the battleship which had now ended. From that moment the capital ships of the naval powers would not be battleships but aircraft carriers.

Since the revelations about the wartime activities of the British code-breakers, the purveyors of 'Ultra' — the name for the secret decrypts of Germany's wartime military traffic — it is reasonable to ask to what extent the hunting and final destruction of *Bismarck* was due to the men and women at Bletchley Park — Station 'X' — where the cryptanalysts worked. Ronald Lewin, in a recent authoritative book, writes:

.... communications to and from the German forces were being transmitted for the first time in *Neptun*, the operational cipher employed on rare occasions for ventures by the enemy's big ships. *Neptun* was beyond Bletchley's immediate capability. The hard fact is that throughout the whole drama no signal to or from *Bismarck* was ever deciphered in time.[8]

It is interesting to speculate as to the possible outcome of Operation Rheinübung (Rhine Exercise), as the Germans called the breakout from Norway of *Bismarck* and *Prinz Eugen*, if the two battleships had been accompanied by one of the two aircraft carriers the German Navy had laid down before the war: *Graf Zeppelin* and *Flugzeugträger B* (thought to be named eventually *Peter Strasser*). *Graf Zeppelin* was launched in 1938 and could have been operational by May 1941. She was to have carried twelve Ju87-Ts (T — Träger, carrier), a navalised version of the Stuka, built specifically for carrier operation; in the event, only a small number of standard Ju87-Bs were converted, being fitted with hooks and a jettisonable undercarriage to facilitate forced landings at sea. Thirty Messerschmitts — Bf109T-1s — were to be embarked; these aircraft were actually constructed, being a variant of the standard Bf109E-1, with a slightly greater wing area, modified flaps and the fuselage stressed to withstand catapult-launching and arrested landings. In addition, the wings could be folded manually. Some seventy were built by the Fiesler Company.

A special unit — Trägergruppe 186 — was formed. It had trained at Kiel-Holtennu in August 1938, but owing to wrangling between the Navy and Göring, who insisted on controlling all the carrier-borne aircraft, work on the ship was stopped, not to be revived until after the British successes in the Mediterranean in 1940. By then it was too late.

However, if *Graf Zeppelin* had been able to sail with *Prinz Eugen* and *Bismarck*, as Admiral Raeder would certainly have ordered, the outcome could have been very different. The Me109-Ts could effectively have defended *Bismarck* from the Swordfish and the carriers *Victorious* and *Ark Royal* might well have been put out of action by the Stukas; even the British battleships would not have been immune from attack. Had that happened, *Bismarck* and *Prinz Eugen* could have raided the convoy routes and the Atlantic almost at will.

In the event *Flugzeugträger B* was never launched and *Graf Zeppelin* never completed. The latter's intended aircraft were issued to land-based units and all had disappeared from the Luftwaffe inventory by 1944.

While the drama of the *Bismarck* was still being played out in the Atlantic, in the Central Mediterranean on 26 May the Fleet Air Arm was launching dawn air strikes from *Formidable* against German airfields on Scarpanto Island, near Crete. The attack was successful but in the afternoon the carrier was attacked by Ju87 and Ju88 dive-bombers, flying from bases in North Africa. Two of *Formidable*'s Fulmars were airborne at the time but could not break up the attack, though four Ju87s were claimed as shot down.

Formidable was hit fore and aft of her armoured flight deck by 1,100lb bombs and, although not so badly damaged as *Illustrious*, *Formidable* had to return to Alexandria, where her repairs were considered to be beyond the dock facility there. So on 24 July 1941 *Formidable*, patched up, passed through the Suez Canal en route to join *Illustrious* at the US Navy Yard at Norfolk, Virginia. In the absence of the two ships there would be no carrier-based operations in the Central or Eastern Mediterranean for the next thirty-eight months. *Formidable*'s aircraft were disembarked to operate from desert airfields, and they gave valuable support to the Eighth Army, effectively attacking enemy shipping and ports.

Ark Royal, after her vital role in the sinking of *Bismarck*, returned with Force 'H' to Gibraltar and was, in company with *Argus*, *Furious* and *Victorious*, engaged in convoy protection and aircraft-ferrying operations to Malta during the summer and autumn of 1941. *Ark Royal* had been bombed by the Luftwaffe and even claimed as sunk by a U-boat; indeed, for some time 'Lord Haw Haw', the alias of William Joyce, the English-speaking commentator on German radio, used to preface his nightly broadcast to Britain with the words: "Mr Churchill, where is the *Ark Royal*?"

The answer, on 13 November 1941, was that the *Ark Royal* was returning from ferrying RAF Hurricanes to Malta when she was torpedoed by *U-81*, which secured a single hit on the starboard side. The sea was dead calm and Gibraltar only 50 miles away. Her crew were taken aboard HMS *Legend*, which came alongside, but owing to design defects and poor damage control, the carrier was to sink while under tow at dawn on 14 November. Only one member of her crew lost his life, but the loss of the ship to a single torpedo was to be the subject of searching inquiry.

The fundamental defect was that, as the uncontrolled flooding in the ship increased her list, the boilers were put out of action; steam pressure failed and, lacking any alternative diesel sets, there was simply no power available to operate the pumps. So inch by inch she foundered, owing to progressive and unchecked flooding — a sad and unnecessary end of a fine ship.

Twenty-four days after the loss of *Ark Royal* in the Mediterranean, in the Pacific, on the other side of the world, there dawned what President Roosevelt was to call "a day that will live in infamy": Sunday, 7 December 1941, the date of the action now simply known as 'Pearl Harbor'. That Sunday in December, without a formal declaration of war, Japanese carrier aircraft delivered a crushing surprise attack on the US Navy's Pacific Fleet in Pearl Harbor on the Island of Oahu, Hawaii.

Owing to an almost unbelievable chain of circumstances, it is true that the attack was a surprise in the sense that the military command in Hawaii was not expecting it. Howewver, in Washington there had been a good deal of political evidence to make the likelihood of a Japanese attack not only possible but probable.

We have already seen how, during the early thirties, the Japanese had developed their Naval Air Arm. that expansion of the Japanese Navy, which had begun in earnest in 1936 with the abrogation of the International Naval Treaties, had continued; China had been invaded in 1937, a National Mobilisation Bill was announced in 1938, and, by September 1940, Germany, Italy and Japan had concluded a three-power pact which was to form the Axis.

During October 1941 the Japanese Prime Minister, Konoye, had tried to negotiate with the United States, which, concerned with Japan's overt military expansionist policy — it had continued with the invasion of Indo-China — had embargoed oil supplies to Japan and had frozen her assets in the United States. Britain had followed suit, the effect of which was seriously to diminish Japan's ability to trade and so reduce her foreign exchange, all of which forced Japan to try to negotiate with the USA for the lifting of the sanctions. Roosevelt, however, declined to negotiate at that time, causing the Japanese Prime Minister to resign, to be replaced by his former Minister of War, General Tojo, who, concerned that dwindling oil stocks would constrict Japanese military ambitions, renewed negotiations with the United States to try to restore the oil supplies. His new Government offered to withdraw Japanese forces from Indo-China, but America now demanded that Japan additionally sever her pact with Germany. At this, Hitler, worried that a peaceful political settlement between the two major powers in the Far East could release the United States Pacific Fleet, enabling it to take part in the European war, assured Japan that, in the event of her declaring war on America, Germany would immediately follow suit. Japan — or rather the Tojo Government — agreed. It was not a difficult decision, for, from mid-1941, Japan had virtually been committed to a war with America and the other Far East colonial powers — the British and the Dutch.

Anticipating war, preliminary Japanese Staff studies had come to the conclusion that the garrisons of Singapore and other British, American and Dutch possessions in the Far East would not present any undue military problems. This was true: they were in reality poorly defended by troops softened by years of garrison duty, many of them living in married quarters or barracks and messes built to generous peacetime standards, and enjoying peacetime routines; the officers in particular were imbued with the ingrained superiority of the western colonial powers towards 'native' or Asian peoples.

The Japanese intention in the Far East was to create what they called the 'Greater Asia Co-Prosperity Sphere' — a euphemism for the total economic and political control of the Dutch East Indies, the Philippines, Indo-China and the Federated Malay States. These possessions were rich in oil, tin and rubber; indeed, one of the first objectives would be the seizing of the oil fields of the Dutch East Indies. These goals would be achieved in a series of rapid amphibious attacks in which the Japanese Imperial Navy would obviously play a key role.

Admiral Isoroku Yamamoto, C-in-C of the Imperial Navy, was a brilliant strategist and — unlike others in the War Cabinet — he realised that any war against the United States would have, of necessity, to be a short one. Japan could not hope to challenge the greatest asset the Americans possessed: their almost unlimited capacity for industrial production. In a war of attrition against the USA Japan must inevitably be the loser. The best that could realistically be hoped for was a lightning and crushing attack by battle-trained Japanese forces to secure as much of her projected Co-Prosperity Sphere, including Dutch oil, as she could, then to resist American counter-attacks long enough to negotiate a political settlement to her advantage. This bold plan seemed feasible: Britain was preoccupied and fully stretched in fighting Germany; the Dutch were weak with their homeland under German occupation; and American public opinion, it was thought, would not tolerate a long-drawn-out distant overseas war which could be costly in American lives and which many would construe as propping up the tottering British Empire with American bayonets.

If the political risks were acceptable, then the military problems were thought to be within the capability of the Japanese forces. The only real danger was the threat posed to the Japanese Navy by the presence of the powerful United States Pacific Fleet based at Pearl Harbor.

The Hawaiian Islands were far out of reach of any Japanese land-based aircraft, but Admiral Yamamoto was one of the few senior naval officers who was a keen advocate of the use of naval air power, and he was convinced that his powerful and highly trained naval air arm, which had perfected its skills for over three years in the war against China, could undertake any tasks required of them. There was also the precedent of Taranto; that British action had greatly impressed the Japanese Staff, who had noted many of its features, including the use the Royal Navy had made of wooden 'air tails' for their airborne torpedoes; these tails broke off on impact with the water but greatly improved the angle at which the torpedo entered the sea, enabling them to operate in a shallow harbour. The Royal Navy had also reduced the safe running distance of their torpedoes so that the attack could be made at point-blank range. The synchronisation of dive-bombing and low-level torpedoing had been successfully employed at Taranto, and the torpedo bombers' run-in had been made at the lowest possible height, which, in the crowded harbour, greatly inhibited the defenders' fire-power because of the obvious danger of hitting adjacent friendly ships in the anchorage, to say nothing of harbour installations and civilian buildings ashore. All this being noted, the Japanese Staff now began to plan a similar blow, on a far larger scale, against the US Pacific Fleet in Pearl Harbor.

The Japanese First Air Fleet, chosen for the operation, began to rehearse the attack in conditions of great secrecy in out of the way inlets and islands around Japan. By November 1941, in the light of extensive trials, the detailed plans for the attack on Pearl Harbor were perfected and accepted by the Naval Staff. The provisional political decision to mount the action was taken on 18 November 1941; anticipating assent, the strike force had left the Home Islands on 10 November for the Kuriles. It was by far the largest carrier-borne force of aircraft yet assembled, embarked on no fewer than six fleet carriers.

Carrier Division 1 consisted of *Akagi*, flying the flag of Vice Admiral Nagumo, commander of the strike force, with twenty-seven A6M2 Zero-Sen (Zeke)[9] fighters, eighteen D3A1 Aichi (Val) dive-bombers and twenty-seven B5N2 Nakajima (Kate) torpedo bombers. *Kaga*, the second carrier of the First Division, embarked twenty-seven Zeros, twenty-six Vals and twenty-six Kates.

Carrier Division 2, led by *Hiryu*, wearing the flag of Rear Admiral Yamaguchi, had twenty-three Zeros, seventeen Vals and eighteen Kates on board. Accompanying her, *Soryu* carried twenty-six Zeros, seventeen Vals and eighteen Kates.

Carrier Division 5 with Rear Admiral Hara, who flew his flag in *Zuikaku*, embarked fourteen Zeros, twenty-five Vals and twenty-seven Kates. The second carrier of Division 5, *Shokaku*, had fifteen Zeros, twenty-six Vals and twenty-seven Kates.

Of the above, forty of the B5N2 Kates were armed with torpedoes (Type 91, 17.7 inch); the remainder carried an improvised AP (armour-piercing) bomb which was actually a 16 inch naval shell with suspension lugs and an air tail welded on. (The Japanese Navy had realised that 16 inch shells are just as effective — and a good deal more accurate — if delivered 300 miles by aircraft rather than hurled from a battleship's gun for 30.) Additional to the six fleet carriers, the strike force had two seaplane-carrying cruisers, *Tone* and *Chikuma*, each with an Aichi E13A (Jake) reconnaissance seaplane, and one Nakajima E8N2 (Dave) biplane spotter.

Including twelve E13A (Jakes) embarked with catapult flights on the battleships, the total air complement that put to sea was 450 operational aircraft. The Japanese fleet of twenty-three vessels included two battle-ships, *Hiei* and *Kirishina*, and a division of destroyers, as well as a number of submarines and midget submarines; though not part of the strike force, these forces would be taking part in the strike, being stationed off Pearl Harbor with the object of sinking any American warships that tried to escape.

Maintaining total radio silence, the Japanese force assembled on 22 November at Hitokappu Bay off Etorofu Island in the southern part of the Kuriles — a long skein of desolate islands that stretch for 600 miles from Hokkaido, the northernmost Home Island of Japan, across the sea of Okhotsk to the Russian peninsula of Kamchatka. The fleet left its anchorages in the Kuriles on 26 November and steamed into an area of the North Pacific little frequented by shipping. On 2 December the Japanese War Cabinet gave its irrevocable assent to the attack on the American base and the prearranged signal — *Niitaka Yama Nobore* (Climb Mount Niitaka) — was transmitted to the strike force, which then steered east into the Pacific to be at a point 200 nautical miles to the north of Hawaii at dawn on 7 December 1941.

The position of the Japanese carriers was unknown to the US Navy, in spite of careful and efficient monitoring of the Japanese Navy radio traffic. The Japanese had used a clever radio deception ploy to mislead the Americans. They changed all their naval call signs — over 20,000 — and telegraphists identified by American listeners to be on certain ships from their distinctive 'fist', or individual way of sending morse, remained ashore, sending bogus messages from radio stations at Kure or similar naval bases, which led American direction-finding stations to conclude that the carriers were still in harbour.

Strict radio silence (something that the Americans usually found impossible to achieve) was enforced on board the actual carriers as they sailed from the home waters in which US Navy Intelligence fondly imagined them to be. Although the Japanese had won this aspect of the radio war, they had, like the Germans, lost another without knowing it.

US Navy cryptanalysts had for some time succeeded in solving most of the Japanese diplomatic and naval codes,[10] including the complex machine cipher, which the Americans knew as 'Purple'; this was based on the German 'Enigma' system which the British had already broken. The decrypts of the Purple cipher were coded 'Magic' and, like the British 'Ultra', were not infallible; there were always some blanks or garble in the intercepts (some ambiguities were of course errors in the original transmission and the intended recipient was probably in as much doubt as to the meaning of parts of the message as the intercepting cryptanalysts).

From several intercepts gathered during November 1941, the US Navy code-breakers learned that instructions on how to destroy 'Purple' machines in Japanese embassies around the world were to be made available from their naval attachés. Specific instructions were sent on 1 December to the Japanese embassies in London, Hong Kong, Singapore and Manila, actually to destroy their cipher machines and codebooks. This led American Intelligence to conclude that a war against Britain and the Dutch was about to break out in the Far East; the Americans therefore decided that, for the moment, the United States was safe. That illusion was dispelled on 5 December when the Japanese Embassy in Washington too was ordered to destroy its cipher equipment. When that message was deciphered the American State Department realised that Japan declaring war against the United States was also a distinctly imminent probability.

In fact Tojo, the Japanese Prime Minister, had instructed his ambassador in Washington to deliver a formal declaration of war on 7 December 1941 at 1300 hours, Washington time, only thirty minutes before the planned attack on Pearl Harbor. Tojo had cynically observed that although the 1907 Hague Convention, of which Japan had been a signatory and which then still governed the rules of warfare, stipulated that prior warning of war must be given, the actual time lag between the declaration and the start of hostilities had not been laid down.

On 6 December, the day before war was to be declared, a thirteen-part polemic, which attempted to justify the Japanese position, was sent in the 'Purple' code to the Washington Embassy; this too was deciphered by US Navy cryptanalysts, but the vital fourteenth part, in which the actual declaration of war was contained, was not transmitted with the others. The nature of the decrypt to hand, however, was sufficient for the US State Department to perceive that war was now inevitable; when President Roosevelt saw the long message that evening, he simply said 'this means war'.

The fourteenth part was transmitted next day, 7 December, and it too had been deciphered by the Americans early that morning. A plain text (in English) was also sent which instructed the Japanese ambassador to deliver the complete fourteen-part note by exactly 1300 hours on that day. Now, 7 December was a Sunday; the Americans were puzzled by the timing which the Japanese ambassador had been specifically instructed to observe, since they were expecting a Japanese attack soon after the formal declaration of war. It should be emphasised that, in all the many (possibly thousands) of messages that the US Navy Intelligence intercepted, not one had mentioned a specific target, but, based on other evidence, the American State Department was anticipating the first strikes to be against either Singapore or the Philippines. What was puzzling them was that 1300 hours Washington time would, in the Far East, be the small hours of the morning and still dark. Classically battles tend to commence at dawn. A naval officer — an ex-aviator in the State Department — realised that there was one US overseas possession where dawn would just be breaking at 1300 hours Washington time: Hawaii.

Immediately it was agreed that a warning of a possible attack should be sent. Now, although Pearl Harbor in Oahu was a great US naval base, the island as a whole was defended by the Army. The possibility of Pearl Harbor being attacked at dawn that day was therefore conveyed to the Army Chief of Staff in Washington, General Marshall, though there was some delay in this as the General that Sunday morning was out exercising his horse. Eventually he was contacted and he authorised the sending of the message. More delay occurred when it was found that the Army communications link to Hawaii was not, for some reason, available that morning. It was unthinkable to ask the Navy to send the message, which was therefore transmitted by ordinary commercial cable. It was delayed still further by the actual attack on Pearl Harbor and was not delivered until some time after the last Japanese aircraft had left.

While all the diplomatic and cryptanalyst activity was going on behind locked doors in Washington, in the morning of 7 December, in the Pacific shortly before dawn, the Japanese carrier force was in position. The commander, Vice Admiral Nagumo, in *Akagi*, dispatched two E13A Jake reconnaissance seaplanes from *Tone* and *Chikuma* to make a search for any US naval presence between the strike force and the Island of Oahu; they found nothing and at 0600 hours Hawaiian time the first strike was ranged and ready for take-off.

From the flight decks of the carriers *Akagi, Kaga, Hiryu* and *Soryu*, 183 aircraft one by one rolled down the flight decks and, cheered by sailors, took off and formed up: forty B5N Kates armed with torpedoes, forty-nine Kates with the improvised but effective AP bombs, fifty-one B3A Vals to dive-bomb airfields, and forty-three Zero fighters to provide air cover and low-level strafing. As the last aircraft of the first strike left its carrier, the machines of the second strike were ranged.

As the strike force was flying south along the 200 nautical mile track to the target, in the Washington Embassy the Japanese ambassador's staff had just decoded the fourteenth and vital part of the message; they had, as instructed, destroyed all 'Purple' machines save the one on which the message had been decoded, which had delayed the decrypt. It was now being laboriously typed in English, first in rough, then as a fair copy, the punctilious Japanese ambassador declining to present the American Secretary of State with anything less than a perfect text.

As the strike force drew nearer to Pearl Harbor, the first shots in the Pacific war were, in point of fact, fired by the Americans without their realising it. A US Navy PBY-5 Catalina of Patrol Squadron 14 on a routine flight off Oahu spotted a suspicious-looking submarine, which dived as the aircraft approached; the Catalina dropped a bomb on the vessel. At the same time a destroyer, USS *Ward*, dropped depth charges on a midget submarine (possibly the same one attacked by the PBY); it was one of five that had just separated from a full-sized mother submarine of the Japanese I class. The time of the attack was logged aboard *Ward* as 0651; the destroyer sent a signal to Naval HQ at Pearl Harbor warning of the presence of unidentified submarines, but this was not decoded until 0730.

That incident was not the only intimation of an attack: an experimental army radar, perched high above Pearl Harbor on a mountain peak, began to indicate a large force of aircraft approaching from the north. The two enlisted men operating the station reported the contact by field telephone, but were brusquely told to close down the station, which they reluctantly did at approximately 0700 hours.

At 0746 the first wave of Japanese aircraft roared over Oahu. Their presence was hardly without warning but came, nevertheless, as a total surprise.

The first bomb dropped hit the seaplane slipway on the southern end of Ford Island at the centre of the harbour at 0755, followed by many others. The exploding bombs and the noise of low-level aircraft engines brought men tumbling out of beds and messes.

As it was a Sunday — a peacetime Sunday at that — everything was standing down or shut. The guns on the battleships tied together in 'Battleship Row' off Ford Island had their covers on, 'ready use' ammunition had not been issued, the magazines were under lock and key. In the confusion that followed there were acts of incredible stupidity alongside sublime courage. 'Book men' refused to issue ammunition and guns without written orders; others broke open arsenals or wrenched machine guns from blazing aircraft, balancing them on fuel drums to fire at the enemy; on open airfields ground crews started aircraft engines while under fire; fighter pilots took off from cratered runways with Japanese fighters strafing the rows of planes which had been parked with commendable precision, wingtip to wingtip, and many of which were now burning fiercely. (The aircraft had been concentrated because of concern about sabotage, there being many ethnic Japanese living on Oahu.)

The worst casualties were among the sailors, asleep below decks in the battleships; most of them never knew what killed them. Within the first minutes of the attack on the seven battleships alongside the dock at Ford Island, *West Virginia* was resting on the bottom, blasted by at least six direct torpedo hits from Kates; *Arizona* had been hit by a torpedo and then her magazines were detonated by an armour-piercing bomb which turned the ship into a blazing hulk in which 1,000 of her crew died, most of them trapped below. *California* took two torpedoes as well as bombs and settled on the harbour bed through uncontrolled flooding. *Oklahoma* was struck by at least five torpedoes, though she rolled over slowly, giving most of her crew time to abandon ship. One ship, *Nevada*, though torpedoed and bombed, managed to slip her mooring and get under way, her depleted engine-room crew remaining at their posts. There were no officers on the bridge, the big ship being conned by Chief Quartermaster Sedberry, who got her from her mooring into the main channel, without tugs or pilots and past the blazing *Arizona*: a very considerable achievement. Two of the battleships, *Maryland* and *Tennessee*, were moored inboard alongside others; they were therefore immune from direct attack by torpedoes but the latter ship was hit by bombs, causing damage to her gun turrets. *Maryland* was hit on her foredeck by an AP bomb which penetrated her hull below the waterline. Apart from the seven ships in 'Battleship Row', the old target ship *Utah* was torpedoed and sunk; *Pennsylvania*, in drydock, was bombed and damaged; and a new cruiser, *Helena*, was torpedoed, which badly damaged her and sank the adjacent minelayer *Oglala*, by underwater mining effect.

Sailors and marines on the ships organised ammunition parties and continued to serve the guns as their ships settled on the bottom. While

(*Above*) An SBD-3 lands aboard *Yorktown* during the Battle of the Coral Sea. The photograph clearly shows the perforated flaps and dive brakes which were characteristic of the Douglas SBDs (*US Navy*); (*below*) USS *Lexington*, soon after she was torpedoed during the Coral Sea action. The 7° list has been corrected by counter flooding, causing the ship to be down by the head, but the AvGas fires had yet to spread (*Capt Perlman via Coral Sea Association*)

(*Above*) An F4F Wildcat is waved away from a landing on *Yorktown* at about the time of the Coral Sea action. F4F pilots felt that their fighters were a match for the Japanese, though the tactics and radio discipline had to be revised (*US Navy*); (*below*) Japanese crewmen on the flagship, the carrier *Akagi*, fighting fires started when the AvGas supply lines were ruptured and the fatal fires began. The ship was later abandoned and scuttled with torpedoes from Japanese destroyers (*US Navy*)

(*Above*) An SB2U-2 Vindicator in drab wartime camouflage photographed in July 1942 while serving with Scouting Squadron 42 aboard CV-4, USS *Ranger*, in the Atlantic (*US Navy*); (*below*) the TBF was to vindicate itself as the most successful torpedo bomber serving with the US and Royal Navies. The photograph shows an early TBF-1 in pre-May 1942 national markings (*Grumman*)

8650

(*Previous page*) The cockpit of the TBF-1 was roomy and afforded the pilot an excellent field of view. Like all aircraft cockpits of whatever nation, it is full of plaques, warning and admonishing the pilot (*Grumman*)

(*Above*) A Sea Hurricane perched on the catapult of a merchant ship; essentially a throwback to the disposable fighters of World War I. From May 1941 to August 1943, these makeshift fighters shot down six Condors, and damaged many others (*IWM*); (*below*) HMS *Biter*, 12,000 tons; a Royal Navy Escort carrier of the *Archer* class, equivalent to the US Navy's *Long Island* class, built on C-3 merchant hulls (*IWM*)

(*Above*) Photographed on 15 October 1942, just two months after being commissioned, the name ship of the *Sangamon* class, USS *Sangamon*, CVE-26, has Grumman TBFs and Douglas SBDs on her flight deck (*US Navy*); (*below*) USS *Prince William*, CVE-31, the eleventh and last of the *Bogue* class of Escort carriers, engaged on ferry duties in August 1943 (*US Navy*)

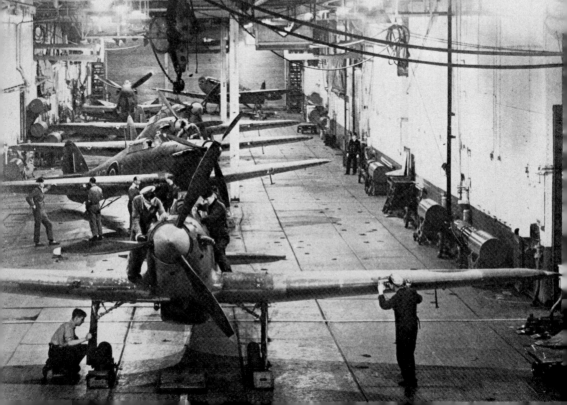

(*Above*) A Sea Hurricane lands on an Escort carrier. Though a landplane conversion, the Sea Hurricane, with its wide-track undercarriage, fared better at sea than the later Seafire (*IWM film still*); (*below*) during the hard-fought convoy battles in the Mediterranean, the ships had only their anti-aircraft armament to rely on. Most RN ships mounted multiple 'pom-poms', here seen in action on an RN carrier with the 'Pedestal' convoy (*IWM film still*)

(*Opposite above*) CVE-98, USS *Kwajalein*, ferrying Vought F4U Corsairs and Grumman F6F Hellcats, on 19 July 1944. Up to 70 aircraft could be ferried on one voyage (*US Navy*); (*opposite below*) the hangar of a Royal Navy CVE. The non-folding Sea Hurricanes were very wasteful of space (*Crown Copyright*)

(*Above*) The second USS *Lexington*, CV-16. She was originally to be named *Hancock*, but assumed the name *Lexington* to commemorate CV-2 when the earlier carrier was sunk on 8 May 1942. The photograph was taken 10 March 1944 and illustrates well the *Essex* class carriers (*US Navy*); (*below*) one of the *Independence* class of light Fleet carriers, CVL-29, USS *Monterey*, photographed 5 June 1943 on a shake-down cruise, shortly after she was commissioned (*US Navy*)

(*Above*) Grumman F6F-3 Hellcats with TBF Avengers below, filmed in 1943; Hellcats were the most successful US-built carrier-based fighters, 5,156 enemy aircraft being destroyed by pilots flying the type (*US Navy*); (*below*) an A6M-3. Zero-Sen fighter photographed after it had been captured by US forces; the A6M-3 was the most numerous sub-type to be embarked on Japanese carriers in 1942 (*US Navy/National Archives*)

1.	Carburetor Protected Air Control (Aux. Stage Only)	17.	Landing Gear Control
2.	Ignition Switch	18.	Altimeter
3.	Clock	19.	Rudder Pedals
4.	Landing Gear Emergency Lowering Control	20.	Airspeed Indicator
5.	Directional Gyro	21.	Gun Charging Controls
6.	Compass	22.	Cockpit Heater Control
7.	Gunsight	23.	Turn and Bank Indicator
8.	Attitude Gyro	24.	Ammunition Rounds Counter
9.	Chartboard Light	25.	Fluorescent Lights Control
10.	Attitude Gyro Caging Knob	26.	Rate of Climb Indicator
11.	Tachometer	27.	Wing Lock Safety Control Handle
12.	Water Quantity Gage—A.D.I. System	28.	Manifold Pressure Gage
13.	Instrument Panel Fluorescent Light	29.	Chartboard
14.	Cylinder Head Temperature Gage	30.	Oil-In Temperaure Gage
15.	Oil Pressure Gage	31.	Fuel Pressure Gage
16.	Landing Gear & Wing Flap Position Indicator	32.	Fuel Quantity Gages

The cockpit of an F6F-3; an illustration from the Pilot's Notes (*Grumman*)

Corsairs were not easy aircraft to land-on: here a Royal Navy Corsair I (*above*), equivalent to the US Navy F4U-4, takes the barrier on the training carrier USS *Makassar Strait* (CVE-91), ending up (*below*) balanced on its nose (*US Navy*)

(*Above*) A Blackburn-built Barracuda II, MD893, displaying Air-to-Surface Vessel (ASV) metric radar antennae on wingtip pylons (*Hawker Siddeley*); (*below*) the 'Beast': the 3-ton Curtiss SB2C Helldiver replaced the Douglas Dauntless in its dive-bomber role from 1944 to the end of the Pacific War (*US Navy*)

The hazards of carrier landings: a Seafire takes the barrier on HMS *Implacable* and sheds her starboard wingtip and a propeller blade (*above*) while another Seafire has taken the barrier on *Implacable* (*below*), its airscrew showing evidence of 'pecking' (*IWM film stills*)

A Japanese Jill torpedo-bomber (*above left*) is hit on the port wing by a 5in AA shell from USS *Yorktown* during the US attack on the Marshall Islands; the outer wing panels fall away with flames erupting from ruptured fuel tanks (*above right*). Still flying on full power, the blazing Japanese bomber plunges past a Hellcat on *Yorktown*'s flight deck and still miraculously level (*below left*) with one wheel hanging down, disintegrates (*below right*) on hitting the sea (*US Navy*)

the battleships, which were the main objective, were being attacked, the dive-bombers and fighters over the airfields and seaplane slipways were finding no lack of targets.

There were over 400 US Navy and Army planes on the island. At Kaneohe seaplane base, the Marine airfield at Ewa, and the Army's Hickam, Wheeler and Bellows Fields, 188 aircraft were completely destroyed, including nearly 50 PBY Catalinas at Kaneohe; in addition, thirty-one aircraft were badly damaged. Several of the attackers were shot down by the few US fighters that managed to get airborne.

At the height of the battle, at 1350 hours Washington time, twenty-five minutes after the attack had begun, the now impeccably typed fourteen-part declaration of war was presented to the Secretary of State, Cordell Hull, who of course was careful to give no indication that he was already aware of its contents. Back at Pearl Harbor, with the war now formally declared, there was only a slight diminution in the fighting between the two strikes, but during the lull fifteen SBD Dauntlesses of Scouting Squadron 6 (VS-6), which had taken off from USS *Enterprise* at roughly the same time as the Japanese attackers, but 200 miles to the west of Pearl Harbor, arrived, bound for Ewa Field; twelve managed to land safely. One SBD was shot down by American fire; its pilot was heard on the radio shouting "Please don't shoot, this is an American plane". The three remaining airborne Dauntlesses went after the attacking Japanese aircraft; one plane collided with a Zero, and the other two were shot down by numerous Zeros, though both pilots managed to bale out.

The Dauntlesses were not the only American aircraft to fly into the battle: twelve USAAF B-17s from California arrived, down to their last drop of fuel. Incredibly only one was shot down, and that by 'friendly' anti-aircraft fire, though another had to make a crash-landing.

The B-17s actually arrived during the second Japanese attack, which consisted of seventy-eight Vals and fifty-four Kates, escorted by thirty-five Zeros. None of these aircraft was armed with torpedoes — only bombs. The second wave was over its targets by 0855. *Nevada*, still under way and heading for the harbour entrance, was struck by dive-bombers and had to be beached to prevent her from foundering, two destroyers in dock were set on fire, and among other ships hit was the repair vessel *Vestal*, which, like *Nevada*, had to be beached. A destroyer, *Shaw*, had its bows blown off and the seaplane tender *Curtiss* was not only bombed but also hit by a crashing Val — to survive both.

By 1000 hours local time the sound of aircraft engines faded, as the last Japanese plane departed: it was all over. In all, 2,335 Americans were dead, nearly half in the *Oklahoma*, and 1,143 were wounded. The

Japanese had lost possibly 100 pilots and twenty-nine aircraft, mainly to ground fire — a loss of slightly over 8 per cent of the attacking force. Although such a loss rate sustained by the RAF and USAAF bombers over Germany in 1944-5 would seem high, for a single raid and considering the results, it was a cheap victory. But was it, in fact, a victory at all?

The first and immediate result of 'Pearl Harbor' was to create, in America, a national feeling of profound revulsion against the Japanese. The war which President Roosevelt declared the next day became highly popular; there were great queues outside the recruiting offices. It also was to become racial in a way that the struggle against Germany was not. The explanation is simple: the Americans were a nation brought up against a background of the code of the Wild West (largely fictitious) and of the ball park, football field and high school and collegiate playing fields, where the rules of fair play are sacrosanct. The 'Japs' had won on a foul; had ambushed and shot their undeclared enemy in the back. "Remember Pearl Harbor" was a slogan to be seen everywhere. The determination to defeat Japan remained green until the holocaust of Hiroshima and Nagasaki and, even then, many Americans (and British) would consider it "served them right" — a sentiment nurtured by the brutal treatment the Japanese would mete out to their western prisoners-of-war.

Considered in purely military terms, the attack on Pearl Harbor had been brilliantly executed by highly trained men. However, although seven old capital ships were out of immediate action, only two — *Arizona* and *Oklahoma* — were total losses; the remainder were eventually raised and repaired. The three carriers that could have been in the harbour were not there. *Saratoga*, CV-3, had just completed a major refit and was still at San Diego; *Lexington*, CV-2, was at sea 425 miles to the south-east off Midway, towards which she was sailing to fly off a Marine Scout bombing squadron; *Enterprise*, as already noted, (it was her fifteen SBDs that arrived during the battle) was 200 miles west of Hawaii, returning from Wake Island after ferrying F4F-3 Wildcats to Marine Fighter Squadron VMF-211 based there.

The damage to the harbour facilities and other shore installations, although it looked impressive enough on the contemporary newsreels, was not significant. Aircraft hangars are soon rebuilt and in war are not really necessary; the aircraft could quickly be replaced; even the drydock was not rendered unserviceable and, perhaps most surprisingly in view of the accuracy of the action, the extensive oil-storage tanks were not hit at all.

In one sense the most obvious success the Japanese enjoyed — the immobilisation of the Pacific Fleet — was really to the advantage of the US Navy in the long run, for it forced the development of the American Fast Carrier Task Force on the Fleet, and the United States Navy carriers were to avenge, in full, that "day that will live in infamy". (Interestingly, it did not perhaps live as long as President Roosevelt imagined. When the Israelis, without any warning, attacked the Egyptian and Jordanian airfields to start the 'Six Days' War' in 1967, it was called by American commentators "a pre-emptive strike" and was widely regarded as a prudent way to start that war.)

The news of the attack on Pearl Harbor reached Winston Churchill informally; he had been dining at Chequers, the country home of the Prime Minister in Buckinghamshire, entertaining American Ambassador Winant and Averell Harriman. The radio was on in the background and Harriman thought he had heard the BBC announcer say something about a Japanese attack; Churchill's butler, who had heard the news below stairs, came in to confirm the story. Churchill at once telephoned Roosevelt and joyfully reported that the President had told him: "We are all in the same boat now."[11]

Hitler, true to his undertaking to the Japanese, declared war on the United States on 11 December, as did Mussolini. The United States now had to wage a two-ocean war in the Atlantic and the Pacific with a large deficit in warships, due to the Japanese attack.

The war was about to become a world-wide conflict. In the Pacific it would be, by its nature, primarily a maritime operation fought on an unimaginably large battlefield, comprising as it did no less than 56 per cent of the total surface of the globe.

Fortunately, the US Navy, concerned at the outbreak of the war in Europe and the probability of America being eventually involved directly against Japan and Germany, had sought an increase in its effective naval air combat strength, which had stood, on 1 July 1939, at five carriers and 1,315 combat aircraft.[12] Congress granted an 11 per cent increase in June 1940 and the following month a further increase of over 70 per cent, which, when eventually implemented, would include eighteen fleet carriers and 15,000 aircraft (these figures were achieved in 1943).

In appealing for the increases, Admiral Stark, then Chief of Naval Operations, pointed out: "From 1921 to 1933, the United States tried the experiment of disarmament in fact and by example. It cost us dearly in relative naval strength — but the greatest loss is time. Dollars cannot buy yesterday. Our present Fleet is strong but not strong enough."[13]

After the Pearl Harbor attack, further tonnage was authorised but the original increase enabled the US Navy to lay down the first of what were arguably the finest carriers to serve during World War II: the *Essex* class of attack carriers. *Essex* herself, CV-9, was laid down at Newport News Yard on 28 April 1940. The ships had a standard displacement of 27,000 tons with a flight deck 866 x 90 feet. The class featured, in addition to two or three conventional lifts on the flight deck, a new departure: a lift on the port side outboard of the flight deck. In essence this was a standard carrier lift with a 60 x 34 foot platform which travelled vertically on the port side of the ship. The advantages of the outboard lift were obvious; if, owing to enemy action, a conventional carrier lift was stuck 'down' (as happened with HMS *Illustrious*), the flight deck would be unusable, whereas the position of the outboard lift had no effect on flight-deck operations. The outboard lift first fitted to *Essex* was so successful that the ship's Commanding Officer, Captain Donald B. Duncan, recommended that an additional lift should be placed on the starboard side of the carriers; however, BuShips turned down this proposition since it would not have allowed the *Essex* class to pass through the Panama Canal. No British-built carrier had an outboard lift, since the side access into the hangar would have compromised the armoured 'box' concept. The Americans, however, considered the reduction in the numbers of aircraft embarked in a fully armoured carrier too high a price to pay. The *Essex* class, with a full width, lightly armoured hangar, could accommodate up to 108 aircraft and a crew complement of about 3,400.

The *Essex* carriers, (in all no fewer than thirty-two were to be ordered during the war years), had a maximum speed of 33 knots — 32 knots could be sustained indefinitely — and a radius of action of 14,100 nautical miles at 20 knots, with facilities for refuelling at sea from naval oilers: the Fleet train technique used by the belligerents in the Pacific which was to enable carrier task forces to remain at sea for an indefinite period. With the imminence of war, eight further *Essex* class carriers, CV-12—19, were ordered on 9 September 1940. Although these carriers were to be built in an incredibly short time — one, CV-13, USS *Franklin*, in only fourteen months — the first, *Essex*, would not be commissioned until 31 December 1942. Before that date there was to be a series of bitter and humiliating defeats for the United States and the British at the hands of the Japanese.

Britain formally declared war on Japan the day following the attack on Pearl Harbor. Japan had declared war on Britain and the United States simultaneously.

In Britain, prior to the actual opening of hostilities in the Far East, the appreciation was that a frontal attack by the Japanese on the strongholds of Singapore and the American base at Manila, in the Philippines, was unlikely, it being considered in London that the base at Singapore was virtually impregnable — which was in reality far from the case. The main threat in the event of war against Japan was thought to be attacks by the Japanese on shipping carrying those vital war materials, oil, tin and rubber.

To create what the Prime Minister, Winston Churchill, called "the vague menace" of powerful naval units at large, "like rogue elephants", the battleships HMS *Prince of Wales* and *Repulse*, with destroyer escorts, were sent to Singapore in November 1941. The new armoured aircraft carrier *Indomitable* was to have sailed in company, but most unfortunately that ship, working up in the Carribbean, had grounded off Jamaica and had been sufficiently damaged to delay her sailing for the Far East.

When, on 8 December 1941 (7 December, Hawaiian dateline), the war against Japan became a reality, the two battleships, with Admiral Sir Tom Phillips flying his flag in *Prince of Wales*, sailed to harass Japan's shipping routes in the South China Sea. On 10 December they were recalled to investigate a report of a supposed Japanese landing at Kuantan, on the east coast of the Federated Malay States. The report was false; no enemy was found but at 1020 hours, Singapore time, Japanese reconnaissance aircraft found the British battleships. Just forty minutes later, land-based Japanese Navy torpedo bombers attacked: *Prince of Wales* was put out of action almost at once by torpedo hits astern which damaged her propellers, and a second wave torpedoed *Repulse*. In spite of heavy anti-aircraft fire from the battleships and the escorting destroyers, Japanese dive-bombers pounded the two ships, and in eleven minutes *Repulse* rolled over and sank. *Prince of Wales* survived until 1320, then she too disappeared into the South China Sea. In all 840 men from the battleships were lost with their Admiral. The Japanese did not bother to attack the destroyers, which picked up 2,081 survivors. The attackers lost just three aircraft.

The bombers were not even escorted by fighters. Had *Indomitable* been with the battleships, as originally planned, they might possibly have survived, at least long enough to withdraw.

The loss of the two battleships was more than the final curtain for the battleship: it was the first intimation that Britain's role as a world-wide naval power was in decline. For the first time in its long and not inglorious history, the Royal Navy, having lost supremacy in a theatre of war, would not regain it. From that time on the Allied ships fighting the war at sea in the Far East would be under the control of the US Navy: British ships and carriers would take part, but only as subordinate units of the American fleets and even then not exactly welcomed by certain US admirals.

The attack on Hawaii and the sinking of the British battleships was only the curtain raiser to an astonishing run of Japanese victories.

The US base at Wake Island had been bombed only a few hours after Pearl Harbor; seven of the twelve F4F-3 Wildcats delivered four days earlier to the Marine Fighter Squadron VMF-211 by the *Enterprise* were destroyed on the airstrip at Wake. The same day, 8 December, by a grim coincidence, at Shanghai the river gunboat USS *Wake*, surprised at the rapidity of events, failed to scuttle and was given up to the Japanese: the only US Navy ship to surrender in World War II.

On 9 December Japanese land forces took Bangkok, in Thailand. Next day the Japanese Army Air Force commenced the bombing of the small US garrison on Guam; two days later the island was captured.

Japanese Army bombers from Formosa destroyed half the USAAF B-17 and P-40 Warhawks based on Clark Field, Manila, in the Philippines. The pattern was to be the same everywhere: Japanese navy or army aircraft bombed and strafed airfields, knocking out aircraft on the ground; the few that became airborne were overwhelmed as the Japanese established total air superiority prior to amphibious landings at widely scattered points, which the depleted Allied navies could do little to oppose.

The defenders on Wake Island fought a desperate rearguard action. The Marine F4F Wildcats of VMF-211 and shore artillery actually drove off a landing on 11 December. The pilots, arming their Wildcats with small bombs, damaged at least two ships of the assault force. Japanese bombers from bases on Kwajalein, in the Marshall Islands, 700 miles away to the south, continued to bomb the American defenders. Navy dive-bombers from the carriers *Soryu* and *Hiryu*, which had refuelled at sea after the Pearl Harbor attack, appeared on 21 December to prepare the way for another invasion attempt. By the next day only two Wildcats were still flyable, and those had been kept airworthy only by cannibalisation. Carrier-borne Zeros shot both down that day and the remaining grounded pilots fought as infantry. They tried desperately to hold the island against the invaders, having been signalled that an American relief force, Task Force 14, which included the *Saratoga*, was steaming from Pearl Harbor; but the force was recalled when less than 500 miles from the beleaguered island. Without air cover, the island fell on 23 December after a sixteen-day siege.

Hongkong fell on Christmas Day 1941 and British forces were soon fighting the advancing Japanese in Burma and Malaya. The European troops were finding the Japanese a difficult enemy to oppose. The British positions were bypassed; the lightly armed and self-sufficient enemy infantry seemed able to fade into the jungles, to reappear miles ahead, having slipped through seemingly impenetrable rain forests and traversed peaks considered unscalable. When engaged, the Japanese soldiers fought with a dedication and ferocity unknown to western armies.

On 8 February 1942 a Japanese Army of 35,000 men, having advanced through Malaya, began the assault on Singapore. In London the island was thought to be capable of withstanding an indefinite siege; after all, it was well supplied and heavily defended by over 80,000 men. Unfortunately the Japanese attacked it from the wrong direction; most of its defences, including heavy artillery, were planned to repel a seaborne assault from the south — not across the narrow Jahore Straits to the north.

After a week of continuous fighting, Singapore's Commander, General Percival, surrendered the island to the Japanese at 2030 hours on Sunday, 15 February 1942. Over 80,000 troops — four divisions — had capitulated, to be marched into a terrible captivity. It was the most humiliating defeat ever sustained by British armed forces. With the fall of Singapore the shadows were lengthening perceptibly over the Empire on which, British schoolchildren had been taught, "the sun never sets".

The relentless Japanese tide flowed on: Burma, Sumatra, Borneo, Celebes, and the rest of the Dutch East Indies. In the Philippines the enemy had landed at nine different invasion points. Manila, the capital, was evacuated and US forces retreated, under their colourful general, Douglas MacArthur, to Luzon and the Bataan Peninsula, and finally the heavily defended base at Corregidor, from which, on 11 March 1942, General MacArthur was ordered to leave. His famous remark, "I shall return", must have seemed at the time mere rhetoric to his successor, General Wainwright, who held Bataan until 9 April when 35,000 American and Filipino troops were marched into captivity. The last US forces fighting in the Philippines were the 11,500 men on Corregidor, who finally surrendered on 6 May 1942. Many of the men captured in the fighting in the Far East were to die while prisoners of the Japanese.

The territorial gains made by the Japanese up to March 1942 covered an area greater than the whole of Europe; from Attu in the Bering Sea, south through the chains of Pacific Islands to Papua and New Guinea, west across the Timor Sea to Java, then north through Malaya, Siam and Burma to Assam and the frontiers of India. These Japanese advances pushed her war zone far beyond the ambitions of the original 'Greater Asia Co-Prosperity Sphere'.

To dismantle this vast stolen empire was to necessitate a long and bitter campaign, involving hundreds of thousands of troops, thousands of aircraft and the largest naval forces ever assembled to fight the greatest sea battles in history.

The Japanese land conquest had been possible only because of their air and sea supremacy; their carriers in particular had contributed to the victory, providing dive-bombers and torpedo aircraft to attack Allied shipping and airfields, and fighters to cover the subsequent amphibious landings. The Allied fleets, in contrast, had achieved little success. Douglas Dauntless SBD dive-bombers from USS *Enterprise* had sighted and sunk the Japanese submarine *I-70* off Oahu two days after the Pearl Harbor attack; but *Saratoga* was torpedoed by a Japanese submarine to the south-west of Oahu and, though able to steam, she had to return to the United States for repairs. *Saratoga* was to be replaced by *Hornet*, CV-8, the last of the *Yorktown* class. The remaining US Fleet carriers were *Lexington* and *Yorktown*.

These carriers were to be involved in several minor actions off the Philippines and the Dutch East Indies. Although they damaged and destroyed a number of enemy aircraft and ships, the raids were, by and large, unco-ordinated and tended to emphasise the US Navy's lack of suitable weapons and tactics; not one of the Japanese carriers was so much as sighted during these operations.

One sad casualty during February was USS *Langley* — the 'covered wagon' and the first US Navy carrier. She was now AV-3, a seaplane carrier, and was engaged in ferry operations taking crated USAAF P-40 fighters to Tjilitjar in Java. On 27 February she was sighted by a reconnaissance seaplane and attacked by a force of Japanese Army G3M (Nell) twin-engined medium bombers and sunk. *Langley* was but one of many Allied vessels sunk in the fighting now generally known as the Battle of the Java Sea, which included the Royal Navy cruiser *Exeter* and the destroyers *Encounter* and *Stronghold*; a US Navy cruiser *Houston*, and an oiler, *Pecos*; the destroyers USS *Ashville* and *Pope*; and some twenty small ships. Most of these vessels were lost in trying to forestall the coming invasion of Java.

The assault on Java was supported by strikes of aircraft from the carriers *Soryu* and *Ryujo*, with land-based bombers from Bali and Timor. The defenders could do little: Java fell on 10 March 1942. After the Java action the Japanese carriers returned to their base in Staring Bay, Celebes, and rearmed for their next move.

With the whole of what is now Malaysia under Japanese control, the powerful strike force under Vice Admiral Nagumo, consisting of five fleet carriers — *Akagi, Shokaku, Zuikaku, Hiryu* and *Soryu* — together with the small carrier *Ryujo*, four battleships and a cruiser division, sailed from Celebes on 26 March into the Indian Ocean to attack British naval bases in Ceylon and shipping off the south-east coast of India, in the Bay of Bengal. The strike force split in two south off the Nicobar Islands, the fleet carriers heading for the British base at Colombo on the west coast of Ceylon and the *Ryujo*, in company with surface ships, searching for shipping off the east coast of India.

The British were warned by intelligence reports, probably due to the activities of cryptanalysts, that an attack on Colombo was imminent. To meet the Japanese, Admiral Sir James Somerville had only a somewhat scratch force, consisting of the modern carriers *Formidable* and *Indomitable*, the old, slow *Hermes*, and five battleships — *Warspite*, in which the Admiral hoisted his flag and four World War I *Revenge* class ships. There were also four cruisers, which alone had the speed required to keep station with the two armoured carriers.

British Intelligence had reported the date of the Japanese attack as 1 April: this proved incorrect. Somerville sailed too soon and anti-climactically the opposing main forces never met — from the Allied point of view, perhaps fortunately. Lacking a sufficient number of oilers, the Royal Navy fleet was compelled to return to its base in the Maldive Islands to refuel; while they were there, four days later than supposed, on 5 April, the Japanese attacked the naval base at Colombo with eighty-seven aircraft from the five carriers. The attack lasted for only thirty minutes. RAF Hurricanes and land-based naval Fulmars intercepted, shooting down six Japanese Vals and one Zero for the loss of sixteen British aircraft.

The shipping losses due to the bombing of the harbour were not heavy — one destroyer and one fleet auxiliary — but the greatest losses were to be at sea. A reconnaissance seaplane from a Japanese cruiser sighted the two British cruisers *Cornwall* and *Dorsetshire* to the south of Ceylon, steaming to rejoin Somerville's fleet. Fifty-three Vals were launched from the carriers *Akagi, Soryu* and *Hiryu* and within twenty minutes both cruisers, lacking air cover, were sunk. The attackers lost one Val.

Following the sinking of the cruisers and the first attack, anticipating another strike, shipping scattered from Colombo. This played into the hands of Rear Admiral Kakuta, commanding the second Japanese force at large in the Indian Ocean, which consisted of four heavy cruisers and the carrier *Ryujo*. On 5 April, the same day as the Colombo strike by

Nagumo's aircraft, Kakuta was searching for shipping in the Bay of Bengal, to the south of Calcutta; one of his Kates found and sank a merchant ship, the first of over 53,000 tons of shipping sunk by the carrier's aircraft alone, which was a good return considering that *Ryujo* was one of the least satisfactory of the pre-war Japanese carriers and at no time during the Bay of Bengal action did *Ryujo* operate more than eighteen B5N Kates. It is an interesting commentary on the efficacy of carriers that the four cruisers sailing with the air group managed to sink only an additional 30,000 tons of shipping among them.

There was to be yet another loss to the Royal Navy off Ceylon. On 9 April Nagumo's carriers launched a strike against the Navy base at Trincomalee on the east coast of the island. An RAF Catalina sighted the force the previous day and thus the base received some fifteen hours' warning of the impending attack; enough to organise the defences and disperse shipping. Sixteen RAF Hurricanes and six Fulmars intercepted the attackers but shot down only one Kate and a Zero for the loss of eight Hurricanes and a Fulmar. However, little damage was done to the port other than the sinking of a merchant ship. But a long-range reconnaissance E13A Jake seaplane catapulted from the Japanese battleship *Haruna* sighted the Royal Navy carrier *Hermes*, in company with other ships, to the south of Trincomalee. Eighty-five of Nagumo's dive-bombers, escorted by Zeros, sank *Hermes*, which had no fighters embarked; in addition, two fleet oilers, a destroyer and a corvette were lost in the action.

The Japanese striking aircraft were intercepted after most of the ships had sunk, by land-based Fulmars, which shot down four of the Vals for the loss of two Fulmars. Nine RAF Blenheim bombers of No. 11 Squadron, a type of plane considered obsolescent in Europe, stalked *Akagi* and the cruiser *Tone*, using cloud cover; they bombed the two ships from medium level but scored no hits. Swarms of A6Ms and D3A Vals rose from the decks of the Japanese carriers, but surprisingly, in view of the limited speed and manoeuvrability of the twin-engined RAF bombers, only five were shot down and even then their air gunners destroyed at least two A6M Zeros and damaged others.

The shipping losses from the incursion into the Indian Ocean by the Japanese carrier forces had been high, but there was perhaps one tiny crumb of comfort to be had: against spirited — though inadequate — fighter defence at Colombo and Trincomalee, the Japanese pilots had not pressed home their attacks with quite the same élan as at Pearl Harbor and other bases in the Far East. Perhaps they were mortal after all.

At the same time as the sorry tale of disaster and defeat in the Far East was filling the papers and newsreels in America and Britain, the US Navy Operations Staff was putting the finishing touches to a bold and brilliant propaganda coup: an air raid on the Japanese capital, Tokyo. Such a raid, if successful, would not only raise public morale at home, it would also have a profound effect on the Japanese, who were enjoying the heady prospect of an early, cheap and total victory over the old Imperial powers. The only difficulty was the rather serious question of distance. Tokyo was approximately 2,200 miles from the US bases on Midway and slightly more from the only alternative — Dutch Harbour in the Aleutian Islands. That distance was beyond the feasible range of US Army bombers then in service, unless a 'one-way' suicide mission be undertaken which was, of course, never even contemplated.

The Navy Staff, however, had a further alternative; it decided that the raid would not be mounted from a land base but from an aircraft carrier. The vessel chosen was the new *Hornet* (CV-8), which had just completed her work-up.

The aircraft selected for the Tokyo raid were North American B-25B Mitchell Army Air Force bombers. (The other type considered, the B-26 Martin Marauder, was rejected in view of its longer take-off run.) The B-25 was an excellent medium bomber, powered by twin Wright Cyclones, offering 1,850hp for take-off. Its normal still air range was 1,275 miles and it could achieve 275mph while carrying a bomb load of 3,000lb. However, its designers had never in their wildest dreams envisaged carrier operations: the B-25's usual combat all-up weight was 33,500lb — a figure twice the weight of the heaviest contemporary naval carrier aircraft.

It was considered that the flight deck of *Hornet* could withstand the loading; the real question was could a bomber normally operating from long paved runways become airborne, at combat weight, in 500 feet? In order to test the feasibility of the plan, Admiral King approached Army Air Force General 'Hap' Arnold, who was enthusiastic and who made a B-25 available. Trials took place from a marked landing strip, supervised by a Navy test pilot, Captain Donald Duncan, on the basis of which the operation seemed just possible. After the tests from a runway, the B-25 was successfully flown off *Hornet* in the Atlantic in February 1942 — a trial conducted in the greatest secrecy.

That a stripped bomber, flown by a highly skilled pilot, could take off from a carrier proved only that a 67 foot wingspan bomber could become airborne from an 80 foot wide flight deck. What of fully loaded operational bombers carrying 2,000lb of bombs and 1,141 US gallons of fuel, flown by Army squadron pilots with no carrier experience whatsoever? The man selected by 'Hap' Arnold to find the answer was Lt Colonel James Doolittle.

At Wright Field sixteen Army B-25B Mitchells were stripped of all non-essential equipment: radios were cut to the minimum; the ventral gun turret was removed, saving 600lb and allowing an additional 50 US gallon fuel tank to be installed in its place; and to save further weight and because of the likelihood of one or more of the bombers being forced down over Japan, the then top secret Norden bombsight was removed and a specially built 'twenty cent' low level sight fitted. Twin dummy .50 guns were added to the tail cones to discourage stern attacks by fighters, in the absence of the ventral turret. Even lightened as far as safety and operational efficiency allowed, the bombers prepared for the unique raid tipped the scales in combat trim at 31,000lb.

Volunteer crews were enrolled for the operation from the USAAF 17 Bombardment Group and the 89 Reconnaissance Squadron, both of which normally operated the B-25s. Though they knew they were expected to fly from a carrier, they were not, at this stage, informed of the target. Training began on a marked runway and the crews found that the lightened B-25s could become airborne in the seemingly impossibly short distance of 500 feet. When taking off from *Hornet*, of course, only the first aircraft would be required to achieve the minimum run, for as each aircraft left the flight deck, the next had its available run increased by some 30 feet; but it would be touch and go, even with *Hornet* sailing at maximum speed.

At the end of March 1942, the preparations being completed, all the sixteen bombers were craned aboard *Hornet* in San Francisco Navy Yard. Seven of the carrier's normal complement of naval aircraft had to be put ashore but the remaining sixty-five were stowed below in the hangar, though some had to be dismantled and slung from the roof beams, thereby increasing the loading on the flight deck above still further.

The impression aboard the carrier and in the Yard, which the Navy did little to dispel, was that *Hornet* was engaged on a ferry trip to the Pacific. The actual plan was for the carrier to get to within 400 miles of Honshu, the main island of Japan, and fly off her bombers, which would attack targets at Kobe, Nagoya, Osaka and Tokyo, thereby achieving the maximum impact on the Japanese population. The military value of the raids, which with sixteen bombers could only be slight, was a secondary consideration. After the attack, the bombers were to fly on to an airstrip at Chuchow, in unoccupied China, over 1,200 miles from Tokyo.

On 1 April 1942 *Hornet*, loaded to her marks, slipped out of San Francisco Bay; she was screened by two cruisers, four destroyers and accompanied by a fleet oiler. When safely at sea, Doolittle briefed his crews as to the true nature of the operation. Not one backed down. It had been planned that one of the B-25s would leave the carrier on a demonstration take-off when 500 miles from San Francisco and fly ashore to encourage the others. Doolittle decided that it would be a waste of an aircraft which could otherwise fly on the raid; the demonstration was cancelled.

The carrier and her escorts sailed to a position north-west of the US Island of Midway — soon to be the scene of one of the great sea battles of World War II — to rendezvous on 13 April with USS *Enterprise* to form Task Force 16, under the flag of Rear Admiral Halsey. The task force sailed west into Japanese-dominated waters towards the planned launch position, 400 miles from the targets, which would give the B-25s just sufficient range to attack and overfly Japan and to land in China, 1,600 miles away from the launching point.

At dawn on 18 April, the day of the raid, when Task Force 16 was still some 700 miles from Toyko, search aircraft from *Enterprise* sighted Japanese naval picket boats. Other enemy patrol ships were located by the carrier's radar; it was certain that they had sighted and reported by radio the presence of the US Task Force. The decision was taken to fly off the B-25s as soon as possible, even though it would add another 300 miles to the flight. On the flight deck Army and Navy men bombed up the aircraft; each was armed with two 500lb bombs and 1,000lb of incendiaries.

Within two and a half hours of the sighting of the patrol pickets (three of which were sunk by combined gunfire from the escorts and bombs from the *Enterprise*'s aircraft), Lt Colonel Doolittle — not a man to lead from the rear — selected take-off flap and opened wide the throttles of the lead B-25. He held the bomber on the brakes to allow the engines to develop full take-off power and then released them. The weather was poor; *Hornet* was steering into a moderate sea and the wind over her flight deck was about 40 knots. With engines flat out, slowly the Mitchell gathered speed, Doolittle following a broad white line painted on the deck well to port as a guide to help pilots squeeze past the carrier's island. The B-25s had tricycle undercarriages — by no means universal in 1942 — and Doolittle had little difficulty in keeping the aircraft straight. 'Goofers' lined every vantage point, though none were watching with more intent than the remaining fifteen crews. They need not have worried: Doolittle's B-25 lifted off like a kite with 100 feet to spare. The others followed. The only casualty was a member of the deck-handling crew who stumbled into a propeller of the last but one Mitchell; he was lucky to lose only an arm.

The average distance to the target was 668 miles, flown mostly at wave-top height. The sixteen bombers reached their assigned targets, which were bombed from 1,500 feet. Tokyo had just ended a practice air raid drill, so there was some confusion when the warning sirens went around noon, the civilian population and the anti-aircraft gunners thinking that it was an extension of the exercise. The flak was sparse and inaccurate: none of the bombers was shot down.

Having bombed the targets without loss, the aircraft flew on to China, but sadly, owing to bad weather and misunderstandings with the Chinese, who thought the bombers were Japanese and promptly extinguished all airfield lights, most crews, running out of fuel, baled out or ditched. Those who landed in unoccupied China eventually, after many adventures, reached home. One crew force-landed near Vladivostok in Russia — nominally an ally — and were promptly interned. Eight men who landed in occupied China were captured by the Japanese: three were executed, one died in captivity, and four survived three years of Japanese imprisonment. Doolittle himself was one of the lucky ones to land in unoccupied China; as he walked away from the wreckage of his B-25, he was convinced that the raid had been a costly failure.

In strictly military terms it was; after all the loss rate was 100 per cent. But the effect on the Japanese was out of all proportion; the military had lost 'face'. More to the point, the authorities realised that most of Japan's cities had a large percentage of wooden, even paper buildings, and further raids could be a very serious menace. The immediate consequence was an increase in defensive fighter and anti-aircraft units around Japanese cities and military bases which tied down men, guns and aircraft. There was, in addition, the effect of the news of the raid on the American public: at last they had a victory to celebrate.

At a press conference President Roosevelt was asked from which base the raid had been flown. "Shangri-La", he replied — an allusion to the mythical Himalayan country of James Hilton's *Lost Horizon*, a novel which had been made into a popular Hollywood film in 1937. The raid was commemorated in 1944 when a US *Essex* class carrier (CV-38) was commissioned as USS *Shangri-La*.

The raid was to cause the Japanese Navy to redeploy their carriers. After the foray into the Indian Ocean in April the Japanese carriers had returned to their anchorages in the Celebes. The force then split up:

Zuikaku and *Shokaku* sailed in the third week of April for the Navy base at Truk, in the Caroline Islands, to prepare for further operations; *Soryu* and *Hiryu* with *Akagi* returned to Japan to re-equip and to be available to cover the approaches to the Home Islands against any repetition of the 'Doolittle' raid.

On arrival at Truk *Zuikaku* and *Shokaku* were joined by a new light carrier *Shoho* (Lucky Phoenix), 11,262 tons standard, which had been converted from the submarine tender *Tsurugisaki* and completed only the previous January. The three carriers were preparing to take part in Operation 'MO': large-scale amphibious landings at Port Moresby, to enable the Japanese Army to occupy the whole of Papua, joining up with their forces already on the other side of the mountain ranges of New Guinea. If successful, the invasion would enable the Japanese to set up major bases in Papua from which the Australian mainland, only 100 miles across the narrow Torres Strait, could be attacked and later possibly invaded.

The orders for the landings which, like many Japanese maritime operations, were exceedingly complex, were sent to Truk and other bases in code by radio. American cryptanalysts had some months previously broken the Japanese naval codes; they could therefore, in a large measure, learn of the intentions of the enemy. Forwarned is forearmed and the C-in-C Pacific Fleet, Admiral Nimitz, could deploy his forces to the best advantage to forestall the landings at Port Moresby.

The Japanese plan was first to establish seaplane bases in the Solomon Islands from which to extend the range of their reconnaissance flights and to search for any enemy warships that could be on the routes which the troop transports and covering forces would have to take to Port Moresby. This was a necessity because Japanese Intelligence knew that the United States Navy had been establishing bases in the New Hebrides and could therefore deploy forces in the Coral Sea between those islands and the eastern seaboard of Australia.

The 'MO' operation began on 30 April 1942, when the Japanese sailed from Truk, the opening move in what was to become known as the Battle of the Coral Sea — the first action between major naval forces in which the opposing fleets would never come into direct contact. The battle was to be fought entirely by carrier-borne aircraft of both sides.

The Japanese naval forces consisted of the fast carriers *Zuikaku* and *Shokaku*, with a combined total of 125 aircraft embarked: A6M Zero fighters, D3A Val dive-bombers and B5N Kate torpedo bombers in almost equal numbers. The two carriers were escorted by three heavy cruisers and six destroyers. The commander of the Japanese strike force was rear Admiral Hara, who hoisted his flag in *Zuikaku*. The fast carrier force was to operate initially from the vicinity of Bougainville, the northern island of the Solomons, providing long-range cover for a secondary assault force which would attack and occupy Tulagi Island in the south of the Solomons group. The object of this minor invasion was the setting up of the reconnaissance seaplane base. That done, the main force would invade Port Moresby. When the transports sailed from Rabaul, the fleet carriers would leave Bougainville to provide long-range cover for the assault fleet.

The landings at Tulagi were to be closely supported by aircraft from the light carrier *Shoho*, with twelve Zeros and nine Kates on board; she was accompanied by four heavy cruisers, three light cruisers and a number of destroyers. Also sailing with the close support force was the seaplane carrier *Kamikawa Maru*, which had on board her operational reconnaissance Nell seaplanes with, additionally, a cargo of Mitsubishi F1M (Pete) biplane single-engined reconnaissance seaplanes, which were to be disembarked at Tulagi to equip the seaplane base.

The landings at Tulagi went according to plan. Little opposition was experienced from the small Australian garrison and the seaplane base had been set up by dusk on 3 May.

To forestall the major landing, the Americans deployed Task Force 17, under Rear Admiral Fletcher. TF17 consisted of the *Yorktown*, which carried twenty F4F-3 Wildcats, thirty-eight Douglas Dauntless SBD-3 dive-bombers and thirteen Douglas Devastator TBD-1 torpedo aircraft. Three heavy cruisers, six destroyers and the fleet oiler *Neosho* completed the Task Force.

Under the overall command of Rear Admiral Fletcher, there was a second Task Force, TF11, consisting of the carrier *Lexington*, two heavy cruisers and seven destroyers. *Lexington* had embarked twenty-two Wildcats, thirty-six Dauntlesses and twelve Devastators.

The two US Task Forces formed up off Espiritu Santo, an island in the New Hebrides, 1,500 miles south-east of Port Moresby, on 1 May and commenced refuelling. The US forces could afford to take their time since the cryptanalysts had provided Admiral Nimitz with the Japanese plans of the whole complex 'MO' operation. The Japanese, on the other hand, were blissfully unaware even of the presence of the two powerful Task Forces; they had assumed that only one enemy carrier, *Yorktown*, was in the South Pacific.

For twenty-four hours the two US Task Forces remained off Espiritu Santo, then TF17 sailed into the Coral Sea during the evening of 2 May, steering north-west to arrive 500 miles south of Tulagi by dawn the next

day. (*Enterprise* and *Hornet*, the only other United States carriers of the
Pacific Fleet, had been delayed by the Doolittle operation and had only
just returned to Pearl Harbor. They took no part in the action.)

On 3 May the Australians at Tulagi reported the invasion of the island
by radio, and TF17, with *Yorktown*, less the oiler *Neosho* and its
attendant destroyer *Sims*, which had been detached, sped towards
Tulagi. On 4 May, when 110 miles from the island, the first strike was
launched, consisting of twenty-eight Dauntless dive-bombers and twelve
Devastator torpedo aircraft. The weather was poor and the attack met
with only moderate success; nevertheless, it came as a total surprise to
the Japanese invaders. Further strikes were launched later that day. The
US carrier pilots were flying their first combat missions and were
inexperienced, though they sank the destroyer *Kikuzuki* and a mine-
sweeper *Tamu Maru*, in addition to several smaller vessels and, perhaps
more importantly, five large four-engined Kawanishi H6K Mavis long-
range reconnaissance flying boats, which had landed following the
seizure of Tulagi. During the strikes the US Navy lost three aircraft.

Task Force 17 then withdrew south. The Japanese could do little to
prevent the *Yorktown* leaving the area; the search operations were
hindered by the loss of the flying boats and the absence of *Shoho*, which,
having covered the Tulagi landing, had left and was refuelling at
Bougainville to the north. The fast Japanese carrier force with *Zuikaku*
and *Shokaku* had sailed from Bougainville and was now 600 miles north-
west of Tulagi, preparing to cover the Port Moresby invasion force,
which included the troop transport that had just sailed from Rabaul in
New Britain, north of Papua.

At dawn on 5 May, the two US Task Forces met 400 miles south-west
of Tulagi off the Louisiades, the archipelago of islands in the Coral Sea
off the eastern tip of Papua. Here the US forces refuelled from oilers and
flew off search patrols to cover the probable approach routes of the
Japanese invasion fleet.

Admiral Fletcher did not at this time know the exact position of the
Japanese forces: in fact the main enemy fast carrier force was steaming
east of San Cristobal Island, off the southern tip of the Solomons. The
Japanese were equally in the dark as to the presence and even the
composition of the US forces. However, TF17 was sighted on 6 May by a
Mavis flying boat but the sighting report was delayed; it was not to be the
only mistake of that day. Seventy miles off Bougainville, four USAAF
land-based B17s sighted and bombed, without effect, what the Army
pilots had identified as an enemy cruiser. In reality it was the *Shoho*,
standing by to join the transports and to provide close air cover for the

Port Moresby invasion fleet, sailing south from Rabaul.

From intelligence reports, Rear Admiral Fletcher was aware of the approximate position of the invasion troop transports and the carrier *Shoho*, though he did not know the position of the other Japanese carriers, the *Zuikaku* and *Shokaku*, which were over 500 miles to the south-west of the invasion fleet. To try to locate the Japanese fast carriers, Fletcher consolidated TF17 and 11 by combining them into a single unit: Task Force 17.5.

At first light on 7 May 1942, *Yorktown* and *Lexington* launched Douglas Dauntless search aircraft to the north-west and the south. At 0815 hours a Dauntless of VS-5 from *Yorktown* signalled "two enemy carriers and four heavy cruisers" 260 miles north-west of the US Task Force. Within one and a half hours the two carriers had ranged, armed and launched between them fifty-one Dauntlesses, twenty-two Devastators and eighteen Wildcats, to attack the Japanese 'carriers'. The US strike force had barely left the flight deck when the scouting SBD landed on *Lexington*. On debriefing the pilot, it was learned that a mistake had been made in the encoding of his sighting report, which should have been read as "two light cruisers and two destroyers".

Fletcher was now in a quandry. He had launched ninety-one aircraft, the major part of his force, against the wrong target; however, he gambled that an enemy carrier must be somewhere in the area reported and the strike was not recalled.

When the scouting SBD from *Lexington* had appeared over the Japanese invasion fleet, it was correctly seen as a prelude to an impending attack; orders were at once given that the vulnerable troop transports should turn about and sail out of the immediate area until the outcome of the expected US carrier aircraft attack was clear. By this time, approximately 0900 hours, the delayed Mavis sighting report of TF17 was decoded and at 1050 *Shoho* turned into wind to launch a strike against it. Nine Kates were ranged about to take off when the first three SBD aircraft of the US strike from *Lexington*, led by Commander Ault, dived through a patrol of A6M Zeros and bombed *Shoho*. One 1,000lb bomb was a very near miss, which blasted five of the Kates overboard and must have inflicted mining effect damage to the ship. Other aircraft from *Lexington* followed, bombing *Shoho* and the escorts, but no further hits were obtained. However, fifteen minutes later, the *Yorktown*'s strike force arrived: at least two 1,000lb bombs from the Dauntlesses were direct hits on the exact centre line of *Shoho*'s flight deck; eleven more bombs hit the stricken ship and an SBD, hit by anti-aircraft fire, crashed on to the side of the carrier's flight deck. The dive-bombers were closely followed by the *Yorktown*'s Devastators, which put seven torpedoes into *Shoho*'s hull.

The weight of explosives delivered on the light carrier would have sunk *Bismarck*; *Shoho* disappeared on this, her first and last operation, in less than ten minutes after the *Yorktown*'s aircraft first attacked. Only three US Navy aircraft failed to return to their carriers.

The sinking of *Shoho*, the first Japanese carrier to be lost during World War II, was welcome news to Admiral Fletcher; but the real targets, *Shokaku* and *Zuikaku*, were as yet not even located, and by the early afternoon of 7 May the weather had deteriorated to the point where further search flights were cancelled. In a sense *Shoho* had been unlucky. The strike forces from the US carriers had been deployed against the Japanese ship by mistake, though, in the event, fortune had favoured Fletcher, who might well have been justified in recalling the attacking aircraft when the mistake in the sighting report was revealed.

Japanese search aircraft also made a mistaken sighting report that same morning. A Kate from one of the two fast carriers had signalled "a carrier and a light cruiser". From the position given it was assumed that this was the US Task force. From *Shokaku* and *Zuikaku* twenty-four Kates and thirty-six Val dive bombers flew to attack the "carrier and cruiser", where were in reality the unfortunate oiler *Neosho* and the attendant destroyer *Sims*, which had earlier been detached from the main force. *Sims* was hit by three bombs and quickly sank; *Neosho* took at least seven direct hits from the Vals and was also near missed by others. Incredibly the oiler did not sink but remained afloat until scuttled by her crew four days later.

When the true identity of the target attacked became known aboard the Japanese carriers, another strike of fifteen Kates and twelve Vals was launched against the American Task Force, the position of which was still only approximately known. In poor weather and failing light the Japanese aircraft flew for some two hours searching in vain for the US carriers. At 1630, running low on fuel, they jettisoned their bombs and torpedoes and headed back to *Shokaku* and *Zuikaku*. On their homeward flight the strike force practically flew over the American ships. They were detected by radar; Wildcats from the *Lexington*'s and *Yorktowns*'s Combat Air Patrols intercepted and shot down nine Japanese aircraft for the loss of two Wildcats. In gathering darkness the F4Fs were recalled to the carriers; the surviving Japanese aircraft, however, which did not have the advantage of radar or radio beacon guidance, had difficulty in locating their own carriers, which were about 50 miles away to the east. Three Japanese pilots circled over *Yorktown*, trying to decide in the darkness if she was Japanese; they were soon disabused of that idea when the American gunners opened up and drove them off. Twenty minutes later a single Japanese aircraft actually made a wheels-down landing approach to *Yorktown* and was shot down by the ship's gunfire. US radar operators watched Japanese aircraft vainly flying round searching for their carriers, dropping one by one into the sea as their fuel ran out. Of the entire Japanese strike force of thirty-seven machines, only four returned safely.

During the night of 7 May 1942, in darkness and heavy tropical rain squalls, the two carrier forces were only 40 miles apart; but neither side wanted a night action, for which the Americans were not even trained. The Japanese steered north; the US Task Force turned south-east and away from the area of bad weather.

The next day, 8 May, dawned clear and bright for the Americans, who had left the cold front during the night. The Japanese, on the other hand, were still in an area of heavy rain squalls. Both sides launched search aircraft as soon as it was light and located each other more or less simultaneously; the two fleets were found to be slightly over 200 miles apart.

Fletcher's Task Force launched a large strike from *Lexington* and *Yorktown*, consisting of a total of eighty-two aircraft: forty-six SBD-3 Dauntless dive-bombers carrying 500lb bombs; twenty-one TBD-1 Devastators armed with the inadequate American airborne Bliss-Laevitt 21 inch torpedo, and fighter cover for the strike in the form of fifteen F4F-3 Wildcats. *Yorktown*'s aircraft were launched first at 0915, followed twenty minutes later by *Lexington*'s. As the American aircraft flew towards the enemy, the Japanese were preparing their strike.

At 1057 *Yorktown*'s air group sighted both the Japanese carriers. The SBD Dauntlesses, protected by broken low cloud, dived on *Zuikaku* and *Shokaku*, missing the former but hitting the latter with two bombs. The first direct hit penetrated the flight deck and exploded in a machinery space below, starting an aviation fuel (AvGas) fire. The second bomb exploded in a workshop. Though the damage to the flight deck was confined to the forward part, *Shokaku* could from now on only recover aircraft, there being insufficient undamaged flight deck to launch.

As the Dauntlesses attacked, nine Devastators released their torpedoes, protected by Wildcat fighters and poor visibility. The previous success of these rather moderate bombers against the *Shoho* had misled the TBD pilots: these Fleet carriers were much larger and faster. The slow American torpedoes were released at far too great a range and no hits were scored.

The second part of the US strike, that from *Lexington*, now arrived on the scene, but, owing to low cloud, had difficulty in locating the targets; also some of the Wildcat fighters had become detached during the flight, so that the attack lacked cohesion. Eventually *Shokaku* — by now well alight — was attacked and bombed by SBDs, though it is doubtful if it was hit, as later claimed. The Devastators' second torpedo attack, like the first, was ineffectual, being again delivered at too great a range; *Shokaku*, damaged as she was, actually outran the torpedoes. *Zuikaku* escaped the attack altogether as the carrier was hidden by rain squalls and low cloud.

While the aircraft from the US Task Forces were attacking the *Shokaku*, thirty-three D3A Val dive-bombers and eighteen B5N Kate torpedo bombers, escorted by a similar number of Zero fighters, were on their way to the US Navy carriers. At 1055 zonal time, *Lexington*'s radar detected the incoming force at maximum range, about 80 miles. Anticipating an attack from the Japanese aircraft, the Task Force had a Combat Air Patrol airborne; unfortunately the F4Fs had been orbiting the fleet for some time and were now too low on fuel to intercept the enemy strike. There was insufficient time to launch a full relief CAP, though nine F4Fs were scrambled to join twenty-three SBD Dauntlesses, which were airborne on anti-submarine patrols or returning from reconnaissance searches. These aircraft, though not primarily fighters, were ordered to augment the Wildcats. The US Task Force had the advantage of fighter control radar, though, since this was the first time it had been used operationally on a US ship, it is hardly surprising that mistakes were made.

The nine Wildcats that had just been launched were climbing, though unfortunately they were vectored in the wrong direction. The Japanese attackers arrived over the Task Force at 1118 hours. The two carriers and their escorts had drawn away from each other as soon as the attackers had been sighted on radar. This dispersal gave the big ships and their screens searoom for evasive manoeuvres, though of course it reduced the effectiveness of massed anti-aircraft fire.

The dive-bombers and torpedo aircraft synchronised their attack, the Kates approaching down sun at wave-top height, the Vals plummeting down in near vertical dives from 17,000 feet out of a clear sky, flashing past the Wildcats still clawing for altitude and badly placed to intercept. However, two of the Kates were shot down by Dauntlesses' twin .50 inch forward-firing guns, before they were in range to drop their torpedoes. Flying through very heavy flak, nine Kates released torpedoes at *Yorktown* but she was able to evade them, though the carrier was hit by a single 500lb bomb which penetrated three decks and started a fire; damage control parties, however, soon had it under control.

Lexington, a much larger and older carrier, was far less handy than *Yorktown*. Six B5Ns attacked her, their experienced pilots making perfect torpedo drops, three angled on either side of the carrier's bows. The Japanese torpedoes ran fast and true; *Lexington* could not comb them all. At 1120 two hit on the port side — one forward, the other amidships. Almost at the same time, while the anti-aircraft gunners were concentrating their fire on the Kates, shooting one of them down, overhead the Val dive-bombers were about to release their bombs, but only two hits were obtained, neither of which did substantial damage, though many near misses shook the ship.

The action was soon over; as the enemy withdrew, at least two further Kates were shot down by the defending Dauntlesses, and others by Wildcats, recovering from the rather inept fighter control. As in any air battle, exaggerated claims were made on both sides, and to this day no one knows the exact number of Japanese aircraft lost. It would seem that at least twenty were destroyed in the immediate area of TF17.5.

As the survivors of both sides flew back to their carriers, it was apparent that *Lexington* was badly hit; she had developed a 7° list. This was corrected by counter-flooding, and although several fires were still burning, they were being brought under control. The carrier's lifts had been jammed, fortunately in the 'up' position, enabling *Lexington* to recover her aircraft, which were now returning from the strike and interception patrols. Six Wildcats, twenty-three Dauntlesses and ten Devastators were landed-on between 1300 and 1345.

The two torpedoes, though they had not inflicted fatal direct damage, had caused the ship's long 888 foot hull to 'whip'; this was what had jammed the lifts but, more seriously, it had also fractured her Avgas aviation fuel lines and highly volatile petrol vapour — that Achilles heel of the early US and Japanese carriers — permeated the ship. Inevitably there was a large explosion followed by fires. The ship's crew fought the new conflagration, but the fires gained ground. Flying operations had to be suspended, *Yorktown* being requested to recover *Lexington*'s nineteen aircraft still airborne. By mid-afternoon the damage control parties reported the fires as being out of control and that the bomb and torpedo magazines were being threatened. Since the danger of the vessel exploding was now a distinct probability, the order to abandon ship was given just after 1700 hours.

Burning fiercely and with 216 of her crew dead and thirty-six aircraft still on board, racked by further AvGas explosions, 'Lady Lex', USS

Lexington, CV-2, commissioned as the first fast US carrier on 16 November 1927, was sunk by torpedoes from a destroyer, USS *Phelps*, at 1853 hours on 8 May 1942 — the first operational US aircraft carrier to be lost.

The Japanese pilots returning to their carriers found *Shokaku* burning, with her flight decks unusable. A number of aircraft, out of fuel, ditched, though others landed on *Zuikaku*, which had escaped the attack; several of *Zuikaku*'s own aircraft had to be pushed over the side to clear the flight deck for the remnants of *Shokaku*'s strike force. According to Japanese sources, the number of operational aircraft remaining after the action was twenty-four Zeros, nine Vals and six Kates, barely a quarter of the number with which the carrier had been embarked prior to the battle.

The Japanese had probably lost forty aircraft in action but the loss of machines was of far less consequence than the loss of their pilots and observers, for these were all experienced, combat-hardened men who had fought in China and the attack on Pearl Harbor. Such men were irreplaceable.

On paper the US losses had been heavier; indeed they had suffered a tactical defeat: one 36,000 ton Fleet carrier, *Lexington*, against *Shoho*, an 11,200 ton light carrier. Including the thirty-six aircraft which went down with *Lexington* and thirty-three others destroyed in action, United States aircraft losses were higher. But the action, though a tactical defeat, was for the US Navy a strategic victory: the Japanese invasion of Port Moresby did not take place; the carriers withdrew to Japan, the troop transports back to Rabaul. Port Moresby was never to fall to the Japanese; attempts to attack it overland from New Guinea failed, Australian troops holding back the enemy.

There was another, less tangible, victory sensed by the Americans after the Coral Sea engagement. Inexperienced in combat as they were, they had taken on the all-conquering Japanese and inflicted severe losses: *Shoho* sunk, *Shokaku* forced to limp back to Japan, where she had to be extensively repaired and was to be effectively out of action for the next crucial three months.

For the first time in warfare at sea, a major action had been fought without a single ship of the ninety-five involved firing a shot at an enemy vessel. The Battle of the Coral Sea confirmed the aircraft carrier beyond any doubt as the capital ship in the Pacific War.

By carrier-borne airpower alone, the Japanese advance, for the first time, had been halted. If it was not the turning of the tide, it was very nearly high water.

As a consequence of the Coral Sea action, Admiral Isoroku Yamamoto came to the conclusion that the US Pacific Fleet would have to be drawn into decisive battle and destroyed as a fighting force. Yamamoto was a realist and not a narrow-minded dogmatist; he knew that once American industrial production got on to a proper war footing, the US Navy would soon be able to overwhelm the Japanese, in sheer numbers of ships and aircraft. It was therefore vital, while the balance of forces still favoured Japan, to destroy the remnants of the US Pacific Fleet which had survived Pearl Harbor, and to consolidate the territorial gains that Japan had acquired since that attack. The Coral Sea action had given the US Pacific Fleet combat experience and confidence. There was little time remaining for a decisive action.

The plan evolved was Operation 'MI' and had as its centre piece the occupation by Japanese troops of the US base of Midway Island, which lay 1,135 miles north-west of Oahu and 2,200 miles from Tokyo. Midway was an unsinkable aircraft carrier which controlled the western approach to the Japanese Home Islands. The object of the projected assault was twofold: to secure a base for Japanese long-range reconnaissance aircraft that could forestall any possible extension of the 'Doolittle' carrier raids on Japan, and to draw the US Pacific Fleet into a decisive action. Yamamoto gambled that the Americans would have to fight to regain the island since, if captured, it would enable the Japanese to threaten Pearl Harbor.

A diversionary force would attack US bases on Kiska, Attu and Dutch Harbour in the Aleutian Islands, 1,500 miles north of Midway, twenty-four hours before the attack on that island, with the twofold object of extending the Japanese perimeter and of drawing away at least part of the slender resources of the US Pacific Fleet from Midway, while the assault force, consisting of four carriers, two battleships, four cruisers, destroyers and troop transports, were taking the island.

The four Japanese carriers were to be used solely in support roles to provide air cover for the Midway landings. Yamamoto had decided that battleships would fight the main sea engagement, which, with the US Pacific Fleet's battleships still sitting on the bottom of Pearl Harbor, or in the Navy Yards being repaired, must have looked on paper a reasonable proposition, especially as at least one of the only two US carriers — which the Japanese Staff assumed to be all that were available in the Pacific — would be sailing to contest the invasion of the Aleutians.

Unfortunately for Yamamoto there were two crucial flaws in the carefully drawn plans. In the first place, the whole 'MI' operation in all its complexity was soon on Admiral Nimitz's desk, thanks to the activities of the US cryptanalysts; the Aleutian operation was immediately seen to be what it was — a trap — and therefore ignored. Furthermore, there would not be two US fleet carriers available, but three.

The date of the Japanese operation against Midway was set at 4 June 1942.[14] On 26 May the only two fully serviceable US fleet carriers, *Enterprise* and *Hornet*, were recalled from the south-west Pacific and entered Pearl Harbor. *Yorktown*, limping back from the Coral Sea, dropped anchor there next day; she was at once placed in the drydock which had survived the December attack and a survey of damage made. The dockyard engineers concluded that the minimum time required to restore the carrier as a fighting unit would be several weeks; they were told by Nimitz that *Yorktown* would be sailing in three *days*.

While the work on *Yorktown* went on round the clock, new fighter aircraft were being prepared for embarkation on the carriers: F4F-4 Wildcats. These aircraft were based on the Royal Navy's Martlet II and had extra armament — six .50 inch guns to the F4F-3's four, an increase in fire-power of 50 per cent. But, more significantly, unlike the F4F-3, which had fixed wings, the F4F-4 Wildcats had folding wings, which permitted almost double the number of fighters to be stowed aboard the carriers. The Task Forces would be able to embark a total of seventy-nine F4F-4s — now the standard US Navy fighter.

The dive-bombers on the carriers were the well-liked and successful SBD-3 Dauntlesses, known to the crews variously as 'Slow but Deadly', 'the Barge', 'the Clunk' and 'the Daunty Lass'. By whatever name, 101 were to sail with the Task Forces from Pearl Harbor. The position about torpedo aircraft was less satisfactory: there was a new Grumman torpedo aircraft on naval charge — the TBF Avenger (Tarpon in RN service) — but this was not yet available in adequate numbers, so forty-two of the obsolete TBD-1 Devastators, armed with their useless torpedoes, had to be embarked.

The three US carriers were part of two Task forces: TF16, comprising *Enterprise* and *Hornet*, with five heavy cruisers, a flak ship and nine destroyers, under Rear Admiral R.A. Spruance (in place of Rear Admiral Halsey who was ill); and TF17, with Rear Admiral Fletcher, who was to fly his flag in *Yorktown*, screened by two heavy cruisers and five destroyers.

The timely warning of the enemy's intentions provided by the code-breakers enabled Midway Island's air cover to be reinforced. By 3 June it comprised VMF-221, a Marine Fighter Air Group of twenty-one F2A-3 Brewster Buffalos and seven F4F-3 Wildcats, and VMFB-241, with eighteen Dauntlesses and sixteen SB2U Vindicators. A naval detachment on the island, VT-8 had six of the new untried Grumman Avengers;

there was also a Patrol Wing of thirty PBY reconnaissance Catalinas. The USAAF 7th Air Force flew in seventeen B-17E Fortresses and four B26 Martin Marauders.

In spite of these preparations, the Japanese forces ranged against the Americans were formidable. Yamamoto had at his disposal nearly 200 vessels, including eleven battleships, eight aircraft carriers, seaplane carriers, nearly 700 aircraft, numerous cruisers, destroyers, fleet oilers and twenty-five submarines.

The deployment of these Japanese forces was typically complex. The first move in what was to become the Battle of Midway began on 25 May when the first of the Aleutian attack forces, consisting of two carriers, *Ryujo* and *Junyo*, with two heavy cruisers and destroyers, sailed from Ominato — a port on the northern tip of Honshu, the main Home Island of Japan.

Junyo was a newly commissioned carrier which had been converted from the passenger liner *Kashihara Maru*, when that vessel was still on the stocks. *Junyo* displaced 24,000 tons and could embark fifty aircraft; she had commissioned only a few weeks before sailing and could hardly have been fully worked up. Also sailing from Ominato for the Aleutians was a screening force with the carrier *Zuiho* and four battleships, three cruisers, three destroyers and oilers.

The main force, under the personal command of Admiral Yamamoto, flying his flag in the battleship *Yamato* — the most powerful in the world — with three other battleships, eleven destroyers and the light carrier *Hosho* with eight Kates embarked, sailed to a position in the Pacific roughly between the Aleutians and Midway, with the object of intercepting the US naval forces confidently expected soon to be steering towards the invasion of the Aleutians and the trap set for them.

The Midway occupation was to be the concern of Vice Admiral Nagumo; his flagship was the carrier *Akagi*, accompanied by *Kaga*, *Hiryu* and *Soryu*, screened by two battleships with cruisers and destroyers. There were also sixteen transports carrying the 2,500 troops that were to occupy and garrison the island.

The Japanese timetable had required a submarine screen to be in position off Pearl Harbor by 3 June to give warning of US fleet movements, particularly of the two carriers. This was suspected by Nimitz, who ordered TF16 to sail before the submarines were on station. On 28 May *Enterprise* and *Hornet* were able to slip unreported from Oahu to take up a waiting station 400 miles north-east of Midway. In the dockyard at Pearl Harbor the seemingly impossible had been achieved and *Yorktown* sailed — with many dockyard civilians still working on board — on 31 May, to embark her aircraft and to join up with the other vessels of TF17. The Task Force then steamed to join TF16 in the late afternoon of 2 June, to await the Japanese.

At dawn the next day a long range PBY Catalina from Midway sighted the Japanese Occupation Fleet steaming 800 miles to the west of the island. Nine USAAF 7th Air Force B17-Es took off in the afternoon, when the convoys were within range, to bomb the vulnerable transports from 12,000 feet. The B17 Fortresses were using the much publicised Norden bombsight, which the American press had reported "could put a bomb in a pickle barrel from 22,000 feet". Be that as it may, they did not manage to put any bombs in the Japanese ships; although the B17 crews had reported two hits, the nearest bomb was 220 yards from a ship.

As the B17s returned, four Navy PBY-5As of the Navy Patrol Wing took off to make a night, radar-guided torpedo attack. The crews had just flown into Midway from Hawaii after a ten hour flight; it was hardly surprising, therefore, that they only managed one hit, on a fleet oiler, *Akebowa Maru*, which though damaged was able to maintain her station.

Meanwhile, 1,500 miles to the north of the Striking Force heading for Midway, aircraft from *Ruyjo* and *Junyo* attacked Dutch Harbour in the Aleutians. Admiral Nimitz, when he became aware of the Japanese intentions, had merely reinforced the Aleutian defences with a cruiser squadron and did not detach any ships from the two Task Forces off Midway, as Yamamoto had hoped. The Japanese Aleutian invasion on 3 June simply confirmed the American cryptanalysts' interceptions: the 'MI' operation was keeping to its meticulous timetable.

At dawn on 4 June the Midway Striking Force was 230 miles north-west of the island. At 0430 over 100 aircraft left the decks of Nagumo's four carriers. They had been airborne for about an hour when a US Navy reconnaissance Catalina found and reported the position of the Japanese force. At a little before 0600 hours the incoming air strike was sighted by another of Midway's patrol aircraft, and, almost simultaneously, by the island's radar, which put the range at 107 miles and closing. Within minutes all the flyable aircraft on Midway were airborne, either to intercept the invaders or to keep out of harm's way.

The fighters which scrambled were twenty F2A-3 Buffalos and six F3F-3 Wildcats of Marine Squadron VMF-221. They intercepted the incoming Japanese strike at 25 miles. In the running fight that immediately developed, two Kates, one Val and two Zeros were shot down. The Marine pilots flying the F4F-3 Wildcats had some chance — only two of them were shot down — but the men in the Buffalos were hopelessly outclassed: thirteen of the twenty F2As intercepting the Japanese aircraft were lost. Of the remainder which managed to return to Midway, most were badly damaged, and of VMF-221's machines, only two F4Fs and two F2As were found to be capable of flying again after the first interception. Had the Japanese known, Midway was now virtually without fighter cover.

Having brushed aside the fighters, thirty-four Vals and thirty-four Kates attacked Midway installations, destroying aviation fuel storage tanks, hangars and the seaplane slipway. The island's anti-aircraft gunners shot down ten of the attackers, which was probably the reason that prompted the strike leader to radio back to his carrier that a second strike would be required to neutralise the still active defences before the landings took place. That report was to be crucial in the carrier battle which was shortly to follow.

While the Japanese aircraft were over Midway, the position on the US carriers was that Rear Admiral Fletcher had been given the PBY sighting report which had placed the enemy carriers at 200 miles north-west of Midway. Fletcher ordered TF16 to launch a strike from *Enterprise* and *Hornet* against the four Japanese carriers which Army bombers and the ships' new untried TBF Avengers from the island were also flying to attack.

As the American carrier aircraft were being ranged for take-off, Nagumo, owing to delay in launching a vital search aircraft from the heavy cruiser *Tone*, was unaware of the presence of the US Task Force.

The first US aircraft sighted by the Japanese ships was the mixed strike from Midway. At a little after 0700 hours four USAAF B26 Martin Marauders and the Navy TBF Avengers dropped torpedoes on *Hiryu* and *Akagi*. Unfortunately for the US pilots, Japanese carrier aircraft returning from the Midway strike had sighted and reported the Americans en route, and the carriers' combat air patrol was ready and waiting. The Zeros fell on the bombers: no torpedo hits were obtained and only two B26s and one badly shot up TBF managed to return to Midway.

The torpedo attack, however, had the effect of dispelling any doubts that Nagumo might have entertained about the necessity of following up the first carrier-borne strike against Midway's airfield. The Japanese aircraft from the Midway strike began landing on their carriers immediately after the unsuccessful US torpedo strike. They were quickly refuelled and bombs for the second attack were brought up from the magazines.

During the first Midway strike, Nagumo, as a precaution against having to deal with a possible US Navy surface vessel attack, had retained some twenty torpedo aircraft aboard his carriers. These were now ordered to be rearmed with bombs to join the others on the second Midway strike; aircraft were struck below to the hangars for this change of armament to be carried out. Owing to the delayed launching of *Tone*'s seaplane, Nagumo at this juncture was still unaware of the presence, 300 miles to the east, of the US Task Force. Fifteen minutes after the order to rearm the torpedo-carrying Kates with bombs was given, *Tone*'s seaplane at last reported the sighting of the American fleet. Incredibly, the crew did not see the carriers, although the three ships were in the middle of launching and forming up some 117 strike aircraft.

The crew of *Tone*'s seaplane was urgently ordered to clarify its report, for Nagumo was now in a very difficult position: he had to make a decision whether to change his orders yet again and replace the Kates' bombs with torpedoes to attack the US Task Force, or leave them bombed up and proceed with the second Midway attack. After some consideration, Nagumo decided that the second strike against Midway be postponed; the aircraft would instead prepare to attack the US fleet. Those Kates which still had their torpedoes would retain them; the rest would carry bombs.

At 0800 hours, in the middle of rearming the aircraft, the Japanese Striking Force was attacked by yet another strike from Midway, this time by Marine SBDs and SB2-U Vindicator dive-bombers, which only near-missed *Hiryu*. USAAF B17s also made another high-level bombing attack but they succeeeded in only near-missing *Soryu*. The high-flying B17s escaped interception but at sea level the Combat Air Patrol Zeros shot down ten of the Marine dive-bombers.

The attacks by the Midway-based aircraft did, however, succeed in delaying the recovery of the remainder of Nagumo's strike aircraft, many of which were running out of fuel and some had to ditch. In all, thirty-six aircraft of the original Midway strike had been lost from one cause or another.

At 0820, as the last of the Marine dive-bombers attacked, Nagumo received from *Tone*'s seaplane the report that a carrier *might* be with the US Task Force. On receipt of this intelligence, Nagumo turned his Striking Force north towards TF16's reported position. On board his carriers there was a fre..izy of activity as aircraft which had just landed-on were rearmed, refuelled and light damage repaired. Down in the hangars, crews heaved heavy torpedoes on their wheeled trolleys into place, hampered by the Pacific swell. Bombs removed from Kates were pushed to the sides of the hangar, as there was no time to return them to the magazines below.

Against this background of distracting toil aboard the Japanese carriers, the first US airborne Strike Force of ninety-eight SBD Dauntlesses was steadily drawing nearer.

The dive-bombers were in two main groups: thirty-three Dauntlesses from 'Bombing Six' (VB-6) and 'Scouting Six' (VS-6) of *Enterprise*, led by the Air Group Commander, Lt Commander Wade McClusky, and thirty-five SBDs from *Hornet* (VS-8 and VB-8) led by Lt Commander Ring. McClusky's group had taken off first and had been circling over the carriers when *Tone*'s search aircraft had been sighted; Fletcher immediately ordered them to fly direct to the targets without waiting for the torpedo TBDs or their fighter escort. *Hornet*'s dive-bombers followed a little later, though with a fighter escort of ten Wildcats.

Nagumo's change of course towards the US fleet was unreported. Consequently, when the two dive-bomber groups arrived over the expected Japanese position, they found nothing and began a search. Ring turned his escorted group south; no enemy ships were sighted. The Wildcats soon ran out of fuel and all ditched; the SBDs, with longer endurance, either made for Midway or returned to *Hornet*. All but two of the thirty-three SBDs landed safely.

Wade McClusky's group, had he turned south with Ring, would undoubtedly have suffered the same fate. Instead he had turned northwest for his search.

Meanwhile the torpedo-armed TBD Devastators from *Enterprise* and *Hornet* had also set course for the enemy. *Hornet*'s group, VT-8, was led by Lt Commander John Waldron, who before take-off had studied the plotting charts on *Hornet*'s navigation bridge and had correctly concluded that Nagumo could be steering for the US Task Force in order to close the range for his strike aircraft. Waldron worked out an interception heading and steered straight for the Japanese fleet. *Enterprise*'s VT-6 torpedo group of fourteen TBDs also flew in the same direction but VT-6's course was somewhat to the south of Waldron's. The Wildcats' cover flew above the *Enterprise* group, and, when attacking, the torpedo commanders would call down the escorts on VHF.

At 0920 hours Waldron sighted the Japanese carriers. It would seem that he called down his escorts on the wrong radio channel; consequently they heard nothing. Thus VT-8 attacked with slow, vulnerable aircraft at wave-top height without any escorting fighters. They were met by scores of Zeros and very heavy anti-aircraft fire; every one of the fifteen, virtually defenceless, Devastators was shot down, many without even releasing their torpedoes. Only one pilot — Ensign Gay, who had been the last to attack — survived.

A second Devastator attack, made by *Yorktown*'s VT-3 and *Enterprise*'s VT-6, led by Lt Commander Eugene Lindsey, did not score any hits either and also suffered catastrophic losses: twenty-two out of twenty-six aircraft. Of forty-one TBDs launched during the Battle of Midway, thirty-nine were shot down, most of the slow torpedo bombers being chopped down by A6M Zeros of the Japanese Combat Air Patrols, without scoring a single hit.

The only US Navy aircraft yet to attack Nagumo's Striking Force were the SBD Dauntlesses of VB-6 and VS-6 from *Enterprise*. This large air group of thirty-seven aircraft had still to locate the enemy. Wade McClusky was, in fact, heading the flight back towards *Enterprise* when he sighted far below a Japanese destroyer (*Arashi*) which was steering north-west at speed. McClusky reasoned that it must be rejoining the main force after being detached for some purpose. He altered his course to that of the destroyer and duly sighted the enemy ships, dead ahead, at 1005. VB-3 with a further eighteen SBDs from *Yorktown* also arrived independently over the Japanese fleet at about the same time. The three squadrons climbed and deployed for a co-ordinated dive-bombing attack.

Thousands of feet below the Dauntlesses, on board the four Japanese carriers, the last disastrous torpedo attack had ended with all but one of VT-3's aircraft shot down. The crews of the Japanese ships were jubilant; they sensed they were about to achieve a great victory; every attack on them, whether from high-level B17s or torpedo strikes, had failed. Over fifty US aircraft had been shot down; not a single ship of the Striking Force had been hit. On the carrier flight decks Kates and Vals were bombed up, ranged and preparing to take off on a strike against the US Task Force. None of the ships in Nagumo's force was fitted with even the most rudimentary radar, and, with the fighters at sea level where they had just trounced the unfortunate and misnamed Devastators, the Dauntless dive-bombers, high above, were able to approach unobserved.

At 1022 Wade McClusky, at 19,000 feet above *Kaga*, pushed over his SBD. He was leading a section of twelve aircraft, each armed with a 1,000 lb bomb. As the Dauntlesses dived nearly vertical over the unsuspecting carrier, two other groups were diving on *Akagi* and *Soryu*.

Kaga took four direct hits from 1,000lb bombs, which penetrated her flight deck and exploded in her hangars. The ship erupted into flames as AvGas from the blazing aircraft and fractured fuel lines fed the fires: *Kaga* was almost immediately abandoned.

Akagi, the flagship, was near-missed, then hit by two 1,000lb bombs which set fire to the AvGas system, the conflagration detonating fused bombs and torpedoes still in the hangar from the hasty rearming following the change of target when the US fleet was reported. The fires gained a rapid hold and Nagumo shifted his flag to the light cruiser *Nagara*.

Soryu, attacked by *Yorktown*'s VB-3, took three direct hits from 1,000lb bombs that detonated in a neat line along her flight deck, which was carrying a full range of aircraft. These immediately caught fire. One bomb exploded in the hangar, blowing out the forward lift and turning the ship into an inferno; the order to abandon ship was piped soon after the hits.

The action had been witnessed from first to last by one American: Ensign George Gay, the only survivor of *Hornet*'s VT-8. He had crashed within the screen and was hiding behind a seat cushion which had floated up from his sinking Devastator. He was picked up by a US PBY thirty hours later and described the Japanese carriers as burning "like blow torches".

The first bomb had hit at 1026, the last at 1030. The SBD dive-bombing attack had, in a little under four minutes, turned not only the Battle for Midway but also the entire Pacific War; such a rapid and total reversal of military fortunes was entirely without parallel. None of the SBDs was lost over the targets, though several ditched later when unable to locate *Enterprise* after the attack.

One Japanese carrier, *Hiryu*, escaped. She had been 2 miles from the others and had not been attacked. After the last of the American planes left the scene, *Hiryu* was the only operational Japanese carrier in the South Pacific. Rear Admiral Yamaguchi on board, unable to communicate with Nagumo on the *Nagara*, ordered an immediate counter strike; as far as he knew there were only two US carriers in the area and they would be busy recovering their own strike aircraft and would not be able to relaunch them for some hours. *Hiryu* had lost a number of aircraft in the Midway strike but managed to launch eighteen Vals, escorted by six Zeros, after a delay in preparing the aircraft, most of which had been in action earlier in the day.

Shadowing aircraft from *Tone* gave the Strike Force an accurate position of Task Force 17. Just before noon the Japanese strike was intercepted by fighters from *Yorktown* about 10 miles from the carrier: nine Japanese aircraft were shot down by the Wildcats and a further three by anti-aircraft fire, but eight of the Vals penetrated the screen and attacked. *Yorktown* was hit by three 500lb bombs: one on the flight deck which set fire to parked aircraft; another detonated below decks, starting fires near the magazines; a third bomb did the most serious damage, striking the funnel and blowing out the fires in five of the ship's boilers. Only a single Val escaped after the attack.

As soon as the enemy aircraft had been sighted, *Yorktown*'s AvGas had been drained and the system filled with Co_2, therefore the immediate danger of aviation fuel explosion was lessened. The fires were brought under control and the boilers flashed up. An hour and a half after the attack *Yorktown* was under way, at reduced speed but operational and far from out of the battle, though with only six flyable Wildcats, which she was able to refuel.

Yamaguchi ordered a second strike against *Yorktown* that afternoon with the only aircraft available: ten Kates and six Zeros. When the incoming attackers were sighted, *Yorktown* suspended refuelling and again drained her AvGas system. Her six Wildcats had therefore to take off with such fuel as they had. One of the F4F pilots was Lt (later Captain) J.P. Adams, who remembers:

> We had only forty gallons of gas apiece, but nonetheless they wanted to get us off to try to oppose the torpedo attack. Lt Thach [later Admiral] and myself and four others manned the planes. By the time we had got off the deck, all the ships in the Fleet were firing. I vividly remember taking off, trying to crank up my wheels and charge the guns, which we had to do manually and then trying to catch the torpedo planes. I did catch one and possibly another.[15]

Yorktown's six fighters shot down five of the torpedo planes but four others got within close dropping range, 500 yards, amid a hail of anti-aircraft fire.

Yorktown's enforced low speed restricted her evasive manoeuvring; nevertheless she avoided two torpedoes but was struck by two others, being hit on the port side. Adams was still airborne: "I saw the torpedoes hit her and there was no way of course to land back aboard and so I ended up landing on the *Enterprise* which was about forty miles away. And very happy to get aboard with few gallons of gasolene left.[16] As her fighters were heading for *Enterprise*, *Yorktown* had lost way due to lack of propulsive power; she had developed a list to port so severe it was feared she was about to capsize. The order to abandon ship was given and her crew were picked up by the destroyers of the screen.

The Kates attacking *Yorktown* had been led by Lt Tomonaga, one of the few to return to *Hiryu*. He reported the hits on the US carrier. Yamaguchi knew that one *Yorktown* class carrier had been hit by the Vals, another of the same class by the Kates: ergo, since only two US carriers were thought by the Japanese to be in the South Pacific, the Task Force should now be without an air strike capacity. (The complication from the Japanese point of view was that all three of the carriers were of the *Yorktown* class; they represented, in point of fact, the entire class.)

Hiryu was to try to launch one final attack to ensure that no US carrier was left afloat. Her aircraft were now mostly survivors of the other Japanese carriers, and as she prepared to launch, Dauntless dive-bombers from the undamaged *Enterprise* and *Hornet* made their attack. *Hiryu* tried desperately to avoid the bombers but was directly hit by at least four bombs and suffered, in addition, four very near misses, which must have damaged her hull; she was soon ablaze with uncontrollable AvGas fires in her hangar.

Hiryu, like the other Japanese carriers, did not sink immediately after the attack. The final fate of the four was as follows: *Soryu*, abandoned and burning, blew up in the late afternoon of 4 June, 190 miles north-west of Midway; a quarter of an hour later, 40 miles away, *Kaga* also blew up, probably due to fires reaching her magazines, and sank; at dawn on 5 June *Akagi* was scuttled by torpedoes from Japanese destroyers, as was *Hiryu*, but the latter survived the torpedoes and an attack by Midway B17s until 0900 hours, when she too sank into 3,000 fathoms of the Pacific Ocean. Admiral Yamaguchi and *Hiryu*'s captain remained on board to go down with the ship.

With the loss of the four Japanese carriers, 2,280 men and 258 aircraft,[17] the Battle of Midway was essentially over. Yamamoto had no alternative but to withdraw; he had no air cover for his battleships — the three carriers which had gone to the Aleutians were too far away to intervene. In withdrawing, the Japanese suffered further losses: the heavy cruiser *Mikuma* was attacked by Midway-based aircraft and damaged, being finally sunk by *Hornet*'s SBDs on 6 June. Another cruiser, *Mogami*, survived three direct hits from 1,000lb bombs but managed to reach Japan, eventually to be rebuilt as a seaplane carrier.

The final loss to the US Task Force was the crippled *Yorktown*, torpedoed by the Japanese submarine *I-68*, which also sank the destroyer *Hamman* as she was standing by the carrier. *Yorktown* finally sank at dawn on 7 June 1942, bringing the total US losses to one fleet carrier, a destroyer, ninety-two carrier aircraft and forty aircraft from Midway.

Admiral Yamamoto must have known, as he steamed away to apologise to his Emperor, that more than a battle had been lost at Midway. From now on Japan would be fighting a defensive war.

Whether Admiral Yamamoto suspected that the Japanese defeat off Midway was due in part to the activities of the American cryptanalysts, history does not record, but, ironically, it was the same code-breakers who were to be instrumental in the death of the Japanese Admiral. On 18 April 1943 Yamamoto was to make an inspection of naval installations on Bougainville in the Solomon Islands. He was to fly in a Sally bomber from Truk. The message containing details of the itinerary was intercepted and decoded. Eighteen Army P-38s were modified overnight to carry long-range fuel tanks, and sixteen of them took off from Henderson Field, Guadalcanal, to intercept. As the P-38s fought off the escorting Zeros, one, flown by Lt Thomas Lanphier, shot down the Sally, which crashed into the jungle-covered island of Bougainville near Kahili. Admiral Isoroku Yamamoto, Commander-in-Chief, Imperial Japanese Navy, was found dead in the wreckage, leaning on his Samurai sword.

The Battle of Midway had been decisive. Although it marked the beginning of the end for the Japanese, that end was still, in the summer of 1942, many a long day's march away.

There were battles of unprecedented ferocity to be fought at sea and from island to island across the Pacific. Many of those actions which now comprise the battle honours of the US Marines and Navy were then the unknown names of tiny coral atolls: one such was Guadalcanal.

Guadalcanal is an island in the Pacific Ocean to the east of the Solomons Archipelago, 1,200 miles north-east of Australia. It was far from a tropical paradise, being in the main a malaria-infested, swampy, fetid jungle, which few white men visited, let alone settled. Circumstances alter cases and Guadalcanal had one feature which in war outweighted its manifest disadvantages: unlike all its neighbours, it had, running along its palm-fringed northern shore, an area of level ground large and firm enough on which to construct an airfield.

From the seaplane base the Japanese had set up at Tulagi in the opening moves of the Battle of the Coral Sea, reconnaissance had been made of the surrounding atolls and islands, and the stretch of level terrain on Guadalcanal was discovered and selected for the construction of an airstrip. Japanese labour battalions were already at work hacking down the jungle on 2 July 1942, when a long-range US reconnaissance B-17, which had been photographing the seaplane base at Tulagi, flew over Guadalcanal and took some photographs at 23,000 feet, as a matter of routine. When the prints were examined the unmistakable runway pattern was seen emerging from the tropical growth on the island.

The American Joint Chiefs of Staff had already decided to attack Tulagi and Santa Cruz as the first objectives in the long amphibious campaign against Japan. As soon as the Japanese army engineers' activities on Guadalcanal were revealed by photo-reconnaissance, it became imperative to deny the airfield to the Japanese; it was therefore decided that a landing be made to capture the strip.

A Task Force, TF61, was formed under Vice Admiral Fletcher, with three carriers to support the landings — *Enterprise* and *Saratoga* (the latter had rejoined the Pacific Fleet from repairs and refitting in the United States) and *Wasp*, which had been transferred from the Atlantic, where she had been engaged with the Royal Navy, ferrying RAF Spitfires to Malta. Between 28 July and 1 August an amphibious exercise to rehearse the landings was held on the Island of Koro, near Fiji, though the operation was hampered to some extent by the necessity of enforcing total radio silence to avoid alerting the Japanese. Nevertheless, valuable experience was gained.

By midnight on 6 August 1942 the Task Force was within 80 miles of Tulagi, unseen by Japanese search aircraft, owing to low cloud and rain showers caused by an unstable warm front. At dawn on 7 August the carriers launched their strike aircraft: Operation Watchtower, the invasion of Guadalcanal, had begun.

Wasp's strike was directed against Tulagi seaplane base; *Saratoga*'s and *Enterprise*'s aircraft attacked defenders on Guadalcanal itself. Following the air strike, 17,000 US Marines landed at Lunga Point and secured the primary objective — the incomplete airstrip, which was named Henderson Field (after a Marine Major who had died defending Midway). It was tenuously held, for the last Japanese soldier on Guadalcanal was not to be killed until 7 February 1943, and between the first landings and that date the US Navy was to fight no fewer than six major naval battles round the Solomon Islands.

The first engagement was the Battle of the Eastern Solomons. Once the US Marines had secured a lodgement on Guadalcanal, the Japanese troops still resisting on the island obviously required both reinforcements and supplies. These came at night from Rabaul in the form of fast destroyer runs which the American sailors nicknamed the 'Tokyo Express'. Admiral Yamamoto, however, was interested in other than simply supplying the Japanese garrison on the island. He decided that the 'Tokyo Express' sailing on 21 August, which included troop transports, would be covered by a heavy naval force, consisting of battleships, heavy cruisers and three aircraft carriers, in order to bring the US Fleet off the Solomons into the decisive naval action which had eluded Yamamoto at Midway.

The plan was inevitably complex: broadly, the light carrier *Ryujo* was to provide close cover for a landing to retake Henderson Field, while the two larger fleet carriers, *Shokaku* and *Zuikaku*, commanded by Admiral Nagumo, would provide a Mobile Force for distant cover of the two heavy battleships and three heavy cruisers with which Yamamoto hoped to destroy the US Fleet Task Force. According to plan, the Japanese Mobile Force sailed from Truk on 21 August for the Solomons.

The broad intentions of the Japanese were known to the Americans, though the intelligence (as so often happened when cryptanalysis was the primary source) was not of the same detailed quality as the previous interceptions, which had been of decisive help to the US Navy in the Midway battle. Alerted by such information as was available, the Americans sighted the enemy invasion force (by a PBY Catalina) when it was 300 miles north of Guadalcanal on 23 August. *Saratoga* launched a strike but failed to locate the transports, owing to a delay in ranging the aircraft, which landed on Henderson Field, where the Navy pilots, unused to the mosquitoes, spent an uncomfortable night. At first light next day PBYs again searched for the Japanese, while *Saratoga*'s aircraft refuelled and thankfully returned to the carrier.

A PBY duly made contact with *Ryujo*, which when sighted had just launched an air strike against Henderson Field; the air group was detected by *Saratoga*'s radar at 100 miles range as they headed for Guadalcanal (where they were to be decimated by Marine fighters). *Saratoga* at once launched a strike of thirty aircraft, which included VT-8 with eight of the new Grumman TBF-1 Avengers. As *Saratoga*'s aircraft flew towards *Ryujo*, *Enterprise*'s search aircraft found the main Mobile Force with *Shokaku* and *Zuikaku*, some 90 miles north-east of *Ryujo*. Thus Admiral Fletcher had repeated the mistake made at the Battle of the Coral Sea: he had launched the major part of his air group against a secondary target. Radio conditions were poor and *Saratoga*'s strike could not be contacted to be recalled. In fact *Saratoga*'s strike leader, Commander Felt, had intercepted the *Enterprise*'s search aircraft's original sighting report and he led his group to the reported position but failed, in bad weather to make contact and so returned to the *Ryujo*, which was attacked soon after 1600 hours. The light carrier was directly hit by four 1,000lb bombs from Dauntless dive-bombers and a single torpedo, the first operational success for a TBF Avenger. *Ryujo* was left burning fiercely, soon to be abandoned by her crew and to sink six hours later.

While *Ryujo* was under attack, Nagumo's flagship *Shokaku*, after being ineffectually attacked by *Enterprise*'s armed search planes, launched, with *Zuikaku*, a counter-strike against *Enterprise*. Fletcher, profiting from the action in the Coral Sea and at Midway, had retained his fighters aboard his two carriers (*Wasp* had been withdrawn to refuel 240 miles to

the south), fifty-three of which were airborne to meet the incoming
Japanese. The fighter control radar on *Enterprise* was now cluttered with
over 200 aircraft returns: fighters from the Combat Air Patrol, anti-
submarine patrols, returning US strike aircraft and the Japanese attackers.
To make matters worse, the radio discipline of the US fighters was non-
existent; the few channels available were jammed by excited pilots
calling to each other as they attacked. Nevertheless, all the Japanese Kate
torpedo aircraft were either shot down or driven off; but, as had
happened when Wade McClusky attacked the Japanese carriers at
Midway, while the defenders were distracted in fighting off the low-
level torpedo aircraft, high above Val dive-bombers were dropping
down on *Enterprise*. An intense anti-aircraft barrage was put up,
shooting down several Vals, but three direct hits were scored on
Enterprise. The bombs passed through the flight deck and started fires
below, but although there were heavy casualties among the crew, the
ship itself was not badly damaged. Her fires were controlled and the
carrier was under way, steaming at 24 knots, within an hour. Her radar
was still operational and detected a second Japanese strike heading for
the ship; inexplicably it turned away and the aircraft commenced
searching for the American Task Force 80 miles south of the actual
position without ever making contact, which was most fortunate, for
Enterprise's steering had developed a fault due to a short circuit caused
by the fire-fighters, which jammed the rudder hard to starboard for forty
minutes, during which time *Enterprise* could only steam in circles.

As the Japanese aircraft searching for the US Task Force faded from
Enterprise's radar, the Battle of the Eastern Solomons was essentially
over. It ended in a victory for the Americans; Yamamoto had not even
been able to force a landing on Guadalcanal, much less the hoped-for
decisive naval action. The US carriers had lost only seven aircraft in
actual combat and three others too badly damaged to be repaired.
Enterprise had suffered seventy-four men killed and about 100 wounded;
the ship was out of action for two months for repairs.

The Japanese losses were severe: Henderson Field, the light carrier
Ryujo, over seventy aircraft with irreplaceable crews. Indeed, her loss of
aircraft and pilots was crippling enough to make it impracticable for
Nagumo to continue the action; he had two undamaged carriers but
insufficient aircraft. The carriers withdrew to Truk.

The invasion convoy, without air cover, continued towards Guadalcanal
but was now confronted by determined opposition from US Marine
aircraft from Henderson Field, which the famed 'Sea Bees' (Naval
Construction Battalion) soon enlarged. USAAF B-17s from Espiritu

Santo at last scored their first hits from high level bombing — the destroyer *Mutsuki*, which they sank. (Perhaps she had a pickle barrel on board.) The combined effect of the land-based aircraft, particularly the Marine SBDs, which could of course make many sorties each day, was such that the Japanese invasion convoy withdrew from Guadalcanal on 25 August 1942.

The Japanese continued the battle, using destroyers in night actions at which, although without radar, they were extraordinarily good. The activities of these destroyers curtailed the development of Henderson Field, since it was repeatedly shelled at night from the sea.

The daylight hours, however, belonged to the side with air superiority — the Americans with their carriers — though that force was to be depleted when, on 31 August, *Saratoga*, patrolling 250 miles south-east of Guadalcanal, was struck by a single torpedo from the Submarine *I-26*. The old ship survived and made Pearl Harbor under her own steam; repairs, however, took six weeks.

Although *Hornet* had replaced the damaged *Enterprise*, the US fleet carriers in the Solomons now only numbered two. *Hornet*, incidently, was narrowly missed by torpedoes from another Japanese submarine, the *I-11*. In the same general area, dubbed 'Torpedo Junction', USS *Wasp* was not to be so lucky; she was hit by three torpedoes from *I-19* on 15 September. Her fuel oil and AvGas tanks detonated and the ship was soon ablaze from stem to stern. After several violent internal explosions, the carrier was abandoned and scuttled by torpedoes from the destroyer USS *Lansdowne*. *Wasp* lost 193 of her crew killed and 366 injured.

Seven miles from *Wasp*, the submarine *I-15* torpedoed the battleship *North Carolina* and the destroyer *O'Brien*. Neither ship, though damaged, sank and both could still steam.

These losses occurred just before the next major naval action in the Solomons: the Battle of Santa Cruz. This action was the result of another attempt by the Japanese to retake Guadalcanal. The single airstrip on the island was of an importance that outweighed the inevitable losses which were sustained by both sides. For this latest action Nagumo had five carriers at his disposal — *Shokaku*, *Zuikaku*, *Zuiho*, *Junyo* and *Hiyo* — with battleships, cruisers and destroyers.

The opposing US fleet comprised Task Force 16, under Rear Admiral Kinkaid, flying his flag in the repaired *Enterprise*, and TF17 with *Hornet*. There were also two battleships, cruisers and destroyers. Although on paper the American fleet was far inferior to the Japanese, the US forces had the use of aircraft based on Henderson Field — the ultimate objective of these Solomons battles.

The Battle of Santa Cruz was fought on 26 October 1942 off the island of that name. It was to result in the loss of USS *Hornet*, torpedoed by two hits from Kates and several direct hits from dive-bombing Vals, though at the cost to the Japanese of most of their attacking aircraft. *Enterprise* was also heavily attacked, but the quality of the Japanese pilots was now not what it had been, and the American carrier was only slightly damaged by three bombs, which nevertheless caused heavy casualties.

In American counter-strikes the Japanese carrier *Zuiho* was hit by SBDs and fires started, which were eventually put out, but the carrier was out of action for the rest of the battle. *Shokaku* was also hit and damaged by SBDs, but not fatally. At the end of the Battle of Santa Cruz the Japanese could consider that they had won the day; with *Hornet* sunk, *Enterprise* damaged, the only operational carriers in the South Pacific were Japanese. But it was a Pyrrhic victory: the land battle of Guadalcanal continued but the vital Henderson Field was still in American hands, with increased numbers of Marine and Army aircraft. The losses sustained by Nagumo's aircrews was unacceptable; most of the veteran pilots who had laid the foundations of the long series of victories, beginning with the war in China, through Pearl Harbor and the days of unbroken successes in the Indian Ocean, the Philippines, Malaya and New Guinea, were now gone. The sinking of *Hornet* was their last action.

The struggle to regain Henderson Field did not end with the action off Santa Cruz, far from it; there was still to be fought the Naval Battle for Guadalcanal 12—15 November, in which *Enterprise*'s aircraft were partially responsible for sinking 89,000 tons of shipping attempting to reinforce the Japanese troops on the island. The final set-piece action was the Battle of Rennell Island, which was fought between 27 and 30 January 1943, when the Japanese finally abandoned Guadalcanal and attempted to withdraw their surviving forces from the disputed island.

The Marines and USAAF units on Henderson Field and carrier-borne aircraft — mainly from *Enterprise* — had accounted for over 1,000 enemy aircraft by 9 February, when the whole island was finally in American hands. Few airfields could have cost more: 24,000 Japanese soldiers, 6,000 US Marines and Army troops killed in the land fighting alone. The US Navy lost two carriers, eight cruisers and fourteen destroyers. It could afford such losses. On 31 December 1942 USS *Essex*, the first of seventeen *Essex* class carriers, was commissioned at Norfolk, Virginia. These 27,000 ton vessels could operate over 100 aircraft and would form the Fast Carrier Force which was to be irresistible in the future actions in the Pacific. By mid-1943 the US Navy alone would have 16,700 aircraft on charge and were training tens of thousands of pilots to fly them. In addition to the huge new fleet carriers, smaller escort and light carriers were being built, some in less than three months. Japan could never hope to equal even a fraction of that industrial potential.

The South Pacific had been the scene of the full flowering of the aircraft carrier, but the Pacific was only one ocean. There was another where a different war was being fought — the Atlantic.

There can have been few battles in history as decisive as the Battle of the Atlantic. It was fought with ruthless dedication from the first day of the European war to the last. Had Britain lost that struggle, as nearly happened in March 1943, then it would have ended the war: without supplies of food and war material from the United States, Britain would have had to sue for peace; there would then have been no base from which American forces could attack Germany.

As mentioned at the beginning of this chapter, the Royal Navy had relied on ASDIC as a counter to enemy submarines. In practice, under wartime operational conditions, it was soon apparent that the effectiveness of ASDIC had to some extent been oversold. In 1941 Britain had lost 2 million tons (432 ships) sunk by U-boats; in 1942 the figure rose to 6 million tons — a rate far beyond the capacity of British shipyards to replace.

With the advent of Air to Surface Vessel Radar (ASV), the most effective counter to U-boats was in the use of aircraft. Coastal Command of the RAF, even when using the early 1941 1½ metre radar, began to inflict serious losses on German U-boats, which tended to shadow convoys they were to attack at night, in daylight on the surface, when they could best be attacked by airborne depth charges.

The difficulty for the RAF was the lack of suitably long-ranged aircraft; even when bases were set up in the west of Ireland, Iceland and North America, there was the so-called 'Black Gap' in mid-Atlantic which the aircraft of 1941-2 could not cover. The solution, so obvious with hindsight (though it was mooted before the war and shelved), was to carry the aircraft along with the convoys.

The first ship-borne aircraft to be available for convoy protection were Fulmars and Hurricanes, which were embarked on HMS *Pegasus*, an old catapult ship. The use envisaged for *Pegasus* and her small number of aircraft — three was the maximum — was really a throwback to the days during World War I when battleships flew Sopwiths off their gun turrets. *Pegasus* substituted the catapult for the turret but the result from the unfortunate pilot's point of view was the same: unless he was lucky enough to be launched within reach of land, he had no alternative other than to ditch in the sea after his one and only flight.

The catapult ship was placed in service initially not to combat U-boats but another deadly convoy predator: the long-range Focke-Wulf Fw200 Condor. These large four-engined aircraft were operated by KG40 from Bordeaux/Merignac in occupied France and Stavanger in Norway. Their endurance gave them a range of nearly 4,000 miles, enabling them to fly to attack convoys in the Atlantic out to 20° West, far out of range of any land-based fighters. Once a convoy was sighted, the Condor could attack it with bombs and machine guns with little risk of being shot down; the anti-aircraft armament on board the few escorting corvettes and destroyers would not be particularly intense and the merchant ships' AA armament would be only twin-mounted .303 inch machine guns. Between August and September 1940 Fw200s sank 90,000 tons of merchant shipping directly, but more deadly was their later use of shadowers for U-boats lying in wait for the convoys. The Condor would signal the sighting, the composition of the convoy, the number and disposition of escorts, the course and speed. Quickly U-boat HQ at Lorient on the French Biscay coast would order a number of U-boats by radio to make a co-ordinated attack; this was known to the Germans as the *Rüdeltaktik* or 'Wolf Pack' to the Allies. The effect on the convoys was appalling; the escorts were overwhelmed and many ships, with their vital cargoes, were sunk.

The Fw200s enjoyed a success disproportionate to their small numbers; yet for a land-based fighter, the Condors were easy targets. In the first place they were not particularly fast — about 240mph maximum — but more to the point, they were basically a conversion from pre-war civil airliner design and consequently, by military standards, frail.

Catapulted fighters were an obvious counter. In addition to *Pegasus*, five merchant ships were converted and commissioned into the Royal Navy; four were used operationally between December 1940 and June 1942. Although their fighters shot down only one Condor, several others were damaged and many forced to abandon their shadowing of the convoys.

Overlapping and outlasting the RN Fighter Catapult Ships were the Catapult Armed Merchant (CAM) Ships, which were just what that name implies: merchant ships with a normal cargo but with the addition of a catapult built over their forecastle, and a Sea Hurricane 1, probably converted from an ex-Battle of Britain veteran, sitting poised ready to take off to intercept the Condors. From May 1941 to August 1943, CAM ships sailed with over 170 convoys, their fighters with RAF pilots being launched operationally eight times, on which sorties six Condors were shot down and others damaged. In all, thirty-five CAM ships were in use during the war.

The success of the CAM ships had a deterrent effect not obvious from the number of aircraft destroyed, and it led to the logical solution of convoy protection, which had been suggested in the early 1930s: small auxiliary carriers. The first of these 'escort' carriers, as they came to be known, was constructed in some haste in July 1941, on the hull of the merchant ship *Empire Audacity*, originally a German prize ship, *Hannover*. Her deckhouses, funnels and bridge were cut down to the level of the hull, which was plated over to make a small (460 foot) flight deck with two arrester wires and a crash barrier. The latter was essential, since HMS *Audacity*, as she was commissioned, had no hangar, her eight Martlet IIs (equivalent to the US F4F-3 Wildcats) being parked at the end of the flight deck where all servicing and maintenance had to be carried out, protected only by the barrier should an aircraft overshoot.

Audacity immediately demonstrated the value of a small carrier in protecting a convoy. She sailed with only four, all on the Gibraltar run; three of those convoys were attacked by the enemy, when *Audacity*'s Martlets took off and shot down five Fw200 search aircraft, damaged another three and chased a fourth away from the convoy. Not content with that excellent debut, the Martlets, though single-seater fighters, also sighted and reported at least nine U-boats shadowing the convoys, sharing with destroyers in the destruction of one of them, *U-131*. During all these operations only one Martlet was lost in action, being shot down by a gunner in one of the Fw200 Condors; another went over the side when attempting to land in very heavy weather. Sadly the ship and her remaining fighters were lost on 21 December 1941 when torpedoed during a night action by *U-741*.

The success of *Audacity* demonstrated the feasibility of 'escort' carriers, and not only against airborne attacks, for even though the Martlets were not in any way equipped for anti-submarine work — they had no radar or observers — yet they had sighted nine U-boats, forcing them to dive. The Admiralty was convinced: the problem was the lack of shipbuilding yard capacity to convert the escort carriers from other merchant ships, which, in any event, were in desperately short supply. The British yards were fully engaged in building and repairing other badly needed warships, but American yards, untroubled by air raids and chronic shortages of manpower and materials, had virtually unlimited resources.

The US Navy had been toying with the idea of light 'pocket' carriers for some years before the war; plans were even drawn up by the Bureau of Construction and Repair for light carriers to be built on merchant-ship hulls, but these were dropped in 1940 when it was considered, wrongly as it was to turn out, that the heavy, fast monoplanes then coming into naval service to replace the older biplanes were making small carriers impractical, owing to the restricted length of their flight decks.

It was Franklin Roosevelt himself who revived the question in October 1940; he suggested that the US Navy convert a merchant ship with a flat deck suitable for operating ten or twelve autogyros, which were the half-way house to the modern helicopter and which the President suggested "could hover ahead of convoys, detect submarines and drop smoke bombs to indicate their location to attacking surface escort craft". (Since, when the United States entered the war against Germany, the US Navy would not at first employ convoys until mounting losses forced them to, one might be forgiven for suspecting 'a Former Naval Person' could have been the moving force behind the President's memorandum.)

The plan was examined but the autogyro proved itself incapable of lifting any sort of worthwhile offensive load, and the first US military helicopter (the Sikorsky R-4) would not enter service until 1944. As an alternative, the Chief of Naval Construction decided that a flight deck which could operate fixed-wing aircraft was essential. After discussions it was agreed, on 6 March 1941, to acquire for the Navy two type C-3 cargo ships — the *Mormacmail* and *Mormacland* — for conversion, which work Roosevelt had ordained must not take longer than three months. The yards just made the deadline, delivering the ships on 2 June. On 1 August *Mormacmail* was commissioned as USS *Long Island*, AVG-1, a flush-deck carrier of 10,000 tons.

Long Island had a flight deck 362 feet long with one lift and a hangar that could accommodate some sixteen aircraft, though her maximum speed of only 17.6 knots dictated aircraft of low wing-loading. The low speed was due to the vessel retaining the original merchant ship's diesel engines, at first thought to be an advantage because of the lack of a funnel, which steam propulsion would of course require. The sister ship, *Mormacland*, was converted at the same time and became the British escort carrier, HMS *Archer*.

USS *Long Island* was extensively used in trials to determine the best configuration for subsequent US escort carriers, with the result that later ships built on C-3 hulls had two lifts installed and the flight deck length increased to 436 x 79 feet. The biggest change, however, was the substitution of a single geared steam turbine for the diesel machinery. Only three carriers were built as the *Long Island* class: *Long Island* herself, the British HMS *Archer* and USS *Charger*. *Long Island* was used operationally for ferrying Marine F4Fs to Guadalcanal; *Charger* was employed as a carrier landing training ship and spend most of the war years in Chesapeake Bay. HMS *Archer* saw little active service, being plagued with machinery defects and eventually laid up, though on 23 May 1943, on one of her only two Atlantic convoys, her Swordfish in mid-Atlantic sank *U-752*, the first U-boat to be destroyed by airborne rockets.

There were to be five main classes of US Navy escort carriers commissioned during the war years. The first three classes were the *Long Island* class (CVE-1), two ships; the *Bogue* class (CVE-6), eleven ships, which were built on C3 merchant hulls but incorporating the improvements found desirable after the *Long Island* trials; and the *Sangamon* class (CVE-26), four of which were converted from *Cimarron* class fleet oilers. Only four were converted because there was a shortage of oilers. *Sangamon* class carriers had a 553 x 53 foot flight deck, a speed of 18 knots and a displacement of 23,200 tons. They could operate two squadrons of aircraft and they were all commissioned in the summer of 1942.

The largest class of CVEs was the *Casablanca* class (CVE-55), the first being commissioned in July 1943. These carriers were not conversions but built from the keel up as escort carriers by the Kaiser yards, which were also turning out the famous 'Liberty' ships on a production-line basis. The entire *Casablanca* class of fifty ships, CVE-55 to CVE-104, was commissioned in exactly one year — July 1943 to July 1944 — almost one a week, a considerable feat of wartime production.

The *Casablanca* CVEs had, like most escort carriers, a vestigial starboard island which contained little more than a navigation and signals bridge; the flight decks were 512 feet long; the maximum speed of *Casablanca* herself on trials was 19.3 knots and the displacement was 9,570 tons. The class could operate about thirty-eight aircraft.

The final wartime constructed CVEs were the *Commencement Bay* class, (CVE-105), the first of which was not commissioned until November 1944. They were the largest of the CVEs — 23,100 tons — with a 557 foot flight deck and a speed of 19 knots. The class numbered nineteen ships, though only ten were commissioned before the end of World War II and few of them saw active service.

The US Navy commissioned seventy-seven CVEs[18] during the war years. A further thirty-eight US-built CVEs were supplied to the Royal Navy, either under the terms of the 'Lend Lease' agreement or ordered directly from American yards. In addition, the Royal Navy commissioned four British-built CVEs: HMS *Activity*, *Nairana*, *Vindex* and *Campania*. These ships were built to full Admiralty standards (which US CVEs were not), with enclosed hangars, steel flight decks, and twin-screw diesels which offered a speed of 19 knots.

In addition to these purpose-built carriers, the Royal Navy provided aircraft for the only civilian carriers to see action during the war years. These were the 'MAC' ships, Merchant Aircraft Carriers which were grain ships or oil tankers partly converted — that is, they had their superstructures removed and a flight deck with two arrester wires, a barrier and a small starboard island fitted. The grain ships also had a single lift to a small hangar which typically accommodated four aircraft. The tankers had no hangar and the aircraft embarked, usually three, were permanently parked and serviced on the flight deck. The MAC ship carried 80 per cent of its normal cargo — oil or grain. MAC ships were operated by their Merchant Navy crews, with a small number of naval ratings, pilots and observers. They normally sailed one or two in a convoy, depending on its size.

The Admiralty and the Ministry of War Transport had opposed the idea initially on the grounds that flying and landing aircraft on a thin flight deck above 10,000 gallons of high octane aviation spirit was dangerous; it was argued that an aircraft crash-landing could penetrate the deck and touch off the cargo. Fleet Air Arm officers managed to disabuse the Director of Naval Construction of this idea.

The first MAC ship to be commissioned was the grain ship *Empire MacAlpine*, 7,950 tons, which when converted had a 422 x 62 foot flight deck and a hangar that could accommodate four Swordfish with folded wings. The ship's armament was one 4 inch gun, four 20mm and two 40 mm Oerlikons. In May 1943 a Swordfish landed-on, the first time a Fleet Air Arm aircraft had alighted on a civilian ship. Swordfish proved ideal for escort carriers; they were simple biplanes with a low stalling speed and viceless flying characteristics. They could become airborne from the short deck in spite of the low speed, typically 12 knots, of the MAC ships. Fitted with centimetric ASV radar, the Swordfish, in spite of its archaic appearance, was a formidable anti-submarine weapon when armed with rocket projectiles or depth charges. All Swordfish embarked aboard MAC ships were operated by 836 Squadron and 860 Squadron of the Fleet Air Arm.

MAC ships were converted in about a third of the time it would have taken to construct an equivalent escort carrier, and of course, in addition, they still carried their vital cargoes. They came to be regarded by the merchant seamen in the ships of the convoys with which they sailed as 'their' carriers and, unlike the formal 'Navy' CVEs, they always stayed with their charges in the centre of the convoy, not being detached to form 'hunter killer' groups, operating over the horizon and out of sight of the merchant ships.

Swordfish from MAC ships made at least twelve attacks on U-boats during 4,177 sorties; the results were uncertain, though it is likely that some of the submarines were damaged. In any event, no ships were lost in any convoy protected by a MAC ship, which was after all the whole point of the vessels' existence. The tragic aspect of the success of these makeshift carriers was that they could have been operational right from the outbreak of the war; had they been, a great many men and ships would have survived the early crippling onslaught of German U-boats in the Atlantic.

To return to the role of the true CVE (the 'Woolworth' carriers as their crews called them), they fell into two broad operational classifications: trade protection duties, that is convoy escorts, a role which gave the small carriers their name and, as the Allies went on the offensive towards the end of 1942, assault carriers, providing air cover for amphibious landings. These latter carriers had rather more elaborate radar, including fighter-control sets; they normally embarked only fighters, though a few assault CVEs were required to operate both fighters and reconnaissance/torpedo bombers. Most CVEs, at one time or another, were involved in aircraft-ferrying operations; indeed some were used solely for that purpose from either the United Kingdom or the USA to distant fronts; up to seventy aircraft could be carried on a ferrying operation.

The role of all the carriers in the European war was either escort duties, assault work or the vital but unglamorous ferrying of aircraft and supplies, mainly to the Mediterranean. The major role of CVEs in the Battle of the Atlantic dated roughly from April 1943 to August 1944. While it would be untrue to say that by the time escort carriers became available for convoy protection the Battle of the Atlantic was won, it was nevertheless at a time when U-boats were being sunk in large numbers, particularly by RAF Coastal Command in the Bay of Biscay, through which the German submarines had to pass to and from their French Atlantic coast bases. The advent of centimetric radar; the breaking of the German U-boats' operational codes and other scientific aids, such as 'Huff Duff' (*H*igh-*F*requency *D*irection-*F*inding); the increasing number

of surface escorts which could be formed into 'hunter killer' groups; the
provision at last of an adequate number of very long-range B-24
Liberator aircraft, which closed the Atlantic 'Gap'; all helped to turn the
tables on the U-boats. But escort carriers played their part too.

USS *Bogue* and HMS *Biter*, two typical CVEs, began convoy protection
in March and April 1943 respectively. The British CVEs usually
embarked nine Swordfish and three Martlets (the US name Wildcat was
not adopted until January 1944); the US carriers were likely to be
equipped with twelve TBF Avengers. By the time the Royal Navy CVEs
were operational, in mid-1943, U-boats had to a large extent been driven
away from the North Atlantic trade routes. Only three U-boats are
known to have been sunk solely by CVE Swordfish, though four others
were sunk by Swordfish with surface vessels also participating. In a sense
these low figures are a tribute to the effectiveness of the CVEs; their
aircraft attacked other U-boats which approached convoys, invariably
driving them off.

The US Navy operated CVEs south of the normal convoy routes off
the Azores, where U-boats were refuelled from the 'Milch Cow' submarine
tankers. These rendezvous were often known as a result of cryptanalysis
at Bletchley Park in England, and every one of the Milch Cows was
eventually sunk. The weather in those southern latitudes was much
better than further north; also the US Navy formed 'hunter killer'
groups with CVEs whose sole purpose was not convoy protection but to
engage in anti-submarine sweeps.The result was that TBF Avengers
from US Navy CVEs sank twenty-eight U-boats in the last eight months
of 1943.

Later in the war Royal Navy CVEs began to escort military supply
convoys to Russia, and if by that time the number of U-boats attacking
convoys in the North Atlantic had dramatically declined, in Arctic
waters this was not so; furthermore the almost continuous daylight of the
Arctic summers enabled the Luftwaffe to mount large-scale bombing
and torpedo strikes from land-based aircraft, mainly Ju88s and He111s,
operating from Norway. CVEs sailing with the Russian convoys embarked,
in addition to Swordfish or Avengers, fighters to protect the strike
aircraft and the ships. Sea Hurricanes were embarked on some CVEs;
these were ex-RAF veterans, often with many operational hours behind
them, but they provided the Fleet Air Arm with the fastest fighters
available. Little modification was needed to convert a Hurricane into a
Sea Hurricane other than local strengthening of the airframe around the
hook suspension, to withstand arrested landings. Sea Hurricanes did not
have folding wings, but fortunately the lifts of the US-built escort

carriers were large enough to accommodate them. By far the best fighter embarked on CVEs was the Grumman Wildcat IV which, though not quite as fast as the Hurricane at altitude, had an excellent low-level performance and was a sturdy aircraft, designed specifically for carrier operations.

The success of the CVEs on the Russian convoys was dramatic, in spite of the appalling weather often encountered: 750 merchant ships sailed in the twenty-seven Russian convoys with CVE protection, which lost only twenty-four ships by enemy attacks. The cost to the Germans was fourteen U-boats, nine of which were sunk directly by CVE aircraft; forty enemy bombers and torpedo aircraft also fell to the F4Fs and Sea Hurricanes from the carriers. Eighty Fleet Air Arm aircraft were lost, mainly in deck-landing accidents caused by the very difficult weather conditions.

Against the epic carrier v. carrier battles in the Pacific, and even the hard-fought convoy protection in the North Atlantic and Arctic, the ferrying of aircraft in the sunlit Mediterranean does not seem unduly arduous. In fact the carriers sailing to provide RAF aircraft for the defence of Malta were subject to some of the most ferocious attacks of the war.

By the end of July 1942 Malta, virtually cut off from supplies by the Luftwaffe, was in a desperate position. Of 275 RAF Spitfires delivered by carriers to the island, only eighty remained operational; the loss rate of RAF fighters in trying to stem the almost nonstop Axis air raids on Malta was running at around three per day. There was also the problem that the island's stores of fuel, food and ammunition were falling below the danger level.

It was decided that the only hope of saving Malta was to sail a convoy from the west into the Mediterranean. Virtually the whole route was within the range of German air bases, through the infamous Narrows between Tunis and Sicily, to the beleaguered island. The operation was code-named 'Pedestal'. The convoy was composed of fourteen modern fast merchant ships, their holds crammed with supplies, defended by two battleships, seven cruisers, twenty destroyers and three fleet carriers — *Victorious*, *Indomitable* and old *Eagle*, with a total of 112 operational aircraft, comprising Sea Hurricanes, Martlets and Fulmar fighters with Albacore strike aircraft. HMS *Furious* was also sailing in a ferry role, with thirty-eight RAF Spitfires which were to be flown from the carrier to Malta by their RAF pilots.

The convoy slipped through the Straits of Gibraltar during the night of 9-10 August 1942. The first attack was in the early afternoon of 11

August. *Furious* had just flown off her RAF Spitfires about 80 miles north of the Algerian coast when *Eagle* was hit by four torpedoes from *U-73*. The old ship, the first carrier with a starboard island, sank within eight minutes. Four of *Eagle*'s Sea Hurricanes had just taken off from the carrier; one of the pilots was Lt Commander Peter Hutton, who said: "I remember clearly looking round to see the ship beginning to list. I was able to see aircraft still on deck slither off over the side and I realised I was not going to land back on her."[19] (Hutton eventually landed on *Indomitable*.)

The loss of *Eagle* was a blow to the Royal Navy; she had in the previous five months flown off over 180 Spitfires to Malta. However, *Eagle* was to be the last RN fleet carrier lost during the war.

To oppose the convoy, the enemy threw in every serviceable bomber available: Ju88s, Italian SM79s and, as in the earlier actions in the Mediterranean, Ju87s were in the forefront of the attack. Thirty of the Stukas were flown by Italian pilots of 102 Gruppo from Pantellaria Island. Other Stukas were twenty-six Luftwaffe machines of I/St.G3, based on Trapani in Sicily.

Enemy air attacks began that evening and continued with mounting ferocity, bombers sinking three merchant ships from the 'Pedestal' convoy, although carrier fighters were airborne almost continuously and succeeded in breaking up most of the attacks. Nevertheless, *Victorious* was directly hit by a bomb on her flight deck, but the armour withstood the hit without affecting the ship's operations.

By 12 August the convoy was in the Sicilian Narrows and, as the fleet carriers detached to sail back to Gibraltar, over 100 enemy aircraft made a co-ordinated attack on them. The Combat Air Patrol fighter pilots were exhausted, having flown and fought all that day; outnumbered, they were overwhelmed. Although *Victorious* escaped further damage, *Indomitable* was attacked by twelve Stukas of I/St.G3 and directly hit on the flight deck by two armour-piercing bombs and near-missed by others, which damaged her hull. The armoured deck withstood the bombs to the extent that the ship was not immobilised, but she could not embark any aircraft. Lt Commander Hutton, who had been flying his ex-*Eagle* Hurricane, was again lucky being airborne when *Indomitable* was hit: ". . . . we were told [by R/T] to land on *Victorious*. There was a problem here in that *Victorious* was not able to strike down non-folding Hurricanes, so these aircraft had to be thrown over the side once the pilots had climbed out of them."[20] *Indomitable*'s damage from the bombs was such that she had to withdraw from operations and sail the sad route to an American Navy yard for repairs.

While the Germans were attacking *Indomitable*, twelve Ju87s of the Italian 102 Gruppo dived on the tanker *Ohio*. Anti-aircraft fire from the destroyer *Ashanti* shot down two of the Stukas, one of which plunged on to the tanker, exploding on her deck, though miraculously she did not catch fire and remained afloat. During the attacks carrier aircraft and anti-aircraft fire from the screen shot down thirty of the attackers.

The convoy sailed on to be ceaselessly attacked by aircraft, torpedo boats and U-boats. The Royal Navy lost two cruisers and another two were badly damaged; of the fourteen merchant ships that left Gibraltar, only five entered Malta's Grand Harbour, including the tanker *Ohio*, loaded with aviation fuel, with the wreckage of the Stuka still on her deck. *Ohio* was kept afloat only by being lashed to two destroyers. As a result of 'Pedestal', 32,000 tons of supplies were delivered, plus the Spitfires flown off from *Furious*.

'Pedestal' was only one of the Malta convoys. Without the carriers, including USS *Wasp*, which had flown off forty-seven Spitfires on 20 April and a similar number on 9 May, the island would have fallen.

Important though Malta was, the major naval operation in the Middle East in 1942 was Operation 'Torch' — a co-ordinated Allied assault on North Africa. The object was to land forces at Casablanca, Oran and Algiers to join up with the British Eighth Army advancing west from El Alamein. The Casablanca landings were to be entirely American. The Oran attack would be made by US troops conveyed by the Royal Navy and covered by combined Anglo-American Air Forces. The third force against Algiers would be composed of troops of both Britain and the US, covered by the Royal Navy and land-based RAF aircraft. The whole complex operation was mounted in just ninety days, which was not long enough: the American CVEs practically sailed from the shipyards that had built them, landing-craft crews were virtually untrained, and most of the carrier pilots lacked any operational experience.

The US invasion force that was to strike Casablanca sailed directly from America; it was the first operation in which CVEs were used in an assault role, as part of Task Force 34, an armada of 102 vessels carrying 35,000 troops. It says a great deal for the deterrent effect of the CVEs that these large forces crossed the Atlantic without loss. The landings were to be made at three places on the coast of Morocco: at Casablanca, Port Lyautey and Safi. The invasion was to be covered by 172 US naval aircraft from four carriers, all but one CVEs. *Sangamon* was to cover the northern landings on Port Lyautey, the old carrier *Ranger* (CV-4) and *Suwannee* to support the centre attack (that on the Casablanca area) and the CVE *Santee* that to the south of Safi. The Task Force also included the CVE *Chenango*, which ferried seventy-six P-40F USAAF Warhawks; these were to fly ashore as soon as an airfield was in US hands.

Prior to the landings, the fire-eating US General Patton, later to be famous for his outspoken criticism of other people's conduct of the war, made one of his earliest recorded pronouncements: "Never in history has the Navy landed an Army at the planned time and place. If you leave us anywhere within 50 miles of Fedhala [the Casablanca invasion beach] and within one week of D-day, I'll go ahead and win." Contrary to Patton's reservations, the Navy, on 8 November 1942, did land the General and his troops at the correct place and on time.

The French Vichy forces fought back, supported by the 15 inch guns of the battleship *Jean Bart*, moored in Casablanca harbour, but the ground forces, the battleship and the rather motley collection of Vichy aircraft — including Boeing P-36 and Dewoitine 520 fighters — could not prevent the landings. By midday, 10 November, Port Lyautey airfield was in US hands and in other sectors the landings were going well. *Chenango* launched her seventy-six P-40s, only one being lost in the process. *Ranger* launched nine SBD Dauntless dive-bombers armed with 1,000lb bombs. Twenty minutes after take-off the strike leader radioed: "No more *Jean Bart*". Later that morning another message was received from *Ranger*'s aircraft flying bombing missions over Casablanca: "Urgent, urgent, cease firing in Casablanca area." US forces were in the town; the Moroccan part of the campaign was virtually over.

The cold light of the post-action analysis revealed a woeful lack of training on the part of the US carrier pilots. The CVE *Santee* suffered most: out of her thirty-one aircraft, twenty-one were lost, only one loss being attributable to enemy action. Of 172 US Navy planes taking part in the landings, forty-four were lost 'operationally'; not one was shot down in an aerial dogfight. The French lost 2,000 men, twenty-six of their aircraft shot down, another 100 or so being destroyed on the ground.

When the ceasefire came (ironically on 11 November, the anniversary of the World War I Armistice), the landings had succeeded, though the cost had been higher than anticipated. Nevertheless, TF34 had given a good account of itself; it was after all the longest — 3,000 miles — amphibious operation in the history of modern warfare. The lessons learnt had been valuable.

The other port landings in the Mediterranean were covered by Force 'H', including HMS *Formidable*, which had on board, in addition to twenty-four Martlets, six Seafire IICs, converted from RAF Spitfires. 'Torch' was to be the first time Seafires would see action. *Victorious*, the other fleet carrier covering, also embarked six Seafires.

The Oran Task Group had three carriers: *Furious*, with twenty-four Seafires, and the CVEs *Biter* and *Dasher*, both US-built and carrying mainly Sea Hurricanes. The Algiers force consisted of the venerable *Argus*, also with Seafires aboard, and the CVE *Avenger*, with Sea Hurricanes.

No ships were lost prior to the landings, which took place just before midnight on 7-8 November. All the British aircraft in the 'Torch' operations had the familiar national roundels replaced by a plain white star, as it was considered that confusion could arise since French aircraft were marked with a red, white and blue roundel, very similar to the British marking, with only the colours reversed. The United States star reduced the possibility of friendly aircraft being shot down.

The airfields at Algiers were captured soon after daylight on 8 November. One of the fields, Blida, was surrendered to one of *Victorious*'s 882 Squadron Martlets; the patrol leader, flying his group over the airfield, saw white flags flying; covered by his flight, he landed and accepted the surrender of the field from its French Commandant. The rest of the flight touched down and remained in occupation until Allied ground troops took over. The other Algiers airfield was captured more conventionally by US combat troops.

Carrier aircraft also flew strikes against objectives in Oran, particularly the naval airfield at Tafaroui. During the fighting Seafires engaged French Dewoitine 520s; in one dogfight, Sub-Lt G.C. Baldwin gained the distinction of being the first Seafire pilot to destroy an enemy aircraft when he shot down a D520. Another Seafire got a D520 over La Senia, one of forty-seven French machines destroyed by *Furious*'s aircraft, mainly by Albacores dive-bombing La Senia airfield. 'Torch' was possibly something of a false dawn for the Seafires, which, being originally designed, like so many Fleet Air Arm aircraft, for the RAF, were not ideal carrier fighters, but on this occasion they performed well, shooting down a number of French fighters.

On the ground the Allied army advanced rapidly and by 10 November Algiers airfields were operational, and the harbour was also in Allied hands to the extent that *Avenger* was able to enter to rectify machinery defects. With the capture of airfields and a harbour, the need for naval air cover was past and the carriers began to withdraw. In so doing *Argus*, the old 'Ditty Box', had a charmed life: at dusk on 10 November she was attacked by fifteen Ju88 dive-bombers, but only one bomb hit her aft, destroying four fighters on board; the carrier remained operational. During the night she was attacked by a German U-boat, but the torpedoes passed harmlessly under her shallow-draught hull. Next day she was able to embark *Avenger*'s Sea Hurricanes.

Avenger, having repaired her defects in Algiers harbour, sailed with one of the returning convoys. On 15 November she was hit by a single torpedo from *U-155* and blew up; there were only seventeen survivors. *Avenger*'s loss caused the Admiralty to revise the AvGas-stowage and fire-protection systems in US-built CVEs. To balance the loss of *Avenger* to some extent, an Albacore from *Formidable* torpedoed *U-331* while she was surfaced north-west of Algiers, and on 21 November another Albacore from *Victorious* sighted and dropped depth charges on *U-517*, sinking the ocean-going Type IXC in mid-Atlantic.

'Torch' as a whole was to prove a disappointing operation as far as the subsequent North African land battles went. But from the combined Allied Navies' point of view, all the landings had demonstrated the ability of carrier-borne aircraft to cover a major invasion until land airfields could be secured.

In the Pacific there was a lull in the carrier operations; both sides, after the severe attrition of the sea battles off Midway, Guadalcanal and Santa Cruz, were re-equipping their naval air arms. The Americans were building carriers at an almost unbelievable rate; not only the small CVEs, but also the 27,000 ton *Essex* class fleet carriers. In just eleven months between 31 December 1942 and 24 November 1943, seven *Essex* class carriers were commissioned: *Essex* (CV-9), *Yorktown* (CV-10), *Intrepid* (CV-11), *Hornet* (CV-12), *Lexington* (CV-16), *Bunker Hill* (CV-17), *Wasp* (CV-18). (Four of the new carriers were given the names of ships sunk in the early Pacific actions: *Hornet*, CV-12, was to have been named *Kearsarge*; *Lexington*, CV-16, was built under the name *Hancock* and the second *Wasp* was to have been *Oriskany*.) In all, twenty *Essex* class or variants were to be commissioned before the end of 1944.

Also commissioned in the first eight months of 1943 was a new class of fast light carriers — nine ships of the 11,000 ton (standard) *Independence* class, CVL-22 to 30. These ships had been ordered in March 1942 and were built on hulls of the US Navy's *Cleveland* class cruisers. The choice of hull was a compromise, being the smallest (622 feet) capable of accommodating a reasonable flight deck and yet fast enough, at 31 knots, to maintain fleet speed. Though much faster than the CVEs, they could operate only a limited number of aircraft — about thirty. The 525 foot flight deck was wooden and unarmoured. A small starboard island offered very little accommodation and the funnel was not part of it; instead the boilers were ducted to four short uptakes running along the starboard side of the ship but cranked clear of the flight deck.

The experience that had been gained in the early battles against the Japanese Navy, together with the reports from the European war, had influenced not only the design of carriers but also that of the aircraft embarked. The F4F-3 Wildcat, though in 1942 undoubtedly the best available Allied carrier fighter, was found to be inferior in some respects to the A6M-2 Zero. The Zero, it is fair to say, came as a profound shock to the Americans, and the British for that matter: its long range, incredible manoeuvrability and heavy armament of two Oerlikon 20mm cannons and two 7.7mm machine guns in the wings made it a formidable opponent. The Zero's Nakajima 'Sakae' 14 cylinder, radial engine of 1,130hp gave a maximum speed of 348mph at 20,000ft (A6M-5), with a ceiling of 36,000ft and a range of 1,130 miles. What had impressed the US carrier pilots was the agility of the Zero; it could turn inside any Allied fighter; indeed pilots were warned not to try to 'dogfight' Zeros, but only to attack them when conditions were favourable, as with a height advantage.

The success of the Zero was such that it was beginning to acquire an almost mystical reputation for invincibility in the eyes of many Allied pilots. To counteract this the US Navy, in 1942, badly wanted an intact Zero to evaluate, but for some time there were only fragments available. On 3 June 1942 a Japanese petty officer pilot took off from the deck of the *Ryuojo* to cover the attack on Dutch Harbour in the Aleutians — part of the Japanese Midway deception plan. His model 21 A6M-2 Zero was practically brand new and was flying on its first operation; on the return flight to the carrier the pilot noticed that his fuel guage was going down at an alarming rate. Suspecting, correctly, that a stray bullet must have holed his tank, he radioed that he was making for a designated emergency strip on one of the many small islands of the Aleutians; he never made it. A month later, an American reconnaissance patrol found the Zero upside down in a swamp on the island of Aktan, with the unfortunate pilot dead with a broken neck.

The capture of that Zero was considered a great prize. It was almost undamaged and was at once shipped to Wright Field, where it was soon repaired, made airworthy and exhaustively evaluated. No doubt there were long technical reports with graphs and pages of figures, but the most succinct comment on the Japanese fighter came from Colonel Heyward of the Wright Field test centre: "The Zero?: it's a light sports plane with a 1,300hp engine." That was the Zero's secret: its unladen weight was only 3,920lb. A British F.11C Seafire of almost exactly the same size turned the scales at 6,200lb. The light weight of the Zero was both a strength and a weakness; the Japanese designers had sacrificed everything to speed and structural strength. It could withstand 12G aerobatics and climb at an angle of 45° at over 3,000 feet a minute; but in aeronautical engineering everything is a compromise based on the simple equation of power plus lift over weight and drag. The Zero's performance was gained at the cost of vulnerability: it had no armour plating for pilot or engine; no self-sealing fuel tanks. The heavier and therefore less agile Allied fighters could withstand a great deal of damage and still fly: the Zero could not. It was an attacking plane for use against weak opposition; it could not absorb punishment. One accurate burst of fire from six .50s and the Zero usually erupted into flames and crumpled up. The tactics to better it were soon evolved: "Do not dogfight Zeros, attack from a dive or in the climb, avoid trying to out turn them in level flight." The US Navy pilots learnt to use the greater weight of their fighters to dive on the Zeros; to get in a short burst of fire then zoom up to regain height. Of course, in the early engagements, especially when most of the Zero pilots had combat experience, that was easier said than done, but as the quality of the Japanese pilots fell off after Midway, it became possible, and increasingly so when the second generation of US carrier fighters came into service.

The first of the new fighters to go into naval service aboard a carrier was the successor to the F4F Wildcat, the F6F Hellcat. This later Grumman design became the outstanding shipboard fighter of World War II: 12,275 would have been built when production ceased in November 1945.

Hellcats were credited with shooting down 5,156 enemy aircraft in air-to-air combat, with a loss ratio of better than 19 to 1. It should be borne in mind that the previous Grumman fighter, the F4F, had been ordered and placed into US Navy service before the Americans entered the war; from reports on air combat in Europe and the experience of the Royal Navy's Fleet Air Arm, operating their essentially similar Martlets, it was apparent that, although a good naval fighter, the Wildcat did not match up to the best land-based European fighters. When the US Navy and Marines began to oppose the Japanese A6M-2 Zeros after December 1941, the F4F's shortcomings were further revealed.

Two new US naval fighters were by then already on the drawing boards: the F6F Hellcat and the Chance Vought F4U Corsair. In fact the prototype of the F4U, the XF4U-1 (BuNo. 1443) had made its first flight before the Hellcat on 29 May 1940, but as a precaution against possible failure of the radically new Vought aircraft, the Grumman Company was authorised to design a development of the Wildcat, which was by then in production. A contract was issued on 30 January 1940 for two prototypes designated XF6F-1 and -3. The US Navy broadly wanted a more powerful version of the F4F Wildcat, built around a 1,600hp fourteen cylinder, Wright Cyclone R-2600 radial. The Grumman design team, led by Leroy R. Grumman and William Schwendler, decided to go further; they even committed the heresy of talking to Navy pilots with combat experience and asking them for their views. The net result was a fighter pilot's aeroplane, with the pilot sitting at the highest point of the rotund fuselage, thereby achieving maximum visibility. What the Hellcat lacked in elegance it more than made up for in performance and rugged reliability.

The first prototype, XF6F-1 (BuNo. 02981), flew on 26 June 1942, just after the Battle of Midway. To gain first-hand impressions from the pilots who had fought in that action, the Grumman President, Leon Swirbul, flew to Pearl Harbor and consulted Lt Commander John Thach, the US Navy's authority on fighter tactics, who had experience of fighting Zeros in Wildcats at Midway. Thach and other pilots were of the opinion that to counter the A6M-2 Zero and its expected later developments, a greater rate of climb and higher level speed than the XF6F-1 offered was essential.

Consequently the second Hellcat prototype, XF6F-3 (BuNo. 02982), was fitted with a Pratt and Whitney R-2800 Double Wasp, an eighteen-cylinder radial engine which offered 2,000hp for take-off — a 25 per cent increase in power over the first prototype. XF6F-3 was test-flown on 30 July 1942. The flight trials were in general satisfactory; though there was some trouble with excessive changes of trim on lowering the flaps and there was some flutter at speeds in excess of 525mph in terminal velocity dives, these faults were soon eradicated without disrupting the production of the 600 which had been ordered 'off the drawing board'. To get the new fighter into squadron service as quickly as possible, the Grumman Company built a Hellcat assembly line at Bethpage, NY, with steel recovered from the derelict New York Second Avenue elevated railway. It is said that as the first production fighters moved down the line, the plant was still being built around them. The first operational F6F-3s were accepted by the US Navy on 16 January 1943, just 6 months after the first prototype had flown.

Service F6F-3s had a maximum speed of 376mph at 22,800 feet, a ceiling of 37,000 feet and a range of 1,085 miles on standard tanks, though this could be increased with a drop tank to 1,620 miles.

The first US Navy Squadron to equip with the Hellcat was VF-9, embarked aboard *Essex*, though the distinction of first using the F6F in combat went to VF-5 on the new *Yorktown* (CV-10). Both these fleet carriers were with *Independence* as part of Task Force 15, which launched strikes on the Japanese installations on Marcus Island on 31 August 1943. It was quite a day for firsts: the beginning of the naval offensive against the Japanese, the first combat flights of the F6Fs, and the first combat missions of the new *Essex* and *Independence* class carriers. Nine strikes were launched against Marcus Island with the F6Fs, escorting TBFs and SBDs, in what were in reality minor actions mounted to provide combat experience for newly trained pilots and aircrew and to perfect the tactics for the coming major offensive against Japan.

By the time the F6F had begun its operational career, its contemporary, the F4U Corsair, was still not considered suitable by the US Navy for carrier service, though the first F4U combat missions had been flown on 14 February 1943 by a land-based Marine squadron, VMF-124, operating from Espiritu Santo. Twelve F4Us formed part of the escort for PB4Ys — the US Navy version of the B24 Liberator — making a daylight strike against enemy shipping at Kahili in southern Bougainville. The attacking force was intercepted by fifty Zeros; in the fighting that followed, the entire top cover of four Army P-38 Lightnings and two P-40 Warhawks was shot down, together with two of the Marine Corsairs, which were flown by inexperienced pilots. The Japanese lost three Zeros, one of which had collided with a Corsair. Following the 'St Valentine's Day Massacre', the Marine Corsair pilots soon gained the upper hand, VMF-124 being credited with sixty-eight Japanese aircraft for the loss of eleven Corsairs and three pilots, on their first combat tour. By the end of 1943 all Marine squadrons in the South Pacific had been re-equipped with F4Us and had destroyed 584 aircraft. The most famous Pacific Marine squadron was VMF-214 — the Black Sheep — commanded by the highest scoring Pacific 'Ace', Major 'Pappy' Boyington, who was personally credited with twenty-eight kills. On one early Corsair mission in September 1943 Boyington shot down five Zeros near Bougainville; he would undoubtedly have added to his total had he not been shot down on 3 January 1944 and spent the rest of the war behind the wire of a POW camp.

The success of the Marine squadrons was establishing the Corsair as a most formidable fighter. Its performance was, by any standards, high: a maximum speed of 415mph at 20,000 feet and a service ceiling of 37,000 feet. But still there was a reluctance to embark the type, although it had originally been ordered as a carrier fighter. The Navy had been impressed when the prototype XF4U-1 had notched up 405mph on 29 March 1940 — the first US fighter to exceed 400mph.

What then were the Corsair's drawbacks? Although it shared the same power plant as the Hellcat, it had a much larger diameter propeller; to enable it to clear the ground and yet have reasonably short undercarriage oleos, the wings of the F4U were cranked, which was to pose certain aerodynamic problems. Also the cockpit gave a rather poor visibility, especially with the tail on the ground. However, carrier trials were conducted on board USS *Sangamon*, CVE-26, in Chesapeake Bay on 25 September 1943. The trials revealed further difficulties: unlike the Hellcat, which warned the pilot of an approaching stall by buffeting, the Corsair stalled without any warning at all, usually dropping the port wing, due to torque reaction, and snapping into a spin which on a landing approach would be fatal. The view from the cockpit when landing on a carrier, using the standard US Navy 'slow and nose high' approach technique, was ten-tenths engine, necessitating landing the F4U from a turn, which, with the aircraft's unpredictable stalling properties, could be dangerous in the hands of average service pilots. Even when safely over the deck, the Corsair tended to float in ground effect: once hooked, the F4U was found to suffer from excessive oleo bounce, causing burst tyres and/or a damaged fuselage.

The stalling problem was alleviated by placing a 6 inch 'spoiler' on the leading edge of the starboard wing outboard of the guns, which caused that wing to stall a little before the port; this corrected the tendency for the aircraft to snap into a spin on stalling. The oleos were modified to reduce the bounce and a longer tail-wheel yoke made the landing attitude easier. By the time these developments had been incorporated, the F4U had mainly been consigned to the Marines, who removed the 'hooks' and locked the wings permanently down. It was not until the end of 1944 that the US Navy decided to embark the Corsairs on carriers.

In mid-1943 the British Fleet Air Arm received the first of 2,012 Corsairs under 'Lend Lease' agreements; these aircraft were equivalent to the US Navy F4U-4. The first RN squadron to equip with the type was 1830, which began training at Quonset, a US Navy base, in June 1943. Owing to the lack of clearance in the British carriers' armoured hangars, 8 inches had to be cut from the wingtips of RN Corsairs to enable the upward folding wings to clear the hangar roofs. This modification caused the stalling speed to increase slightly, but it was considered by test pilots to improve the rate of roll. In any event, the distinction of first operating the Corsair from carriers went to the Royal Navy, when, in April 1944, No. 1834 Squadron from HMS *Victorious* provided cover for Barracuda torpedo bombers when they attacked the German battleship *Tirpitz*, then lying in Kaafjord in northern Norway.

The attack, which achieved fourteen direct bomb hits, immobilised the ship for three months and she never became operational again, being finally sunk by 12,000lb bombs for RAF Lancaster bombers.

Soon after the *Tirpitz* attack, Barracudas were in action in the Pacific. But before returning to that conflict, it is worth mentioning other second generation carrier aircraft that would be used in the war against Japan. The SBD Dauntless would slowly be supplanted by the Curtiss SB2C Helldiver, the second of that company's designs to bear the name. The name the US Navy pilots bestowed was less warlike but apposite: 'The Beast'. The name was partly a reference to the SB2C's 3 ton dead weight, but mainly because of its difficult flying characteristics: early versions had a tendency to shed their wings — hardly a recommendation for a dive-bomber. This weakness was never really cured and restrictions were placed on the aircraft, which could never throughout its service career dive as steeply as the lately lamented SBD. The biggest problem with the Helldivers was simply producing them: it was to take nearly four years to get them from prototype to operational service aboard the first carrier to embark them — *Yorktown*. Fortunately, in view of the type's shortcomings, by the time it eventually got into action, the Japanese Navy's fighter aircraft were a shadow of what they had been in the early Dauntless days. The Royal Navy received twenty-six Helldivers (JW100 to JW125) but did not use the type operationally.

If the Helldiver was something of a disappointment, the same could be said of the British Seafire. The last wartime variant, the fully wing-folding Seafire III, would be used in the Pacific in the closing stages of the war. But the Seafire was not really a successful naval fighter; it had, after all, been adapted from the original RAF Spitfire, which had been conceived as a high-performance, high-altitude interceptor — a role it had performed to perfection during the Battle of Britain. When used on carriers, the naval variant suffered from several drawbacks: in the first place it was always restricted in endurance to only about an hour on internal tanks; it had to be landed aboard with care, there being little latitude between stalling and landing too fast and floating over the arrester wires into the barrier. RN Seafire pilots remember well the three 'P's: Pecking, Pintling and Puckering. 'Pecking' was the result of the tail being thrown up when arrested too fast, causing the 10 foot propeller to 'peck' the flight deck; if, as was often the case, the airscrew was wooden, it was irreparably damaged.

During the landings at Salerno in August 1943, there was practically no wind and forty-three Seafires out of the 105 embarked were lost in accidents, and many more made unserviceable due to 'pecking'. The spares embarked were quickly used up; so serious was the shortage of propellers that eventually the less damaged ones simply had their blades 'cropped' to remove the splintered tips. Surprisingly it was found that at least 2 inches could be removed without any noticeable difference in performance.

'Pintling' was caused by the rather fragile undercarriage, which had been designed to be used on smooth grass airfields or paved runways, suffering from the inevitable shock of landings; these could be very heavy, which caused the oleo pintles, on which the main undercarriage leg pivoted on retraction, to be bent or sheered. The result was that if the oleo did not collapse immediately, which was often the case, the next time the undercarriage was retracted it would not extend, which usually meant the unfortunate pilot having to bale out. 'Puckering' was the result of a heavy landing forcing the tail down with sufficient violence to bend the fuselage just in front of the tail assembly, the aircraft invariably being written off.

Defenders of the Seafire will point out that it was the fighter's misfortune seemingly always to be involved in operations in areas of light winds; that is not really a tenable excuse, for winds had ceased to be a reason for naval operational shortcomings once steam became the universal propulsive power. There is no denying that once airborne the Seafire was a superb interceptor fighter, though its lack of endurance was always a worry for its pilots. One of the most successful was Commander Dick Reynolds, who said of the Seafire's limited range:

> In the Home Fleet when we were protecting the Russian convoys this did not really matter; we were there to defend the Fleet and that we could do adequately, but when we got to the Far East we were restricted, at the very maximum, to one hour and thirty minutes airborne .... as you can imagine, it is very uncomfortable when you know you have only got ten gallons of fuel left to get back on board, which happened fairly often.[21]

Commander Reynolds succeeded in shooting down four enemy aircraft while flying Seafires, including two Zeros, and sharing in the destruction of another. In all, Seafires from FAA carriers shot down thirty-seven enemy aircraft for the loss of only eight Seafires and one loaned RAF Spitfire.

The last British designed and produced naval aircraft to enter service with the fleet during the war was the Fairey Firefly. It had been conceived as long ago as 1939 as a Fulmar replacement, but, like that aircraft, it was a two-seater fighter; it was therefore a large machine with a wingspan of 45 feet, powered by a Rolls-Royce Griffon, liquid-cooled engine of 1,730hp. This proved inadequate for the weight and size of the

Firefly, which could achieve only 316mph at 14,000 feet. Although slow for a fighter, the Firefly nevertheless was surprisingly manoeuvrable with excellent handling; it also possessed a good endurance, offering a range of 1,300 miles, making it a useful escort and strike fighter — a role which was to be the Firefly's first combat mission when 1770 Squadron from *Indefatigable* flew on the *Tirpitz* attack in July 1944. Fireflys were also to gain distinction when with the British Pacific Fleet.

In the Pacific during 1944 the Japanese Navy would be fighting firstly a defensive withdrawal, then a retreating battle for survival as it was overwhelmed by the output from the shipyards, aircraft factories and training establishments of the United States, which became an unstemmable tide. On 1 July 1944 the US Navy had thirteen fleet carriers, nine light *Independence* class CVLs, and sixty-three escort carriers, in addition to oilers, repair ships and seaplane carriers. On charge were 34,071 aircraft, of which 22,116 were classified as combat types. The pilots and aircrew were, without doubt, the best trained of any of the belligerents; most pilots had at least 400 hours' flight time in their logbooks before even being posted to a carrier. The US Naval Staff was perfecting the concept of the 'Fast Carrier Task Force', an independent group which could refuel and resupply at sea, remaining operational more or less indefinitely.

In contrast to the Americans' plenty, the Japanese did not begin to make good the loss of their fleet carriers until 1943, when the 13,600 ton seaplane tenders *Chitose* and *Chiyoda* were converted as 29 knot aircraft carriers capable of operating thirty aircraft each. The two ships had flight decks of 590 feet, which were devoid of any island. *Chiyoda* was commissioned on 31 October 1943; *Chitose* the next day.

The Japanese Navy had earlier approved a programme of construction that was to have produced six large fleet carriers of the 29,300 ton *Taiho* class; this class was envisaged just after the start of the European war as an answer to the heavily armoured carriers of the Royal Navy. *Taiho* was laid down in 1941; though possibly initiated by the British *Illustrious* class, she was not as well armoured, only the flight deck itself being of 3 inch armour. *Taiho* (the name means 'Great Phoenix') was a very fast ship — over 33 knots — and was designed to embark eighty-four aircraft, but since not all Japanese carrier aircraft had folding wings, the actual number operated was nearer sixty. She was a conventional starboard island ship; the boiler uptakes were fed to a single large funnel cranked out to starboard away from the island, apart from which she looked remarkably like the British *Illustrious* class. *Taiho* commissioned in March 1944 and was the only one of the proposed class to be completed.

Among other Japanese carrier construction projects which would not be completed before 1944 was the ambitious *Shinano* of no less than 64,800 tons standard displacement, the largest carrier built outside the United States. The hull was laid down as a battleship of the *Yamato* class (67,935 tons). After the Battle of Midway it was decided to complete the ship as a carrier, not to operate directly with the fleet, but to provide a floating reserve of operational aircraft and repair and maintenance facilities — a role she was destined never to fulfil, for *Shinano*, completed on 19 November 1944, was sunk on passage to Kure ten days later by the US submarine *Archerfish*.

As far as aircraft were concerned, the Japanese found it impossible to replace the A6M Zero in significant numbers, though several promising designs were in production. The A6M-2 was slowly supplanted by the A6M-5, or Zeke 52, which was a development of the earlier Zero. Potentially the most effective new carrier aircraft was probably the Nakajima B6N2 Jill torpedo bomber which in 1944 replaced the aging Kate; only 770 Jills, however, were completed and, owing to inadequate pilot training, its main operational use was to be in the forthcoming Kamikaze attacks.

The allied strategy to defeat Japan was complex and to some extent compromised by dificulties, not only between the Allied Chiefs of Staff, but also between the US Army and Navy. General Douglas MacArthur wanted to advance through the chains of Pacific islands, thus making the eventual conquest of Japan essentially a soldiers' war, although naturally the Navy would be required to undertake a large number of amphibious landings. The island-hopping through the Pacific would form one pincer, the other being a great land assault through China via Burma. Admiral Nimitz, C-in-C of the Pacific Fleet, on the other hand, saw the Pacific War as a purely naval matter; he argued that instead of the costly occupation of every fortified island and atoll, these should be bypassed to wither on the vine, while a powerful Task Force based on fast carrier groups would, in a series of decisive actions, leap-frog through the south Pacific to the Japanese Home Islands. That strategy would have reduced the campaign in Burma to a sideshow, with the consequent savings of thousands of lives and a great deal of supplies and equipment.

The naval plan was not in its entirety accepted, partly because of the powerful influence that the Chinese leader General Chiang Kai-Shek (or more specifically his wife and her brother) had on the American President and opinion in Washington. The result was a land campaign mounted initially to open the Burma road to China, along which men and supplies would flow. The Burma campaign, one of the great successes of the British army, under the unassuming but very able General Slim, eventually forced the Japanese army out of that country, using the same tactics that had driven the British out in 1941.

In the eventual advance in the south Pacific a compromise was reached between a straightforward naval war, driving directly towards Japan, and the slow conquest of islands. The new strategy envisaged taking only the main Japanese bases in the central Pacific in a series of amphibious assaults. MacArthur insisted, however, that the Philippines be occupied, presumably to enable him to make good his much publicised declaration, "I shall return". That apart, the retaking of the Philippines would, MacArthur argued, enable major bases to be set up to be used as a springboard for the final assault on Japan, linking up with the expected Chinese advances.

The Japanese Staff anticipated the attack in the Pacific. Its Navy was still confident enough to believe that in any major sea battle with the US Navy, it could use its battleship strength to defeat the Americans.

The Pacific advance was to be led by the US Fifth Fleet, commanded by Vice Admiral R.A. Spruance, which consisted of the two large amphibious Attack Forces 52 and 53. Each of these self-contained groups had battleships, transport, assault craft and CVEs for close air support. The large fleet carriers were formed into four separate Fast Carrier Groups which combined as Task Force 50 to provide distant heavy cover.

As part of the planning, in October 1943, the US Navy issued detailed orders as to the aircraft complement to be embarked on the carriers. *Essex* class ships were to have their fighter Air Groups increased to thirty-six aircraft, with the addition of thirty-six bombers and eighteen torpedo machines. At the same time the CVLs were authorised to carry twelve fighters, nine bombers and nine torpedo planes. The CVL complement was to be revised in November 1943 to twenty-four fighters and nine torpedo machines — figures that were to remain until the end of the Pacific War.

During October preliminary strikes were mounted against the Japanese-held Wake Island and Rabaul; on the second Rabaul strike on 11 November, Helldivers were used for the first time. In these strikes valuable experience in handling groups of carriers was gained in time for the first major amphibious landings, which were to be made between 18 and 20 October to occupy the Gilbert Islands.

Six fleet carriers and five light carriers, plus two CVEs engaged in ferrying forty-four F6Fs for the island's garrison air force, comprised the air component of Task Force 50, with a total of 440 fighters and 390

This Firefly of HMS *Implacable*'s complement is suffering the eventual fate of many badly damaged carrier aircraft. Stripped of re-usable equipment, it is slung from a crane and unceremoniously dumped over the side into the depths of the Pacific Ocean (*IWM film still*)

(*Above*) HMS *Indefatigable*: it was from this carrier that Fireflys first took off in action against the German battleship *Tirpitz*, in July 1944 (*David Brown*); (*below*) the final form of World War II US Navy nightfighter radar was AN/APS-6 (Army Navy/Airborne Pulsed Search), carried in a pod on the starboard wing, here shown on a Grumman F6F-5N (*Grumman*); (*below right*) an OS2U Kingfisher spotter over Angaur Island, in the Palau group, on 17 August 1944, as landing craft head for the beaches (*US Navy*)

(*Above*) In the first planned Kamikaze attacks of the Pacific War, at Leyte Gulf, this Escort carrier, *St Lo* (CVE-63), was sunk on 25 October 1944 (*US Navy*); (*below*) USS *Bunker Hill* was hit by two Kamikazes when operating off Okinawa on 11 May 1945. Large AvGas fires started, but her crew fought the blaze and saved the ship, though at a cost of nearly 400 dead (*US Navy*)

(*Above*) Pilots being briefed for a mission aboard a US Escort carrier in the Pacific. All aircrews carried personal arms which can be seen on the man on the left (*US Navy*); (*below*) the outcome of the US Navy's angled-deck trials was the reconstruction of surviving wartime *Essex* class carriers, under the designation 'Project 27C'; USS *Essex* (CV-9) became CVS-9 in March 1959 and is seen a year later with an F4D approaching (*US Navy*)

HMS *Hermes*, 28,500 tons. The ship served as a true fixed-wing carrier until 1972, when she was converted to her present (1980) role — a commando ship, operating helicopters and V/STOL Sea Harriers (*MOD (RN) Crown Copyright*)

(*Opposite top*) *Ark Royal* landing-on a De Havilland Sea Vixen, possibly the prototype XF828, which took part in deck-landing trials on 5 April 1956 (*British Aerospace*); (*opposite centre*) using the Mirror Landing Sight, a pilot was able to effect the flat approach to the flight deck of the new generation of jet fighters. The photograph depicts an early experimental sight (*Royal Aircraft Establishment, Farnborough*); (*opposite below*) a De Havilland Sea Vixen of the RN Test Squadron has the bridle of a steam catapult attached. The safety chocks are hydraulic and will be lowered just before take-off (*British Aerospace*)

(*Above*) The first US carrier to have a 'steam slingshot' (a British innovation) was a reconstructed *Essex* class, USS *Hancock* (CVA-19), photographed off San Diego in February 1975 (*US Navy*); (*below*) a Marine F4U-4B of the Marine VMF-214, 'Black Sheep' Squadron, about to take off from USS *Sicily* (CVE-118), on a Korean strike (*US Navy*)

(*Above*) The Korean War saw the active-service debut of another piston-engined aircraft — the Douglas Skyraider, here demonstrating the heavy offensive load it could carry: a torpedo, two 500lb bombs and twelve 5in rockets (*US Navy*); (*below*) the Hawker Sea Fury was the last piston-engined fighter to serve with the Fleet Air Arm, equipping several squadrons from 1947 until 1954 and was the main RN fighter during the Korean War (*British Aerospace*)

(*Above*) This photograph is believed to depict the US Navy's first helicopter, a Sikorsky NNS-1 Hoverfly, Bu No 46445, obtained from the US Army Air Corps, undergoing sea trials in early 1943 (*US Navy*); (*below*) the power of the steam catapult on the *Midway* class is evident from this 1962 photograph which shows a 33,900lb F3H-2, McDonnell Demon, being launched from USS *Coral Sea* (CVA-43) at anchor at Yokosuka, Japan (*US Navy*)

(*Above*) The De Havilland Sea Venom was the first all-weather jet naval fighter on charge to the Royal Navy. XG612, illustrated, is an F.A.W.-21 (*British Aerospace*); (*below*) the Grumman F-14 Tomcat, a multi-role fighter with three primary functions: (1) an air-superiority fighter (2) flying carrier Combat Air Patrols (CAP) to protect the carrier Task Force (3) attacking tactical ground targets (*Grumman*)

The largest warship in the world — the 95,000 ton, nuclear-powered USS *Nimitz*, CVN-68; built at a cost believed to be in excess of $1,000 million (£434 million), these carriers require a crew of around 6,000 trained men (*US Navy*)

(*Above*) A close-up of the missile tubes of *Kiev* photographed from an RAF Canberra in the Mediterranean, after the Russian ship had passed through the Bosphorus from the Black Sea in July 1976 (*MOD* (*RAF*) *Crown Copyright*); (*below*) a US Marine Harrier demonstrates its short take-off run from a US carrier. The US Marines are investigating the 'ski jump', which will confer a range/payload advantage when Harriers are operated from short flight decks (*British Aerospace*)

The 'ski jump' was first tested at the Royal Aircraft Establishment, Bedford. Here the Flight Systems Squadron's T2 Harrier, XW 175, is about to leave the test ramp (*British Aerospace*)

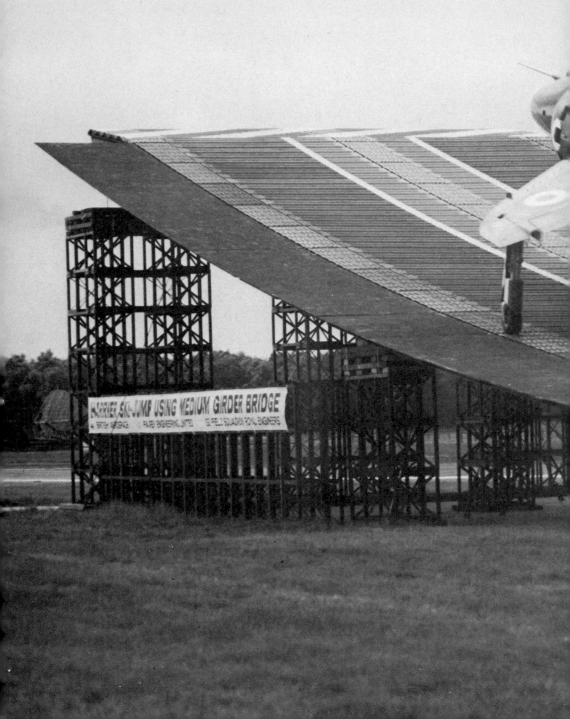

(*Above*) The US Marines were the first to demonstrate the possibility of using a VTOL aircraft in a maritime role. In 1967 came the XC-142A, a tilting-wing aircraft, shown here lifting off from the helicopter pad of USS *Ogden* (*US Navy*); (*below*) a 15,000-ton sea-control ship of US Navy design under construction (1980) for the Spanish navy; she is designed to carry Harriers (Matadors) (*British Aerospace*)

bombers embarked. The operation opened on 18-19 November 1943 with *Essex*, *Bunker Hill* and *Independence* flying off preliminary strikes to soften up the defences on Tarawa and Makin Atolls, prior to their assault by Marines and the Army on 20 November. *Saratoga*, with the escort carrier *Princetown*, attacked Abemama on 21 November, then flew cover for the troops fighting ashore, for four days. During the bitterly contested but successful landings, additionally the US Navy deployed light escort carriers flying anti-submarine and combat air patrols in the invasion area and 'on call' close support missions for the ground forces.

After the islands were secured on 24 November, one carrier group remained in the area for another week as a precaution against a possible Japanese counter-attack, which did not materialise. The two ferry CVEs, *Barnes* and *Nassau*, flew off VF-1 to land its Hellcats on Tarawa airstrip. During the Gilberts' campaign, the cost in carriers was *Liscome Bay* (CVE-56) sunk by Japanese submarine on 24 November, and the *Independence* (CVL-22) damaged by air attack on 20 November.

It was during the Gilbert Island landings that the first night interception by an aircraft from a US carrier was attempted. Two F6F Hellcats from *Enterprise*, led by the Air Group Commander Lt Commander E.H. (Butch) O'Hare, flew with a radar-equipped TBF Avenger which was to vector the fighters on to the enemy aircraft until visual contact was obtained. On the first flight, on the night of 25 November, no contact was made, but on the second sortie, the next night, the Avenger secured a strong radar contact. In the action which followed it was the TBF which succeeded in shooting down a twin-engined Mitsubishi G4M Betty in flames — the first enemy aircraft shot down in a night air-to-air engagement by a US carrier-borne plane. Unfortunately, O'Hare was also shot down, a tragic loss of one of the US Navy's foremost fighter pilots. (His memorial is O'Hare airport, Chicago.)

The next major operation for the Fifth Fleet was the occupation of the Marshall Islands: Kwajalein, Majuro, Roi and Namur. Simultaneously, covering operations were provided by the great mobility of the fast carriers, enabling them to mount attacks on the Japanese naval base on Truk Atoll in the Caroline Islands, 950 miles away. The first strike on Truk on 17 February 1944 was from three fast carriers which were part of Task force 58; in a two-day attack the carriers launched 1,250 combat sorties against the supposedly impregnable base. Japanese radar had alerted the defenders and about forty-five Zeros and Rufes — the seaplane variants of the Zeros — were waiting for the Americans. In the dogfights which followed F6F Hellcats proved that at last the Zero had met its match; they did so by shooting down over fifty Japanese aircraft.

TBF Avengers dropping fragmentation bombs accounted for another 110 or so on the ground. Thirty-seven Japanese ships were also claimed as sunk and base installations damaged. The American losses amounted to four Hellcats and nine Avengers.

It was during the action against Truk that the US Navy carried out its first night-bombing attack, when VT-10 from *Enterprise* flew off twelve radar-equipped TBF-1Cs, which made low level attacks on the harbour, claiming direct hits on several ships. The Japanese countered with a night strike on the US Task Force, using B5N Kates. Radar-fitted night-fighter F6F-3Ns from *Yorktown* were launched, but the carrier's fighting direction was inadequate and one Kate managed to get into an attacking position to hit *Intrepid* (CV-11) with a single aerial torpedo, without, however, inflicting fatal damage.

After the successful operations against Truk, the US Navy was so replete with carriers and aircraft that it could afford to detach USS *Saratoga* with a three-destroyer screen to join the Royal Navy's Eastern Fleet in the Indian Ocean, in attacks on Japanese-held territory. There had been some anxiety amongst the Allied Naval Staff that the Japanese might make another major carrier offensive in the Indian Ocean when it was learned from intelligence sources that five Japanese battleships and three fleet carriers had arrived at the naval dockyard in Singapore: in fact they were simply availing themselves of the excellent ex-British naval facilities there, having been driven from Truk by the US Navy attack.

Another pointer to a possible offensive in the Indian Ocean occurred at the end of March 1944, when three Japanese cruisers had sunk merchant shipping off the Andaman Islands. HMS *Illustrious* sailed from Ceylon to try to intercept the Japanese ships but failed to make contact. *Illustrious* then joined up with *Saratoga* and her escorts, and the ships of both navies returned to the Royal Navy Eastern Fleet base at Trincomalee in Ceylon. After spending two weeks in exercises to enable the different operating techniques and communication systems to be rationalised, the Allied force, comprising HMS *Illustrious* and USS *Saratoga*, with twenty-four other warships, sailed to make a major strike on the Japanese harbour at Sabang, on the northern tip of Sumatra, in what had been the Dutch East Indies.

The aircraft were launched from a point 100 miles from the target at dawn on 19 April 1944, *Illustrious* providing seventeen Barracudas, escorted by thirteen Corsair IIs of the Fleet Air Arm; the largest contribution, *Saratoga*'s, was eighteen TBF-1 Avengers and eighteen SBD-5 Dauntlesses, covered by twenty-four F6F Hellcats. (The US Navy still had not cleared its F4U Corsairs for carrier operations.)

The combined strike was well co-ordinated and achieved complete surprise; no fighter opposition was encountered and flak over the target started only after the raid was well under way. The amount of shipping in Sabang harbour was disappointing; consequently only one freighter was sunk, though another had to be beached, but the port installations, including oil-storage tanks, were heavily damaged. The fighters, having no airborne opposition to deal with, strafed airfields, destroying over twenty-four Japanese aircraft on the ground. One F6F was shot down by ground fire, but its American pilot was rescued by a RN submarine. Later, three G4M Bettys attempted to counter-attack the Allied Fleet, but in the very changed circumstances of 1944 all were shot down by F6Fs from *Saratoga*'s Combat Air Patrol.

After the Sabang operation *Saratoga* was due to return to the USA for a refit, but it was decided that, en route, she and *Illustrious* would attack the port of Surabaja in Java, *Saratoga*'s Avengers and Dauntlesses being escorted by RN Corsairs. After the Surabaja operation on 17 May, the results of which, due to lack of targets, were disappointing, *Saratoga* finally sailed for Pearl Harbor on the way to the United States.

The next month, June 1944, was to see the opening of the second front — the invasion of Normandy, D-Day — which tended to obscure two very significant victories over the Japanese in the Pacific which the US carriers of Task Force 58 (Vice Admiral M.A. Mitscher) were to achieve. These were the occupation of the Marianas and the Battle of the Philippine Sea.

The significance to the Japanese of the major Mariana Islands — Saipan, Tinian and Guam — was that they formed the outer perimeter of the Home Islands' defences, being only 1,200 miles from Tokyo. Any attack on the Marianas would therefore be countered by the full Japanese Combined Fleet, under its new C-in-C Admiral Toyoda, who had replaced the late Admiral Yamamoto. Toyoda was to direct operations from his Naval HQ in Tokyo, the commander at sea being Vice Admiral Ozawa.

The Marianas campaign began on 11 June 1944. Like all the other major actions in the Pacific War, it was mainly to be mounted by aircraft carrier groups. Task Force 58 was centred on seven large carriers and eight light carriers with nearly 1,000 aircraft embarked; they were opposed by fewer than 500 Japanese aircraft in the Marianas, though there were others available on airfields in the Carolines and, by staging on Iwo Jima, from Japan itself.

The US Task Force, TF58, sailed from the newly won anchorages at Majuro, in the Marshall Islands, on 6 June 1944, at which time, on the other side of the world, D-Day was dawning. On 11 June TF58 launched a sweep of 211 F6F Hellcats on Saipan, Tinian and Guam; the American fighter pilots were by now seasoned and confident in their ability to take on whatever the Japanese put up against them. Their confidence was justified: during the first three days of strikes preparatory to the invasion of Saipan, 150 Japanese aircraft are believed to have been shot down in aerial combat. The landing, supported by close carrier air cover, took place on 15 June 1944. As US Marines were consolidating their lodgements after fierce fighting, the Japanese Combined Fleet, which had sailed from its base at Tawitawi in the Sulu Archipelago as soon as the American attack had been reported, was by 18 June 400 miles to the west of TF58 and steaming to attack in three main groups.

The Van Force (Vice Admiral Kurita), though it included two light carriers, *Chitose* and *Chiyoda*, was principally a battleship group with the mighty 68,000 ton *Yamato* and *Musashi*, with the *Haruna* and *Kongo*, screened by four cruisers and nine destroyers. Vice Admiral Ozawa, the C-in-C, was flying his flag in *Taiho*, commissioned only the previous March; his Force 'A' was composed of the veterans *Shokaku* and *Zuikaku*, with three cruisers and nine destroyers. Force 'B' had two fleet carriers, *Junyo* and *Hiyo*, with the light carrier *Ryuho*, accompanied by the battleship *Nagato* and the cruiser *Mogami* (which had survived Midway) and ten destroyers.

The Japanese fleet looked a great deal more effective on paper than in reality, for the key to the coming battle would be airpower. Ozawa's carriers had only about 430 planes embarked, and many of these were to be flown by inexperienced pilots fresh from training. However, the Japanese Naval Staff had gambled on Ozawa being able to use aircraft from the islands to supplement the rather slender seaborne resources.

On 16 June the presence of the Japanese fleet was made known to the C-in-C US Fifth Fleet, Admiral Raymond Spruance, aboard his flagship, the cruiser *Indianapolis*, when a sighting report from an American submarine patrolling off San Bernadino Strait was received. Considering the situation, Spruance decided that the forthcoming Battle of the Philippine Sea could not be joined before 19 June, and, rather like Drake finishing his game of bowls on Plymouth Hoe after the sighting of the Spanish Armada, he declined to sail to meet the Japanese but remained 100 miles off Saipan to cover the amphibious forces still fighting ashore.

Task Force 58 was divided into four main carrier groups: Task Group 58.1 (Rear Admiral J.J. Clark), with the *Essex* class fleet carriers *Hornet* (CV-12) and *Yorktown* (CV-10) and the light carriers *Belleau Wood*, *Bataan* and four cruisers; Task Group 58.2 (Rear Admiral A.E.

Montgomery), with *Bunker Hill* (CV-17) and *Wasp* (CV-18) and the light carriers *Monterey* and *Cabot*, screened by the light cruisers *Santa Fé*, *Mobile* and *Biloxi* and twelve destroyers; Task Group 58.3 (Rear Admiral J.W. Reeves) comprising two fleet carriers, *Enterprise* (CV-6) and *Lexington* (CV-16), with the light carriers *San Jacinto* and *Princeton* and five cruisers, including the anti-aircraft cruiser *Cleveland*, and thirteen destroyers; and the final Carrier Group, 58.4 (Rear Admiral W.M. Harrill), consisting of *Essex* (CV-9) and the escort carriers *Langley* (CVL-27) and *Cowpens*, together with anti-aircraft cruisers and fourteen destroyers.

By the afternoon of 18 June Spruance had disposed three of his four carrier groups 15 miles apart in a line running north-south. To the westward, also at 15 miles distance, there was a large battleship group, TG 58.7, with nine battleships, four cruisers and a screen of thirteen destroyers; this non-carrier group had TG 58.4 stationed 12 miles away to afford air cover.

Steaming eastwards towards the US forces, Ozawa's groups were formed into a large triangle, with the Van Force of battleships leading. The hope being that the Van, some distance ahead of the Japanese carriers, would attract the main American battleships, leaving the carriers exposed.

As the distance between the two fleets decreased, the initial advantage went to the Japanese admiral, whose superior long-range reconnaissance sighted three of Spruance's groups at a range of 400 miles, just within his aircraft striking capability, though as night was now falling the attack was postponed until first light next day. Although the position of the US Task Force was now known to Ozawa, Spruance's search aircraft had yet to make contact. However, during the night, direction-finding stations in Hawaii had intercepted heavy radio traffic from Ozawa's fleet which had indicated its position as being 350 miles from Task Force 58. Though urged, principally by Vice Admiral Mitscher, to close the gap between the opposing fleets and to launch a dawn strike, the cautious Spruance declined to sail from within covering distance of Saipan, taking the view that the continued support of the US ground forces outweighed any other consideration. Dawn searches from the US carriers were still unable to make contact with Japanese ships and, although a long-range naval PBY had sighted them, the report it sent was delayed in reaching Spruance, arriving too late to be of tactical use.

At 0530 hours on 19 June 1944 Japanese shadowers regained contact with the US carrier groups; one of the shadowers, a Judy (Yokosuka D4Y Suisei) based on Guam, was shot down by a Hellcat from *Monterey*.

It was the first enemy aircraft to be destroyed during the battle, though not before it had transmitted a sighting report, on receiving which Ozawa immediately ordered his first air strike on the US Task Force.

As Ozawa's flagship, the newly commissioned carrier *Taiho*, turned into wind to enable her aircraft to take off, she was sighted by the US submarine *Albacore* (Commander James Blanchard, USN) which fired a salvo of torpedoes at the Japanese carrier, one at least of which struck the ship. To the disappointment of *Albacore*'s crew, it appeared that the armoured carrier had shrugged off the hit; even on board little damage was apparent, but the strongest ship will 'whip' after a heavy underwater explosion. *Taiho* was no exception and an undetected crack had started in one of the bulk AvGas stowage tanks. For the moment all seemed well and the carrier continued to operate her aircraft.

As the Japanese strike aircraft were flying towards TF58, the US carriers were embarking F6Fs which were returning from dawn strikes they had made over enemy-held airfields on Guam. Japanese aircraft were discovered to be arriving from the Carolines to reinforce the air garrison. In just over an hour's fighting thirty-five A6M Zeros and several bombers were destroyed. The Hellcats found fewer aircraft on Guam than Intelligence had led them to expect, so, lacking targets, at about 1000 hours they were recalled to their carriers.

As the US fighters were landing on, the battleship *Alabama* reported a radar contact which was correctly interpreted as an incoming Japanese strike, then about 150 miles from the American carriers. The radar gave sufficient warning for the US Task Force to have 200 F6Fs airborne, stacked between sea level and 30,000ft, ready to intercept; to keep the carriers' decks clear for fighter replenishment, bombers and torpedo aircraft were launched and ordered to orbit well out of the fighting area. The air battle which followed, the biggest single air combat in history, was to be called 'The Marianas Turkey Shoot' by the American fighter pilots: 300 F6Fs, refuelling and rearming in rotation, kept up the day-long battle as succeeding waves of Japanese aircraft attacked the fleet. The result at the end of the day was that, out of a total of 373 Japanese aircraft launched from carriers, 243 had been shot down and over thirty severely damaged. The Americans had lost a total of only thirty aircraft, six of these in accidents. Pilot losses were twenty-seven. The Japanese had lost 400 aircrew. The only positive results to show for this crushing defeat was slight damage to the carriers *Wasp* and *Bunker Hill*, though the surviving Japanese pilots, on returning to their carriers, had reported to Ozawa that many of the American ships had been left blazing and sinking.

The day was not yet over, however; the US submarine *Cavalla* (Commander Herman Kossler, USN), which had been shadowing Ozawa's Group 'A' for hours, was rewarded for her persistence when the 29,800 ton carrier *Shokaku* turned into wind to recover aircraft returning from the air battle; in so doing she was unwittingly placed in an ideal attacking position. *Cavalla* fired a salvo of six torpedoes; hits turned the carrier at once into an inferno of AvGas fires. For three hours her crew tried to fight the blaze, but a final enormous explosion rent her hull and the *Shokaku*, veteran of Pearl Harbor and of every other Pacific carrier engagement with the exception of Midway, sank.

At about the same time, on board Ozawa's flagship *Taiho*, there was growing anxiety: as a result of the *Albacore*'s torpedo hit, the carrier was permeated by heavy AvGas fumes. Her captain gave the order to turn the ship's ventilation system on in an attempt to clear the vessel. The result was predictable: the fans spread the lethal gas and it was inevitably ignited. The explosion which followed set the vessel on fire from stem to stern and fractured the hull. *Taiho* was obviously lost; Ozawa transferred his flag to the cruiser *Haguro* to watch the newest Japanese carrier capsize and sink, taking with her 1,650 of her crew. The two carriers foundered with twenty-two aircraft still on board, thus increasing the Japanese losses for the day to 265 carrier-based aircraft destroyed in action or in accidents, plus another fifty or so over Guam and about fifty on the ground — a total of over 365.

During the night of 19-20 June 1944, as the exultant American pilots celebrated as best they could in the bone-dry US Navy wardrooms, even the ever-cautious Spruance felt able to go over to the offensive: leaving one group (TG58.4) off Saipan to cover the ground forces, he set off westwards to seek Ozawa. So far no US carrier strike had found the Japanese: throughout the battle on 19 June, the only strikes against the Japanese ships had been by US submarines. While Spruance was chasing the Japanese fleet through the night, Ozawa was under the mistaken impression that, in spite of having only about 100 serviceable planes aboard his ships, many of his aircraft, after attacking the US Task Force, had succeeded in reaching Guam and would thus be available, refuelled and rearmed, to support a counter-strike the next day. He was further persuaded by surviving pilots' reports that the US fleet had suffered crippling losses; thus he rejected the very sound advice of the far less gullible Vice Admiral Kurita, who had prudently advocated breaking off the action and returning to Japan with such aircraft and ships as were still operational, to fight on another, distant day.

It was not until late afternoon on 20 June that a search aircraft from

Enterprise first sighted the Japanese ships, just within striking range. Spruance made an uncharacteristic decision to launch an immediate strike, although by the time the attacking aircraft were ranged and ready to launch it was barely half an hour before sunset and the attack was to be made at the limit of the strike force's endurance.

The sun was well down as fifty-four TBF Avengers and seventy-seven Helldivers, escorted by eighty-five F6F Hellcats, took off from the US carriers. The strike was intercepted by forty A6M Zeros — probably the total available — which fought tenaciously to defend their ships. Six Hellcats, four Avengers and ten Helldivers were shot down for the loss of twenty-four Zeros, but the defenders were overwhelmed and several Avengers got within torpedo range of *Hiyo*, which was sunk. *Zuikaku* was hit by bombs from Helldivers, being badly damaged by a consequent serious fire which her crew managed to contain. *Chiyoda* was also dive-bombed, set on fire and her flight deck ruptured, though she too managed to remain afloat. *Ryuho* was less badly hit.

After the attack the US strike force turned and set course for their ships, but a freshening wind was now blowing against them; one by one the aircraft ran out of fuel and ditched. The remainder, flying slowly with engines running on the leanest possible mixture, got back to their ships as the last trace of daylight faded. The US Task Force, against all the rules and standing orders, fired star shells and flares, turned on navigation lights and even searchlights, but many of the exhausted pilots, untrained in night operations, crashed on landing. Of the few who did get down safely, many had their aircraft heaved over the side to make room for others approaching on their last drop of fuel. Altogether eighty aircraft and forty-nine aircrew were lost through ditching or deck-landing accidents; 160 men were picked up from the sea the next day by destroyers or submarines searching along the track taken by the returning aircraft.

While the drama of the night recovery was being played out on the US ships, Ozawa, with a total of only thirty-five flyable aircraft on board his shattered carriers, accepted defeat; he signalled his resignation to Admiral Toyoda in Tokyo and turned for Okinawa. The Battle of the Philippine Sea had been a victory for the Americans. The last of the Japanese pilots thoroughly trained to operate from carriers had gone; they simply could not be replaced in time to affect the now inevitable outcome of the Pacific War. The aircrew losses suffered in 'The Marianas Turkey Shoot' forced the Japanese into a bizarre alternative: Kamikazes. They were to appear in the next and virtually final Pacific naval action, the Battle of Leyte Gulf.

The retreat of Ozawa's fleet after the Philippine Sea action enabled the US Task Forces to mount widespread covering air strikes in support of the land forces. Saipan fell to the Americans on 9 July; Guam and Tinian were then invaded. The landings were successful enough to enable the Task Group to extend its air strikes to Japanese bases on Yap, Palau and Ulithi in the Western Caroline Islands. Guam was securely in US hands on 10 August 1944, and by that time carrier aircraft in these operations had accounted for 110,000 tons of enemy shipping and 1,223 Japanese army and naval aircraft.

So total was the American victory that the cabinet of General Tojo resigned and the US Chiefs of Staff, influenced by the admirals, came to the conclusion that the way was now open for a direct assault on Formosa (Taiwan), only 650 miles from the southern Home Island of Japan itself, there to establish B-29 Superfortress bases from which could begin strategic bombing of Japanese industry and the launching of a series of amphibious landings to take the key Ryukyu island of Okinawa, from which the final assault on Japan could be made. With the weakened state of the Japanese Navy it is probable that the undertaking would have succeeded. That plan, if adopted, would of course have bypassed the whole of the Philippines; this was totally unacceptable to General Douglas MacArthur, who had sworn to return and to liberate the Filippinos.

In July 1944 a meeting was held at Pearl Harbor between Roosevelt, MacArthur and Nimitz. MacArthur managed to convince Roosevelt that the taking of the Philippines would be a far less costly undertaking than Formosa, aside from any question of the ethics in honouring pledges previously given to liberate the Philippines. Reluctantly Nimitz agreed to what was something of a compromise: Luzon would be invaded and occupied; from there Iwo Jima would be taken, further bases set up, and then Okinawa in the Ryukyu Islands would supplant Formosa as the main base from which to attack the Japanese Home Islands. During the discussions the question of the possible use of the atomic bomb was not raised; indeed it is probable that neither Nimitz nor MacArthur even knew of its existence, so secret was the 'Manhattan Project'. On 4 October 1944 it was decided that the operation to occupy the Philippines would begin with the invasion of the island of Leyte on which MacArthur would set up a huge army, naval and air base, from which to launch a large-scale invasion to recapture Luzon and the remainder of the Philippines.

Prior to the landings on Leyte, the large US naval Task Force 58 was now under the command of Admiral Halsey and had been redesignated

Task Force 38; this was in accordance with a US Navy Staff decision whereby each Admiral — Spruance or Halsey — alternated; while one was in active command, the other was at Pearl Harbor with his staff planning the next major operation. Thus the Fifth Fleet, under Spruance, became the Third Fleet under Halsey, though Vice Admiral Mitscher remained in command at sea.

The opening moves of the Leyte campaign began on 10 October 1944 when seventeen carriers of TF38 launched massive air strikes to neutralise Japanese airfields on Luzon and, further afield, on Formosa, Okinawa and other Ryukyu islands. In five days, during these strikes, US Navy pilots were credited with destroying 438 enemy aircraft in aerial combat and a further 366 on the ground, for the loss of forty-eight Hellcats.[22]

The aircraft which the US pilots had encountered were mainly Zeros, but there were also two new types. One, the Japanese Army Air Force Nakajima Ki84 Hayate (Frank), was powered by a 1,900hp Nakajima Ha.45/1 Type 4 Radial; this 400mph fighter, in the hands of an experienced pilot, could outfly the F6F-5 Hellcat and, even when flown by pilots of moderate ability, was still a formidable opponent, possessing, unlike the Zero, adequate pilot and fuel protection. The other new fighter encountered was the naval Kawanishi N1K2-J Shiden (Violet Lightning) — Allied code-name George. This too was powered by a 1,900 Nakajima radial and was possibly the best all-round Japanese naval fighter of the war in 1944, but it was often flown by under-trained pilots and mounting production difficulties restricted the number of new aircraft available. The only counter at this stage to the multiple air strikes made on the Japanese airfields took the form of dusk air raids made on TF38 by massed G4M Army Bettys, which attempted to torpedo the fleet carrier *Franklin*; when the torpedoes all missed, one G4M tried to crash on to the carrier but it bounced off the *Franklin*'s deck into the sea. In another strike a single Japanese aircraft torpedoed the cruiser *Houston*, which was damaged and later was attacked and hit again, but just survived. In the twilight and dusk raids, forty Japanese attackers were shot down, but as usual the surviving pilots returning to base made exaggerated claims of ships and carriers sunk or left burning and heavily damaged.

The US air strike made on the Japanese airfields had cleared the way for the landings by the US Sixth Army, which were to be at two points on Leyte Island — Tacloban and Dulag — and were set for 20 October 1944. As in previous Pacific amphibious landings, carriers were to provide close cover for the troops; they were eighteen CVEs designated Task Group 77.4 (Vice Admiral T.L. Sprague), which contributed eighty-five

Hellcats, 219 F4F (FM-2) Wildcats (which were still embarked in CVEs) and 199 TBF Avengers. The initial landings were, surprisingly, only lightly opposed, and as the fighters ranged further afield to destroy remaining island-based Japanese aircraft, General Douglas MacArthur waded ashore to declare to the fortuitously present newsreel cameras: "People of the Philippines, I have returned."

For some time prior to the Leyte landings it was not clear to the Japanese Naval Staff if the Philippines or Formosa was the principal American objective. When the Leyte operation clarified the position, Admiral Toyoda had to decide whether to keep his fleet intact in home waters to defend Japan itself, or to take the risk of exposing his weakened resources in an all-out attack on the Americans off Leyte. The latter plan was adopted. To achieve the hoped-for decisive victory, the Japanese fleet was divided into three main groups which would be used in a typically complex Japanese strategy. Group 'A' under Vice Admiral Kurita, flying his flag in the cruiser *Atago*, with a force of five battleships, including the *Yamato* and *Musashi*, and nine cruisers, screened by fifteen destroyers and light cruisers, would sail on 22 October from Brunei in North Borneo for Leyte via the South China Sea, Mindoro Strait and the Sibuyan Sea, through San Bernadino Strait, rounding Samar, to approach Leyte from the north by 25 October. Force 'C' (Vice Admiral Nishimura), with the two remaining battleships, the cruiser *Mogami* and four destroyers, also sailed from Brunei on 22 October to pass through the Sulu Sea and Surigao Strait to attack the US landing forces off Leyte from the south, also on 25 October.

To reduce the opposition these two groups would undoubtedly have to face, particularly from US carrier aircraft, it was planned by Toyoda's staff that a Japanese carrier group under Vice Admiral Ozawa (his resignation offered after the Philippine Sea débâcle had not been accepted) would sail from the Inland Sea to a position 300 miles west of Luzon, to lure TF38 away from Leyte at the precise moment the other Japanese forces were in an attacking position. Ozawa had under his flag the carriers *Zuikaku*, *Zuiho*, *Chitose* and *Chiyoda* and, partly as an expedient and partly as a stratagem, the battleships *Ise* and *Hyuga*, whose after turrets had been removed and a short flight deck substituted: the wheel had turned full circle, for they therefore exactly mirrored the world's first aircraft carrier, HMS *Furious*, which had proved in 1918 that such a deck was unusable for landing, owing to turbulence. In practice this did not matter, since none of the unfortunate pilots embarked in any of Ozawa's carriers had received deck-landing training; they were expected to land ashore in the unlikely event of surviving their first combat. These makeshift air groups had been hastily formed to replace the losses sustained in the Battle of the Philippine Sea and were the survivors of the US air strikes over Formosa and Okinawa prior to the Leyte landings. Consequently only about 100 aircraft in all were embarked aboard the Japanese carriers.

Also sailing from the Inland Sea was the Second Striking Force (Vice Admiral Shima), which consisted of two cruisers, a light cruiser and four destroyers. This fast group was to sail through the Sulu Sea to join up with Nishimura's Force 'C'.

The stage was now set for what the US Naval historian S.E. Morison described as "the greatest naval battle in history".[23] It was to involve 282 warships (thirty-eight more than at Jutland).

The Battle for Leyte Gulf began badly for Japan. At dawn on 23 October two US submarines, *Darter* and *Dace*, were lying surfaced at the southern entrance of the Palawan Passage, close enough for their captains to talk with loud hailers, breaking the endless tedium of a long spell of patrol duty. The conference was abruptly terminated when the submarines' radar sets painted multiple echoes from an approaching fleet, at a closing range of 15 miles. *Darter* immediately transmitted a sighting report which Admiral Halsey received at 0620 hours.

The two submarines then dived and manoeuvred into attacking positions. The enemy ships were in fact Vice Admiral Kurita's Force 'A', zigzagging through the shoal waters off Palawan. At a range of 1,000 yards *Darter* (Commander David H. McClintock) fired a full salvo of six torpedoes from his bow tubes at Kurita's flagship, the cruiser *Atago*. The time was 0632. Four of the torpedoes hit, setting the *Atago* ablaze. *Darter* then fired four more torpedoes at the *Takao*, securing two hits. The other submarine, *Dace* (Commander Claggett), hit the heavy cruiser *Maya* with four torpedoes.

Within twenty minutes of the attack *Atago* and *Maya* were sunk and *Takao* was limping back to Brunei escorted by two destroyers. Vice Admiral Kurita actually had to be picked out of the water by a destroyer and did not get aboard to hoist his flag in *Yamato* until late afternoon. It was not an auspicious start to the Japanese operation, for it now had to be assumed that the Americans were aware of the presence of all the major Japanese naval units; in point of fact it was not until mid-afternoon of 24 October that Halsey discovered the whereabouts of Force 'C' and the Second Striking Force; the existence of Ozawa's carriers was as yet unknown.

Following the submarine attacks, a Helldiver from USS *Enterprise* sighted Kurita's Group, steering east into Tablas Strait for the Sibuyan

Sea; at 0746 the SB2C signalled "Enemy in Sight". Halsey, from his flagship *New Jersey*, at once ordered a strike. Part of his Task Force was at that moment fighting off an attack from what were assumed to be shore-based aircraft from Luzon (in fact some of the aircraft attacking were from Ozawa's undetected carriers). Only twelve Helldivers, and twelve TBF Avengers armed with torpedoes, escorted by twenty-one F6Fs were launched at 0910; as these aircraft left the flight decks of *Intrepid* and *Cabot* and set course for Kurita's ships, the other carriers of TF38 were still under heavy attack. The strike was, however, delivered by unskilled pilots and no hits were obtained on the American ships. Hellcats from *Essex* shot down nine torpedo planes and fourteen Zeros. Fighters from *Lexington*, and the light carriers *Langley* and *Princeton*, disposed of at least fifty-two Japanese aircraft. Not one co-ordinated attack was made on any of the American carriers, which had used heavy rain showers to provide a measure of cover. Among the inexperienced Japanese pilots, however, there was at least one veteran. This man, flying a Judy, cleverly used the rain clouds for cover and dived undetected over the *Princeton*, dropping a 550lb bomb from 1,000 feet, which penetrated the flight deck and exploded in the hangar, destroying aircraft and setting their fuel ablaze, which detonated armed torpedoes still slung beneath aircraft. For two hours the *Princeton*'s crew fought the fires, aided by the cruiser *Birmingham*, which came alongside to assist. At that moment the *Princeton*'s torpedo store exploded; huge metal fragments scythed down 224 men on the cruiser's deck and seriously injured a further 480. The *Birmingham* backed away and *Princeton*'s surviving crew were ordered to abandon ship, which was scuttled half an hour later.

However, the Japanese attack did not prevent the launching of further US air strikes against Kurita's battleship force, even though, lacking a carrier escort, every ship in it had its anti-aircraft defences increased to a maximum practicable. Even the great 18.1 inch naval guns on the battleships *Yamato* and *Musashi* were brought in to be used for anti-aircraft splash fire.

The first wave of US aircraft arrived over the Japanese battleships at 1026 to open one of the many engagements off Leyte Gulf: the Battle of the Sibuyan Sea. The US carrier pilots attacked Kurita's armada in the Tablas Strait. The weather was perfect: a cloudless blue sky, which soon became blotched by the black puffs of the intensive AA barrage as Avengers and Helldivers attacked the huge battleships *Musashi* and *Yamato*. Two Avengers were shot down; the remaining aircraft launched torpedoes and bombs. The heavy cruiser *Myoko* was torpedoed, badly damaged and was ordered to return to Brunei. *Musashi* and *Yamato* both took several hits from torpedoes and bombs but their massive hulls seemed impervious and they sailed impassively on.

During the day TF38, with the *Intrepid, Cabot, Franklin, Essex* and *Enterprise*, launched further strikes against the battleships in the Tablas Strait. Avengers put eight torpedoes into *Musashi*, causing her to drop astern from Kurita's group; the huge battleship was then so badly damaged that she was no longer operational and was therefore ordered to withdraw and make for Brunei. Further co-ordinated strikes were made and the retreating *Musashi* took at least ten more torpedoes (some sources put the figure at nineteen); even after that the warship remained afloat but, four hours later, rolled over and sank into the Sibuyan Sea with over 1,000 of her crew still aboard, without ever firing her monster 18 inch guns at an enemy ship. Eighteen of the US carrier aircraft were lost to anti-aircraft fire during the attacks.

In the absence of the expected Japanese land-based air cover from Luzon or Ozawa's carriers, Kurita decided to reverse his course from the confines of the Tablas Strait into the Sibuyan Sea as night was now falling. The news of the large battle-fleet's withdrawal reached Admiral Halsey at the same time as the sighting of Ozawa's decoy carriers away to the north-east. The lure of the Japanese carriers proved irresistible, particularly so as Halsey assumed that Kurita was withdrawing because of massive damage to his fleet; returning pilots had reported "at least four and possibly five battleships torpedoed and bombed, one probably sunk, a minimum of three heavy cruisers torpedoed and bombed". In the light of such reports Halsey believed that Kurita's force was no longer a menace and, as an ex-aviator, he had scant respect for battleships. Japanese carriers, on the other hand, were a different proposition. Eager to destroy the last of them — Pearl Harbor's memory was still green — Halsey at 2000 hours decided to steam north with three carrier groups of TF38 to attack Ozawa's ships. The bait had been taken.

Halsey departed with practically the whole of the American Third Fleet: he had no intention of repeating, as he saw it, the mistake made by Spruance after the Battle of the Philippine Sea, in allowing Japanese carriers to escape. The result of Halsey's dash north was to leave nothing guarding the San Bernadino Strait; the only US naval force left in the vicinity of Leyte was the Seventh Fleet under Vice Admiral T.C. Kinkaid, mainly support units comprising old battleships, including Pearl Harbor survivors and escort carriers which had been covering the Leyte landings. Kinkaid was now in a vulnerable position, for, with the appearance of Kurita's ships in the Tablas Strait, he had concentrated most of his available fleet to meet the Japanese battleships, assuming that the battleships of Halsey's forces would still be covering the northern end of the San Bernadino Strait. Unknown to Kinkaid, however, Halsey had taken all his ships to meet Ozawa, a fact that was not made clear to Kinkaid; nor of course could he know that Kurita, who had lost only three ships, had withdrawn merely to get out of range of the US carrier aircraft during daylight and would shortly return.

During the night of 24-25 October Kinkaid mistakenly assumed that the only Japanese ships remaining anywhere near Leyte were Force 'C', and the Second Strike Group; these had been sighted during the afternoon of 24 October, steaming through the Sulu Sea for Surigao Strait and presumably Leyte Gulf. Kinkaid decided to attack this group.

The night action which followed, the Battle of Surigao Strait, was one of the few in the Pacific conducted by surface vessels without aircraft being involved. The superior radar of the US ships under Rear Admiral Olendorf enabled them to rout the Japanese battle-fleet before daylight. Two Japanese battleships and three destroyers were sunk and other ships badly damaged, including the cruisers *Mogami* and *Nachi*, which had collided. The only US ship damaged in the Surigao Strait action was the destroyer *Albert W. Grant*, which, having hit the Japanese battleship *Yamashiro* with torpedoes, had then come under accidental American fire, with heavy damage resulting.

At dawn on 25 October, aircraft from the Seventh Fleet's CVE groups operating off Leyte searched for Japanese stragglers from the night action. Avengers found the cruiser *Mogami* barely under way in the Mindanao Sea: the torpedo attack they delivered inflicted further damage; her crew abandoned her and a final torpedo sank this much attacked vessel.

It seemed to Kinkaid that the US forces had achieved another total naval victory. The night action had shattered Nishimura's and Shima's forces; the survivors were now limping back to base following, it was assumed, Kurita's routed fleet. The Vice Admiral signalled his congratulations to the victors. Hardly had this been acknowledged when the radio operators on his Seventh Fleet Flagship, *Wasatch*, received the disturbing news that US escort carriers operating cover for the beach-heads on Leyte had just sighted the fighting tops of Japanese battleships approaching over the horizon. The next act in the complex Battle for Leyte Gulf was about to begin: the Battle off Samar.

The battleships were, of course, those of Kurita's fleet so prematurely written off by Halsey. After withdrawing from the Sibuyan Sea the previous day, Kurita had, during the hours of darkness, slipped unopposed through San Bernadino Strait while the big night battle was being fought 200 miles to the south. By daylight, Kurita's very considerable fleet, including the *Yamato* and three other battleships, had rounded Samar and had only a group of 18 knot, under-armoured CVEs between himself and MacArthur's massed and vulnerable invasion fleet. The weather was perfect for gunnery: a flat calm with unlimited visibility. At 0659 hours the first Japanese salvoes were fired on the group of escort carriers: *Fanshaw Bay*, flying the flag of Rear Admiral Clifton Sprague, with *St Lo*, *White Plains*, *Kalinin Bay*, *Kitkun Bay* and *Gambier Bay*. Sprague, in the face of overwhelmingly superior enemy forces, at once took the only possible course of action; he ordered his ships to withdraw at flank speed, covered by smokescreens from the destroyer escort, and flew off all his Avenger aircraft that were serviceable. Sprague then signalled Kinkaid for assistance, but his Seventh Fleet ships were up to three hours' fast steaming away, following the night action in Surigao Strait. Kinkaid at 0830 sent an appeal to Halsey: "Urgently need fast battleships Leyte Gulf at once." The message was repeated at 0900 — this time in the clear, to try to avoid any delay or ambiguity that coding might cause. The answer was that Halsey's carriers of TG38.1 would help but were too far away to intervene in time to affect the expected outcome of what looked like being a very short, one-sided engagement.

What actually happened is now legendary. Sprague ordered his three screening destroyers, *Hoel*, *Johnston* and *Heermann*, to attack the vastly superior Japanese fleet with torpedoes. The captain of the nearest destroyer, *Johnston*, Commander Ernest Evans, at once gallantly headed for the enemy. The scratch force of Avengers, which Sprague had flown off his carriers, were able to distract the Japanese guns to some extent and managed to hit the cruiser *Suzuya*, causing her to drop out of line; bombs also struck the cruiser *Kumawo*. Under cover of the Avengers' attack, USS *Johnston* got to within 10,000 yards to launch ten torpedoes; the *Kumawo* was hit and set on fire, dropping out of the action. The other US destroyers joined in the attack, including the light escort ship *Samuel B. Roberts*. Having expended all their torpedoes — and the *Roberts* had only three — the US ships then fired their 5 inch guns at the Japanese battleships; the return fire of armour-piercing shells from 14 inch guns actually went straight through the *Johnston*, to explode on contact with the sea. The destroyer was fearfully damaged but still managed to fire her remaining five torpedoes at the heavy cruiser *Haguro*; although they missed her, in manoeuvring to avoid them, the *Yamato* and *Nagato* became detached from the main fleet. Other torpedo attacks from the selfless destroyers caused many of the Japanese ships to steer away to comb torpedo tracks while all the time the CVEs were drawing further away.

At 0855, USS *Johnston*, engaged at point-blank range by two heavy cruisers, her engines wrecked, her captain and most of his crew dead, sank. *Samuel B. Roberts* took a full salvo of 14 inch shells from *Kongo*, sinking an hour later, as did *Hoel*. *Heermann* which was badly damaged and three other destroyers, *Butler*, *Dennis* and *Raymond*, all took part in the action but miraculously none of these was sunk. The 5 inch gunfire from the destroyers was having some effect on the Japanese cruisers, but, at 18,000 yards, the heavy guns of the battleships were now ranged on the CVEs. The first to be hit was the *Kalinin Bay*; she was damaged but remained afloat and steaming. The *St Lo* and *Fanshaw Bay* were desperately manoeuvring in a forest of tall shell splashes; *St Lo* was hit by 8 inch shells, but though damaged, like *Kalinin Bay*, was able to steam at speed. *Gambier Bay*, after avoiding shells for twenty minutes was badly hit by fire from *Chikuma*, which closed to 10,000 yards; other Japanese cruisers shifted fire on to the crippled carrier. At 0907 *Gambier Bay* (CVE 73) capsized and sank.

The Avengers, launched in haste as soon as Kurita's force appeared, had been armed for quite different missions; some had depth charges, others rockets, but some were carrying bombs. They all attacked the Japanese ships, crippling the *Suzuya* and hitting the cruiser *Haguro*. These attacks, together with the destroyers' action, all helped to delay the onslaught on the carriers; this delay was to prove priceless. Kurita, on board *Yamato*, which was well in the rear of the action, having had to manoeuvre to avoid *Johnston*'s torpedoes, had no clear picture of the action; he became convinced that the CVEs his forward elements were engaging were in fact Halsey's Third Fleet carriers and that the motley collection of Avengers dropping everything from depth charges to signal flares were the van of a full-scale attack. He gave the order at 0925 for his ships to withdraw.

As the Japanese ships turned away, the astonished Americans simply could not believe what they saw. The carriers could now make repairs and land-on their aircraft, in the knowledge that within a few hours reinforcements from TG38 should be within striking range.

Having withdrawn, Kurita pondered his next move for three hours, then turned towards San Bernadino Strait at 1230. Half an hour later he was being attacked by waves of aircraft from *Wasp*, *Hornet* and *Hancock* of TG38.1. The relieved CVEs regrouped and congratulated themselves on what appeared to be an almost divine deliverance.

Divinity in a very different form was now about to manifest itself on the escort carriers of Rear Admiral Sprague. The CVE *Santee* was with the southernmost group; she had just rearmed her aircraft, which were about to take off again to attack Kurita's ships, when a Zero carrying a 550lb bomb dived out of a thin cloud, crashing through the wooden flight deck to explode inside the hangar. A fire at once broke out, which damage control was fighting when the ship reeled to another tremendous explosion; she had been torpedoed by the submarine *I-56*. Incredibly the *Santee* not only remained afloat but later that day was capable of operating her aircraft. The escort carrier *Suwannee* also survived a crashing Zero on her flight deck without sustaining fatal damage, which, considering that both carriers were built on merchant ships' hulls, was a tribute to their builders.

The northern CVE group, hardly recovered from the surprise of the withdrawal of Kurita's battleships, also were attacked by bomb-carrying Zeros. *Kitkun Bay* was lucky: a Zero hit her flight deck and bounced off to explode harmlessly in the sea. Anti-aircraft gunners got two others making for the CVEs *Fanshaw Bay* and *White Plains*. The *St Lo* was hit, though; a Zero tore into her hangar and detonated in a torpedo storage space, the resulting explosion ripping the lightly-built ship apart, and sinking her within thirty minutes. Yet another CVE, *Kalinin Bay*, survived two hits, and after that the sky remained clear until the end of that crowded day.

The Zeros crashing on the carriers were of course Kamikazes. They were the result of Vice Admiral Takijiro Onishi, Commander of the First Air Fleet, based on Clark Field, near Manila, combining medieval mysticism with twentieth-century pragmatism. Onishi, newly appointed to the command, realised that his so-called First Air Fleet consisted in the main of outclassed aircraft flown by half-trained recruits whose chances of returning from air strikes against the US carriers were virtually nil. Since they were practically certain to die, why not take a carrier with them, by crashing on to its flight deck with a 550lb bomb?

Addressing his men, Onishi, after stressing the desperate nature of the war now being fought off Leyte and the threat it posed to Japan itself, then called for volunteers to form a 'Special Attack Unit'; the pilots volunteered to a man. The name 'Kamikaze', which has since passed into the English language, was a reference to a 'Divine Wind' which had in 1281 scattered the fleet of Kubla Khan, which was about to invade Japan.

From Onishi's point of view it had been a successful début: one CVE sunk and three badly damaged for the expenditure of eight obsolescent Zeros — and of course their pilots. Considering the fifty or so aircraft destroyed to sink the *Princeton*, it probably made sense to Onishi and his Samurai-imbued men. Thus a new phase in the Pacific War had begun. Although Sprague's hard-pressed CVEs were the first to suffer the Kamikaze attacks, they would by no means be the last.

The final act of the Leyte struggle was to be the last carrier v. carrier action in the Pacific: the Battle off Cape Engano. This was fought on the same day — 25 October — as the astonishing events off Samar, which had so very nearly ended in disaster for the Americans. The two actions were, of course, linked, as indeed were all the phases of the Leyte engagements.

The whole point of Ozawa's presence off Cape Engano was to lure Halsey away from Leyte to enable Kurita's battleships to attack the landing fleet. The Japanese had succeeded in this beyond their most optimistic hopes, for at dawn on 25 October, as Kurita's battle-fleet was rounding Samar, Halsey's search aircraft were 350 miles to the north, trying to regain contact with Ozawa's carriers, which had been lost during the night. As the search aircraft fanned out north-west and north, on the decks of the American carriers a maximum strength strike was ranged; 180 aircraft divided almost equally into Helldiver bombers, Avengers with torpedoes and Hellcats to provide the cover. At 0540 the first machines left the flight decks; the sky was clear and a fresh trade wind dappled the unbelievably blue ocean with white horses. Flight after flight took off and formed into attack groups, orbiting about 70 miles ahead of their carriers to await the reports from the search planes.

The first "Enemy in Sight" signal came at 0710 and the Air Groups set course for Ozawa's fleet. At that moment, 350 miles away to the south, Rear Admiral Sprague's CVEs were desperately using rain squalls to survive the massed gunnery of Kurita's battleships. On receiving news of this distant action, it must have seemed to Ozawa that the complex Japanese strategy was about to be fulfilled, for his carrier force, which Halsey was so vigorously pursuing, was an empty threat, a lure; there were no more than twenty-nine aircraft, only nineteen of them Zero fighters, left on board his four carriers and all of them flown by inexperienced pilots — inexperienced not only in combat, but in basic flying.

As the US search planes flew around shadowing his fleet, Ozawa, in *Zuikaku*, prepared to meet the expected strike and to sell his sacrifical fleet as dearly as possible. He did not have very long to wait. At 0810 hours the US attackers were sighted. The Japanese carriers increased their speed to 24 knots and steered north, drawing the US Task Force further away from Samar, where Ozawa imagined Kurita would now be engaged in a decisive onslaught on MacArthur's thinly defended invasion fleet.

As the US strike approached, from the decks of the Japanese carriers the pathetic handful of aircraft took off; the Zeros attempted an interception but were no match for the US fighter pilots. Just how many of the Japanese aircraft were shot down there and then no one really knows; some may have managed to escape to land ashore, but within minutes of the attack starting there was no further aerial opposition, though heavy and light anti-aircraft fire was intense, claiming several American aircraft.

The first ship attacked was the light carrier *Chitose*; Helldivers hit her so severely that she was soon lying dead in the water. Next the air group turned on *Zuiho*, which managed to avoid two torpedoes but not a Helldiver's bomb. *Zuikaku*, the flagship, was torpedoed to a standstill and lost all her communications, which caused Ozawa to shift to the cruiser *Oyoda*. In that first strike, in addition to the hits on carriers, the destroyer *Akitsuki* was sunk.

For Admiral Halsey, reports from his Air Groups must have made sweet reading; at last the Japanese carriers were cornered and this time they would not — as in the Philippine Sea action — escape. Halsey was first and foremost an attacking admiral and his hatred of the 'Japs' was very real. The fact that Ozawa's force contained as it did *Zuikaku*, the last of the Pearl Harbor attackers, gave Halsey particular satisfaction. He should perhaps have reflected on the old Italian maxim, 'Revenge is a dish best taken cold'.

As Ozawa's ships were under attack, Halsey received the first chilling signal from Kinkaid that Sprague's CVEs were under attack from a powerful Japanese battle-fleet and that immediate aid was required. The fact that one of the messages had been sent in plain language lent weight to its urgency. Kinkaid, in further messages, requested — implored might not be too strong an interpretation — that "Lee proceed top speed to cover Leyte . . . ."

This signal made Halsey realise for the first time that Kinkaid was under the impression that the battleships of TG34.1 (Vice Admiral W.A. Lee) were still guarding the San Bernadino Strait. They were not. All six of them — including the *New Jersey*, Halsey's flagship — were at that moment in the van of TF38's carriers 400 miles to the north and steaming to intercept Ozawa's fleet, eager to use their new 16 inch radar-laid guns.

As the signals from Kinkaid continued with mounting urgency, Halsey's main reaction was one of frustration and anger. Why should he release his ships at the very moment they were poised to deliver a decisive blow at the last remaining Japanese carriers just below the horizon? How could Kinkaid claim that the survivors of Kurita's routed and shattered fleet posed a dangerous threat? Halsey's only concession to the Seventh Fleet's entreaty was to send to Sprague's aid Task Group 38.1 (Vice Admiral J.S. McCain), comprising *Wasp*, *Hornet* and *Hancock*, with the light carriers *Monterey* and *Cowpens*, which had been detached to refuel 400 miles east of Leyte and had consequently been the only carrier group not to sail in pursuit of Ozawa.

Lee's ships increased speed to 25 knots and sped on towards the Japanese carriers. A second air strike was launched from Halsey's carriers. This wave set the light carrier *Chiyoda* ablaze and disabled her engines; two light cruisers were also hit, one of them, *Tama*, falling behind the main fleet trailing oil. Meanwhile, one of the victims of the first strike, the light carrier *Chitose*, had sunk while under tow.

Ozawa's ships, of which several were damaged, were now spread out over 20 miles of ocean. Before a third air strike could be launched, however, Halsey, on direct orders from Nimitz, was goaded into detaching Lee's six battleships, including the *New Jersey*, accompanied by TG38.2 (Rear Admiral Bogan) with the carriers *Intrepid*, *Cabot* and *Independence* for air cover. This group broke off the pursuit of the Japanese carriers, turned 180° and steamed at full speed to intervene in the battle off Samar. Thus the fiery American admiral was virtually forced to repeat Spruance's much criticised action in dividing his forces just when a decisive victory was within his grasp.

In strictly military terms and aside from any other consideration, Ozawa's carriers were in reality a spent force; carriers without aircraft are as guns without ammunition, and the remaining US ships in pursuit would be entirely adequate to finish them off. At 1410 *Zuikaku* ended an operational career that had begun at Pearl Harbor, when Avengers from *Lexington* hit her with three torpedoes and she rolled over and sank. *Zuiho* followed her to the bottom of the Pacific about an hour later. By then Ozawa knew that Kurita's sortie had failed at the moment when a great Japanese victory was not only possible but probable; the sacrifice of the last carriers was to be in vain.

Halsey was to be disappointed too; the Task Groups arrived too late for a night action in the San Bernadino Strait. However, Kurita's ships were pursued next day through the Mindoro Sea, but only the cruiser *Noshiro* was sunk. The battleships *Yamato*, *Kongo* and *Nagato*, with several cruisers and destroyers, escaped back to Japan. It did not really matter; at the conclusion of the Battle for Leyte Gulf, the enemy had lost twenty-six warships, including four carriers (a total of 300,000 tons), and

1,046 aircraft. The Imperial Japanese Navy had ceased to exist as a cohesive fighting force. The escaped battleships, bereft of air cover, would be impotent; never again would the United States Navy be challenged by Japanese ships at sea.

The Pacific War, though, was far from over; it was to continue for another ten months and to witness some of the bitterest fighting, as the Japanese Army was driven from Burma, through the Philippines, from large islands and minute atolls, across the Pacific. Everywhere their soldiers fought with a fanatical bravery, few surrendering. The losses to the US ground forces were very high.

The US Navy carriers supported the numerous amphibious landings, though now opposed only by Japanese land-based aircraft, mainly on the Kamikaze missions. The greatest losses at sea had been three destroyers foundering and four light carriers and several other ships damaged in a severe hurricane on 18 December; but neither Kamikaze nor typhoons could stem the American advance.

In January 1945 Luzon was invaded. During that operation, between 3 and 22 January, Kamikazes were to prove a serious menace and in unprecedented attacks they damaged several ships, including the CVEs *Ommeney Bay*, *Manila Bay*, *Savo Island*, *Kadashan Bay*, *Kitkun Bay* and *Salamau*.

During the land fighting the size of the island attacked in no way lessened the severity of the opposition. The US Air Force had urgently required an intermediate base between Tokyo and Tinian, in the Marianas, to enable their valuable B-29s, facing a 1,500 mile journey home, to set down if damaged over Japan. The island selected was a volcanic island in the Bonin Group: Iwo Jima, which was roughly halfway from Tinian to Tokyo. It was to take US Marines a month, and 5,500 men killed in some of the bitterest fighting in the Pacific, to take the island. (Joe Rosenthal's classic photograph of the Marines raising the Stars and Stripes on Mount Suribachi is one of the great war pictures, and a bronze based on it forms the Marine War Memorial in the Arlington Military Cemetery.) The first damaged B-29 landed on Iwo Jima's runway before it was even completed: by the end of the war 2,400 B-29s had used the emergency field.

During the Iwo landing, Kamikazes again attacked and sank the escort carrier *Bismarck Sea* and damaged the veteran *Saratoga*, which, not for the first time in her long career, had to return to the United States for repairs.

Kamikazes would form a more serious threat during the last, and largest, amphibious campaign fought by the US Navy's carriers: the invasion of Okinawa. Task Force 58 (Vice Admiral M.A. Mitscher) began the assault on 18 March with ten fleet and six light carriers, flying neutralisation attacks on Kyushu in Japan, and destroying 482 enemy aircraft. The Task Force had then covered the Okinawa assault on 23 March. Kamikazes from numerous bases on Formosa made concentrated strikes in seven distinct raids with Aeros and Baka flying bombs, which were carried under G4M Bettys. Although TF58 lost no carriers sunk, eight fleet carriers and one CVE were hit, some more than once, and damaged.

During the Okinawa campaign carrier air support was on a very large scale. Carrier aircraft flew over 40,000 sorties, destroying 2,516 enemy aircraft and attacking enemy positions ashore with 8,500 tons of bombs and more than 50,000 air-to-ground rockets. One fleet carrier, USS *Essex*, created a record for remaining on continuous active service for seventy-nine consecutive days.

Operating south of Okinawa was Task Force 57 of the Royal Navy (Admiral Sir Bernard Rawlings), formed round the armoured carriers *Indomitable*, *Victorious*, *Illustrious* and *Indefatigable*, with escorts of cruisers and destroyers. The participation of the British Fleet in the Pacific had been opposed by some American admirals; but they were overruled by Roosevelt. One can see their point of view; the Pacific naval war had been and was continuing to be an American affair. Senior American officers suspected the motives of the British appearing at the eleventh hour as being largely political. There were other valid practical objections; in the first place, the nearest Royal Navy base was Sydney, Australia, 3,580 miles away (though that problem was solved by using CVEs to ferry replacement aircraft and spares). British carriers were smaller and slower than the US fleet carriers, with a reduced aircraft complement, and they were less skilled at replenishment at sea; the ships of the Royal Navy had never been designed to operate for indefinite periods away from base, particularly in the tropics. In European waters they were seldom away from port for more than a week or so. All the British communication and radio equipment, coding systems, signals techniques, had to be changed. Many of the aircraft embarked on the RN carriers were unknown types to the Americans, and recognition flights had to be arranged. Spares were a problem; even the US aircraft aboard the British ships were often of a different mark from the standard US types and therefore required special spares.

One by one the difficulties were overcome, with generous help from the Americans. Seafires, for example, were fitted with drop tanks designed for USAAF P-40s, to try to increase their chronically short endurance. (The one aspect of the participation of the British which was whole-heartedly endorsed by the parched men on the American ships was the availability of spirits, the unoffical rate being 4 gallons of ice cream to a bottle of Scotch.)

The first operation of the Royal Navy's TF57, which, although under overall American command, operated independently of the US fleet, was on 26 March, when the first strike against Japanese airfields on Sakishima Gunto, in the Ryukyu Islands, was mounted. Avengers armed with 4 x 500lb bombs hit runways and hangars while Fireflys strafed the airfields and installations; Hellcats and Corsairs escorted the strikes. It was difficult to assess the results of this and subsequent strikes; airfields are not easy targets to destroy, particularly in the Pacific, where the runways were simply packed coral — a material in inexhaustible supply. The Royal Navy continued with strikes against enemy airfields, fuel supplies, railways and communications; all four RN carriers were struck by Kamikazes but their armoured decks were strong enough to withstand the bombs and all four ships remained operational, to earn a special commendation from Admiral Halsey.

While the Okinawa battle was still being fought, there occurred the last action of the Japanese Navy: the Battle of the East China Seas. It was a one-way sortie of the battleship *Yamato*, for she had insufficient fuel to return and, though escorted by a handful of destroyers, had no air cover whatsoever. Thus the biggest battleship in the world paid the inevitable price, being attacked by wave upon wave of US carrier aircraft, which put at least twelve torpedoes into her hull. No ship, not even the 68,000 ton *Yamato*, could withstand such punishment; on 17 March 1945, at 1423 hours, she rolled over and sank — the very end of the battleship era. Carrier planes from the US and Royal Navy roamed over the Tokyo plains and the Inland Sea. In these final strikes of World War II carrier aircraft destroyed 1,223 enemy aircraft, but owing to chronic fuel and pilot shortages in Japan, over 1,000 of these were on the ground.

On 6 August 1945 a B-29 Superfortress *Enola Gay* — the name of the pilot's mother — took off from Tinian in the Marianas and set course for the Japanese city of Hiroshima . . . . The atomic bomb she dropped destroyed 5 square miles of built-up area, killing 71,000 people instantaneously. No offer of surrender came from the Japanese Government and on 9 August another B-29 dropped an atomic bomb on Nagasaki, killing a further 80,000 people. Emperor Hiro Hito intervened and on 14 August Japan agreed to unconditional surrender. Next day all operations in the Pacific ceased.

Nineteen days later, on board the battleship USS *Missouri*, lying in Tokyo Bay, General Douglas MacArthur put his signature to the instrument of unconditional surrender. It was 2 September 1945, six years and one day from the time German Stukas had dropped the first bombs of World War II on Poland.

Although MacArthur accepted the Japanese surrender on board a battleship, it should have been on a carrier, for they had borne the brunt of the war in the Pacific. Carriers had also fought in every ocean of the world, from escort duties in the Arctic to hard-fought convoys in the Mediterranean, and from the Battle of the Atlantic, across the Indian Ocean to the immense Pacific. The Royal Navy had operated forty-eight carriers, of which eight had been sunk. The US Navy had commissioned 150, of which twelve were lost through enemy action. The Japanese alone had lost 15,000 aircraft and 174 ships to Navy and Marine air attacks. Of the twenty-four carriers of the Imperial Navy completed during the war years, twenty had been sunk; the only one left completely intact was the very first: *Hosho*.

Soon the carriers and the ships sailed home and the hard-won airfields abandoned on the Pacific islands were reclaimed by the inexorable jungle. Most of the British and Americans who served aboard the carriers were returning to civilian life, thankful to be still alive; a good many of them were not — they lay with their sunken ships and aircraft. It had been a long and bitter struggle and the survivors could well have echoed Kipling's 'Song of the Dead':

> If blood be the price of admiralty,
> Lord God, we ha' paid in full!

5

Into the Nuclear Age

When World War II finally ended with the defeat of Japan, the aircraft carrier was the undoubted capital ship of both of the victorious Navies. The Royal Navy, in six years of unremitting warfare, had commissioned forty-eight, having begun the war with seven. The United States Navy had in commission or under construction twenty-six *Essex* class, eight *Independence* class, seventy-eight escort carriers and three of the new *Midway* class.

For a short time following the cessation of hostilities many of these vessels sailed from all over the world back to Britain or the United States to be sold, placed in reserve fleets, or simply scrapped. Most of Britain's large fleet of escort carriers were returned to America, which had built them under the terms of the Lend-Lease agreements; not only were the ships themselves returned but most of the 737 US aircraft embarked on them — over half the strength of the Fleet Air Arm. To be strictly accurate, many of these American types were simply pushed over the side, for the United States did not want them, and if the Royal Navy had retained them, they would have had to be paid for. Hundreds of Corsairs, Hellcats and Avengers, particularly those from the Pacific, were therefore dumped. Those that were returned, including thousands of the US Navy's own aircraft, were soon scrapped.

The facts were that no one really knew the future role to be played by aircraft carriers in the very changed circumstances of the post-war world. Nuclear weapons and jet aircraft had arrived and, although the latter had played only a small part in the European war, obviously jets were to be the warplanes of the future.

Apart from the ships and aircraft, there was an immediate and dramatic reduction in both navies in the 'hostilities only' men; the vast majority of these temporary sailors stepped ashore for good. The run-down of naval air strength can be gauged from the following figures. They are for the US Navy, simply because its figures are readily available:[1]

Carriers in service 1 July 1945:   20 *Essex* class (CVs)
                                       8 *Independence* class (CVLs)
                                       70 Escort (CVEs)
                          Total   98

Aircraft on hand 1 July 1945:   40,912 (Total US Navy all types)

Carriers in service 1 July 1946:   12 CVs
                                       1 CVL
                                       10 CVEs
                          Total   23

Aircraft on hand 1 July 1946:   24,232 (many of which were held in storage).

Although the remaining Allied carriers and aircraft were adequate for the immediate post-war needs, mainly in support roles for occupation forces, naval staffs were appraising the future role of naval aviation. The fully armoured fleet carriers which the Royal Navy had operated had proved their value, certainly in the Pacific, where the four British ships had suffered little serious damage from the Kamikazes, a fact which had begun to influence US naval architects even before the war ended; the successors to the American *Essex* class, which, though excellent carriers, had proved vulnerable to that extreme form of attack, were to be three armoured *Midway* class ships, CVBs: *Midway* (CVB-41), *Franklin D. Roosevelt* (CVB-42 and originally to be named *Coral Sea* but on the death of the President renamed), and the eventual *Coral Sea* (CVB-43). The first two were commissioned in 1945, the third in 1947. These carriers were of 45,000 tons displacement and incorporated many of the lessons acquired in the Pacific. The decks were fully armoured and the ships heavily compartmented to avoid a repetition of the wartime losses from single torpedo hits due to progressive flooding. Although the *Midway* class ships were the first post-war US carriers to be commissioned, the first two had been laid down during 1943.

The *Midway* class comprised very large carriers, only just capable of passing through the Panama Canal. Their overall length was 968 feet, the flight deck being 136 feet wide. During her trials *Midway* achieved 33 knots, driven by four propellers with a shaft horsepower of 212,000. During a shakedown cruise in November 1945 she embarked fifty-seven F4U-4 Corsairs, fifty-nine SB2C-4E Helldivers and four F6F Hellcats — 120 aircraft, though the ship could accommodate a total of 137 if required.

The Royal Navy had, during the war years, to a large extent relied on the shipyards of the United States to provide the escort carriers which

were essential for convoy protection, especially in the Atlantic, across
which of course all the supplies for the American forces in Europe had to
pass. No fleet carriers were made available to the Royal Navy from the
US; the *Illustrious* class of six vessels had been constructed in British
yards. They were to have been followed by three 45,000 ton carriers of
the *Gibraltar* class, which was to have comprised *Gibraltar*, *New Zealand*
and *Malta*. These would have been roughly equivalent to the US Navy's
Midway class. In addition, there were four *Audacious* class carriers of
34,000 tons and eight light carriers of the *Hermes* class, 18,300 tons; six
Majestic class of 14,000 tons and a further ten light fleet carriers
displacing 13,190 tons, comprising the *Colossus* class, completed the
ambitious programme.

The massive *Gibraltar* class never left the drawing boards; of the
Audacious class, only two were laid down, to become *Eagle* and *Ark
Royal*, which were commissioned in 1951 and 1955 respectively. Four of
the *Hermes* class were completed between 1954 and 1955: *Centaur*,
Albion, *Bulwark*, and *Hermes* herself, though she was not commissioned
until 1959. The six *Majestic* carriers were laid down, though only five
were completed; the hull of the sixth was launched but not fitted out.
The ten *Colossus* light fleet carriers were constructed and *Colossus*,
Venerable and *Vengeance* arrived in the Pacific just too late to see action
against the Japanese. Post-war only four of the class — *Glory*, *Ocean*,
Theseus and *Triumph* — served with the Royal Navy (the others were
sold abroad to the navies of the Netherlands, France, Brazil and the
Argentine; two built as RN maintenance carriers were scrapped).

The four *Colossus* Royal Navy carriers, despite the class name, were
small by wartime standards but were to form the backbone of the Fleet
Air Arm's post-war activities. The ships were not armoured; indeed the
hulls were built to mercantile standards. Interestingly enough, although
these small vessels displaced 10,000 tons less than the heavily armoured
British *Illustrious* fleet carriers, they were little shorter and could operate
thirty-seven aircraft, against the original thirty-three envisaged for the
larger ships — though this figure was later (1945) increased to fifty-four.

It was one of the *Colossus* class — HMS *Ocean* — that had the distinction of landing-on the first pure jet aircraft. The jet was the third prototype (LZ 551) de Havilland Sea Vampire 1, piloted by Lt Commander E.M. (Winkle) Brown, who landed safely aboard *Ocean* on 3 December 1945. During two days of intensive trials, fifteen take-offs and landings were made, proving to the assembled high-ranking naval officers that operating jet aircraft from carriers was a feasible proposition. To keep the record straight, although the Vampire was the first true jet to land on a carrier, a Ryan Fireball (XFR-1), which was a hybrid machine, being powered by both a piston and a jet engine, landed on board USS *Wake Island* on 6 November 1945, forcibly jet-powered due to the piston engine failing. (Sixty-six of these curious aircraft were delivered to the US Navy but by the summer of 1947, all had been withdrawn.)

Jet-aircraft operation from carriers was to pose many difficult problems. Some of these were solved by Britain, which, though already in decline as a world naval power, nevertheless was to make some fundamental contributions to the post-war development of the aircraft carrier.

The obvious increments in weight and wing-loading of military aircraft, especially the jets then in prospect, were raising questions as to whether the existing flight-deck operations could accommodate them. The technique of the LSO, arrester wires and crash barriers were, after all, some thirty years old. Jets not only land faster than piston-engined aircraft, but the early types suffered from a pronounced lag in throttle response which, together with a landing speed 60mph or so faster than the wartime fighters, was to make the LSO's task begin to approach the limits of human reaction.

Thus the LSO, or Batsman, gave way to a system of lights reflected by a concave mirror. The Mirror Landing Sight, as it was called, was invented by a British naval officer, Commander H.C.N. Goodhart. This new landing aid consisted basically of a number of white lights shining inwardly on the curved surface of an oblong concave mirror; the pilot flew his aircraft so as to keep the apparently horizontal line of the white lights aligned with green datum lights placed at the side of the mirror. The mirror could be adjusted to give the required angle of approach for the particular aircraft type landing and it was stabilised to compensate for the roll of the ship. The old 'nose high, power on' steep descent, followed by an engine cut and flare-out of the piston-engined aircraft was superseded in jets by a much flatter powered approach, virtually down to the flight deck arrester wires. Jets, lacking bulky engines in the nose, offered a much better forward view than piston aircraft, which at least enabled the pilot to see the mirror lights easily throughout the approach and landing, but the jet's high stalling speed required continuous monitoring of the airspeed. The problem of these conflicting requirements was solved by giving the pilot a continuous aural indication of his air speed by a system of coded tones in his earphones, so that he did not have to take his eyes off the mirror, enabling him to fly the highly accurate approach necessary.

The next development, also British, was in some ways the most fundamental and yet the simplest: the angled or axial flight deck. From the earliest days in USS *Langley*, when the crash barrier was created to protect aircraft in the forward deck park, the layout of all carriers' flight decks had changed not at all in essentials. The prospect of heavy and very expensive jet aircraft missing the arrester wires and hurtling into the barriers (or over the side) made some new departure an urgent necessity.

The blindingly simple solution was conceived by Captain D.R.F. Campbell, RN, who pointed out that if the flight deck were angled to port about 10° relative to the ship's centre line, the forward part of the flight deck could be used safely for parking. Any pilot missing the arrester wires would simply open the throttles and overshoot. The bane of generations of carrier pilots, the crash barrier, was no more.

Like so many British inventions, the axial deck, as it was at first called, was put into practice initially by the United States Navy. Between 26 and 29 May 1952, pilots tested the concept aboard USS *Midway*. The modifications to the ship were minimal, since they involved only paintbrushes; the new 'axial deck' was simply painted on the existing one, the ship steaming to give a 'felt wind' down the offset deck. The experiments, conducted by US Navy test pilots and *Midway*'s normal aircrew, flying both piston and jet aircraft, were in every way satisfactory. The angled deck was soon to become standard. The only question raised was "why on earth was it not thought of thirty years earlier?" The answer, had anyone cared to give it, was simply that it had been — during the Great War when the guns of battleships were angled to allow the Sopwith 1½ Strutters to take off from the turrets — but no one thought to translate the basic idea to the first carriers when they appeared.

A further benefit from the adoption of the angled deck was the ability to launch and recover aircraft at one and the same time; the forward part of the flight deck was available to catapult aircraft off over the ship's bows while the angled flight deck was being used for recovery. The catapults then in service were in themselves a new development and were yet another British — or perhaps it would be more accurate to say a European — concept: the steam catapult.

Catapults had been used since the earliest days of naval aviation — they were incidentally originally called, by the Royal Navy, 'accelerators', since that was literally what they did. Usually these devices were either powered hydraulically or by compressed air (unbelievably, 15 inch gun charges were also used for a time). As aircraft weights and stalling speeds rose, it was becoming increasingly difficult to achieve enough acceleration in the short run available with hydraulic/compressed air catapults to launch jet aircraft safely. A Royal Naval Volunteer Reserve Officer, Commander C.C. Mitchell, OBE, suggested that since aircraft carriers were steam-powered, it should be possible to utilise this very convenient source of energy to power aircraft's catapults. The idea was adopted and a long process of development undertaken as a joint Royal Navy and Royal Aircraft Establishment project.

Reduced to its essentials, the steam catapult consists of a long tube with a slot cut into the top which is mounted flush with the carrier's flight deck; the slot is closed by a flexible covering which is inside the tube. The aircraft being launched is attached to a bridle or strop, the other end of which is fastened to a disposable shuttle that acts as a piston and is fitted within the tube. Steam under high pressure from the ship's boilers is introduced, pushing the piston along the tube at a very high rate, pulling the aircraft with it. The flexible cover allows the bridle to travel along the length of the tube, steam pressure sealing it once the piston and bridle have passed a given point. In practice the sealing is effective and insufficient steam escapes to detract from the performance of the catapult. The tube at the far end is open, allowing the shuttle to project out of the pipe; the bridle falls away as the accelerating aircraft it was pulling overtakes it. Each launch of course requires a new shuttle and bridle, as these fall into the sea and are lost — a small price to pay for a very efficient system which, in its fully developed form, can launch any aircraft ever likely to be embarked on a carrier. The rate of launching is also very high — as fast as aircraft can be placed in position and connected to the bridle.

The invention of the steam catapult is claimed, and usually acknowledged, as British, and it was certainly developed in England for carrier use; but in examining its provenance, it must be pointed out that the Germans really got in first — in 1944. The infamous V1s — the pilotless 'Buzz Bombs', 'Doodlebugs' or, to use their German identity, the FZG76s — were launched from a 125 feet long steam catapult, using exactly the same concept of the slotted tube, piston and bridle. The steam was generated rather elegantly: not having large boilers in the woods of the Pas de Calais, the Germans produced steam chemically by decomposing hydrogen peroxide with potassium permanganate; 220lb of hydrogen peroxide was required for each launch, the steam pressure generated launching the V1 at 340 feet per second, giving it sufficient initial speed to enable its low-powered 660lb thrust pulse jet engine to climb the missile away to fly to London.

The mirror sight, angled deck and steam catapult were all adopted by the US Navy and incorporated with other new ideas in a long series of rebuilding and modernisation programmes to existing carriers under the designations Projects 27A, 27C and 110. Project 27A was instigated on 4 June 1947 and initially called for a modernisation of the wartime *Essex* class carriers to enable them to operate jet aircraft up to 40,000lb weight. The first of nine carriers to be so refitted was *Oriskany* (CV-34). After the successful angled-deck trials on USS *Midway*, the *Essex* class modernisation plans were extended to include Project 27C, which called for the reconstruction of the carriers to incorporate a true angled deck.

Project 27C was also to include the British steam catapult, which had first been fitted to the RN carrier HMS *Perseus* and demonstrated to the US Navy at the Navy Yard, Philadelphia, and at sea during the spring of 1952. The tests so impressed the US Navy that it was decided to adopt it there and then, a report in the authoritative official publication, *Naval Aviation News*, commenting:

> .... while the amount of steam required for sustained operation is large, tests have shown that the boilers can meet the demand without interfering with the ship's operations .... The new catapult fared so well during the tests that the Navy has already begun an investigation into the adaptability of it to their new [59,900 ton] carrier USS *Forrestal*, which is now under construction.[2]

Forrestal (CVA-59) was laid down on 14 July 1952. While it was building, USS *Hancock* (CV-19) became the first US carrier to include the American 'steam slingshot' — the C-11 catapult which was tested on 1 June 1954, launching an S2F-1 twin-engined Grumman Tracker. During the successful tests at sea aboard *Hancock*, 254 launches were made with a Tracker and other naval aircraft, including jets.

Eventually nine of the wartime *Essex* CVs were remodelled with steam catapults and angled decks which were standardised at 10.5° to port. The original central elevator of the *Essex* carriers was removed and replaced by side elevators. The first carrier converted to an angled deck, USS *Antietam*, was extensively tested between 29 December 1952 and 1 July 1953, during which time 4,107 launches and landings were made, including night operations, without any major accident or even minor incidents that could be attributed to the new flight deck, which was permanently to alter the silhouette of aircraft carriers and which all future vessels would incorporate. Project 27C, which originally was intended only to convert *Essex* class ships, was extended to the newer *Midway* class as Project 110. The term 'angled deck' was officially recognised by the Chief of Naval Operations when it was directed on 24 February 1955 that 'angled deck' was the term to be used, replacing 'axial', 'canted', 'slanted', 'slewed' and other names.

Although the angled deck was a British concept, the Royal Navy first tested one when USS *Antietam* came to Portsmouth and RN jets were able to land aboard her angled deck. Prior to that, HMS *Triumph* had a painted mock-up and a number of 'touch and go' landings were made, no full-stop landings being possible since the arrester wires were disconnected. HMS *Hermes*, *Centaur*, *Bulwark* and *Albion* were altered during construction to incorporate an angled deck.

The new techniques of carrier operation and the provision of the later generation of aircraft to replace the wartime types was naturally to take some time. Meanwhile there were, on both sides of the Atlantic, vociferous critics questioning — from various motives — the continued need for any aircraft carriers, which it was claimed were an outmoded, very expensive luxury. Russia, it was pointed out, had no aircraft carriers of any description. The arguments, however, were quickly stilled when on 25 June 1950 the army of North Korea crossed the 38th parallel to invade the Republic of South Korea. Within a week, following a UN resolution, a seven-nation force was on its way to Korea. Among the ships deployed were two carriers: the USS *Valley Forge* and the British light fleet carrier HMS *Triumph*.

On 3 July the two ships, operating in the Yellow Sea, launched strikes against ground targets in and around Pyongyang, north-west of Seoul. Thus, just five years after the ending of the Pacific War, British and US carriers were again operating together in the Far East.

The Korean War was fought at a time of transition as far as the naval air arms were concerned. None of the carriers to be involved had the new angled deck or steam catapults; the aircraft embarked were a mixture of old and new. Only the US Navy had operational jets available and these were supplanted by wartime F4U-4s and a piston-engined attack bomber, the Douglas Skyraider.

The Royal Navy operated, in rotation, four carriers off Korea: HMS *Triumph*, *Theseus*, *Glory* and *Unicorn*. The last, although ostensibly a repair and maintenance carrier, indulged in some shore bombardment. The Australian carrier HMAS *Sydney* also formed part of the UN fleet.

The aircraft which flew from the British carriers were all piston-engined; the Royal Navy's first jet, the Supermarine Attacker, was not available in time. Seafires and the Navy's last piston-engined fighter, the Hawker Sea Fury, provided the cover for Fireflys, which acted as ground attack aircraft, though the Sea Furys were also pressed into service as dive-bombers — a role they performed exceptionally well. The naval aircraft from both the US and British carriers were used mainly in an 'interdiction' role: supporting ground forces. Korea was a difficult country for air support, being mountainous and offering few major targets. During the three years of the Korean struggle there was only one major landing on the World War II scale — that at Inchon — which carrier aircraft fully supported. For the remainder of the conflict, the aircraft roamed over the barren countryside attacking targets of opportunity.

One US commander put the conduct of that air war well: ".... a day to day routine where stamina replaces glamour and persistence is pitted against oriental perseverance." That observation is well supported by the statistics. British and Australian aircrews had flown 23,000 sorties by the time the truce was signed in Panmunjom on 27 July 1953. The figures for the US Navy and Marines are illuminating: though Korea was, in comparison with World War II, a small affair; during the three years hostilities lasted, the US Navy, which never operated more than four fleet carriers together at any one time, flew 276,000 combat sorties and dropped 177,000 tons of bombs, 74,000 tons more than dropped by that Navy during World War II. Had they flown a further 7,000 sorties, they would even have equalled the wartime number flown in all theatres by the US Navy.[3]

The Korean War ended with a truce, not an armistice, and the world became irrevocably polarised into East and West; any naïve hopes that the post-war world would simply be a better version of the pre-war one were dashed. The continuation of the seemingly endless 'cold war' and the intractable attitude of the Russians had the effect of prompting the western Governments to make available funds to maintain their military strength at a high level of readiness.

The five years following the Korean War were perhaps the most turbulent ones for Allied naval aviation. The wartime carriers were brought up to date, as previously discussed; new ships were laid down; a new generation of jet aircraft was built; and a new generation of pilots was trained to fly them, as the regular wartime aircrews moved away from the flight decks to assume command and Flag rank.

Britain's Fleet Air Arm seemed set fair for an indefinite existence. An excellent jet fighter, the Hawker Sea Hawk, was in service, and in 1955 the First Lord of the Admiralty, J.P.L. Thomas, stated unequivocally: ". . . . the carrier has a future as firm as any airfield." (No one questioned the firmness of the airfields.)

There were then commissioned the first post-war fleet carriers — *Ark Royal* and three carriers of the *Hermes* class, the 27,800 ton *Centaur*, *Albion* and *Bulwark* — and, if by the standards of the day, these ships were considered of moderate size, it should be remembered that they approximated to the highly successful wartime *Essex* class. In spite of the new ships and the roseate future promised by the politicians, the final decline of the Fleet Air Arm was about to commence.

Not many people in Britain would be able to tell one anything about 'Operation Musketeer'; if, on the other hand, 'Suez' were mentioned, the reaction would be immediate, for Musketeer and the Suez intervention were one and the same thing, and one most Britons would rather forget. Leaving aside the political rights and wrongs, the object of that ill-starred adventure was ostensibly to stop the Israelis and Egyptians from fighting; the real object was to secure the Suez Canal, which the Egyptian President, Gamal Nasser, had just had the temerity to nationalise. From an operational standpoint, because the distance from British Mediterranean bases in Cyprus and Malta was too great, the strike against the Egyptians was to be mounted from aircraft carriers. It was to be an Anglo-French operation involving seven carriers. *Eagle, Bulwark* and *Albion* were to provide the strike aircraft — turbo-prop Westland Wyverns, pure jet Sea Venoms and Sea Hawks, and piston-engined Douglas Skyraiders. The two French light carriers taking part, *Arromanches* and *Lafayette*, embarked F4U-7 Corsairs and Grumman Avengers.

The first carrier strikes were launched on 31 October 1956, backed up by Cyprus-based RAF Canberra bombers. The attacks continued for four days until Anglo-French paratroopers were dropped on 5 November. *Theseus* and *Ocean* then used helicopters to land Royal Marine Commandos with the object of securing the Canal. It was, from a military point of view, a copybook operation, and poignantly the last major operation that the Royal Navy would ever mount. With Britain reviled and deserted by her major ally, the United States, and threatened by the Soviet Union with atomic retribution, a ceasefire took effect from midnight on 6 November, UN troops taking over soon after.

If the sinking of the *Prince of Wales* and *Repulse* in December 1941 had signalled the beginning of the decline of Britain as a world naval power, the outcome of 'Suez' plumbed the very depths. It began a retreat from international power politics which accelerated over the next decade. The Royal Navy, for centuries the instrument of British foreign policy, was, however unfairly, regarded abroad and by many at home as having acted improperly, and would never again act independently of NATO except to honour, by invitation, defence obligations pledged to former protectorates.

With Britain's role in the world contracting, its one-time huge global empire dwindled, as country after country was granted self-rule; and, in the face of mounting economic difficulties at home, the RAF was cut back, beginning with the 1957 White Paper on Defence, in which Duncan Sandys argued that manned aircraft were obsolete in the intercontinental atomic missile age. The Royal Navy carrier fleet was also reduced: HMS *Glory*, *Ocean* and *Theseus*, together with the maintenance carrier *Unicorn*, were withdrawn and disposed of. *Bulwark* was retained to be rebuilt as a helicopter support ship, and went in 1961 to the aid of the Sheikdom of Kuwait, which was being threatened by her powerful neighbour Iraq.

Helicopters were by then the most numerous aircraft in the Fleet Air Arm's inventory, in the form of the jet-engined Wessex, a licence-built version of the American Sikorsky S58. After the Kuwait incident, helicopters from *Bulwark* and a second commando carrier *Albion* flew in support of troops in the newly created Federation of Malaysia, when Indonesia tried to occupy part of North Borneo. These and other operations were on a small scale and made in response to treaty obligations; though vital to the people involved, they hardly justified the growing expense of maintaining a large fixed-wing carrier force.

During the sixties the fixed-wing aircraft embarked in the remaining fleet carriers *Victorious*, *Eagle* and *Ark Royal*, were paradoxically the best ever to see service with the Fleet Air Arm. The de Havilland Sea Vixen, Supermarine Scimitar, the Blackburn Buccaneer and Fairey Gannet were aircraft designed specifically for shipboard operations and not adaptations of landplanes, as was in the the history of the Fleet Air Arm so often the case.

In 1962, to celebrate fifty years of the Fleet Air Arm, there was a flying review before HRH the Duke of Edinburgh, when the new jets were demonstrated along with a British air-to-air guided missile — Firestreak. That same year it was announced that the final designs of the Navy's new fleet super-carrier, known as CVA-01, were nearing completion and that she would be built and commissioned in 1971 with a number of the 138 American McDonnell F-4K Phantoms then ordered, embarked. In 1962 the FAA seemed to have indefinite future existence. It was not to be.

Although politicians blandly made reassuring statements to the effect that carriers would continue to form the backbone of the Navy throughout the 1970s, behind the scenes the old internecine struggle between the RAF and the Fleet Air Arm was renewed as both fought for existence. The British annual Defence Budget was rising towards £2,000 million, and it was plain that swingeing cuts would soon have to be made.

The result of numerous government inquiries and sub-committees, and despite the resignations of a Navy Minister and the First Sea Lord, was that CVA-01 was cancelled in 1966. The old, old argument was put forward that shore-based — RAF — aircraft could undertake the necessary strike and reconnaissance roles of the fleet, leaving the Navy with an anti-submarine capability, using helicopters embarked on surface vessels. The long shadow of Trenchard still stalked the Whitehall corridors. What followed was inevitable.

Victorious was scrapped in 1967 and *Eagle*, requiring an estimated £50 million refit, followed her to the breakers in 1972. The aircraft from these ships, plus many of the new Phantoms, were given to the RAF, who found their wing-folding capability useful in the small dispersal hangars which had become fashionable in NATO air forces after the Egyptian air force had been destroyed in the first hour of the 'Six Day War' by pre-emptive Israeli strikes. *Ark Royal* continued in service until 1979, enjoying a last-minute fame when the ship starred in the very popular BBC TV series 'Sailor', which was filmed aboard during her last commission.

Long before that, the fixed-wing pilots of the FAA had seen very clearly the writing on the wall and, like their fathers before them in 1918, had been faced with the choice of transferring to the RAF, joining a non-flying branch of the Navy, or simply leaving the Service. Few fixed-wing men were attracted to the doubtful joys of flying helicopters.

There was only one tiny glimmer of hope. On 8 February 1963 Hawker's chief test pilot, Bill Bedford, took off from the Company's snow-bound airfield at Dunsfold, flying the first prototype (XP831) of the revolutionary Hawker P1127 vertical take-off jet. He headed south and performed a controlled, powered descent on to the after end of *Ark Royal*'s flight deck. The ramifications of that landing we will examine shortly after discussing the post-Korea developments in the United States Navy.

The Navy of the United States enjoyed one supreme advantage: it was a major part of the military forces of the richest nation on earth. With the end of the Korean conflict, the US Navy proceeded with the modernisation

plans of the wartime *Essex* carriers, though these were soon downgraded
from their pre-eminent position as fleet carriers. The designation 'CV'
was amended in 1953 to 'CVS': Anti-Submarine Support Aircraft
Carriers. The *Midway* class, then the largest carriers in commission,
became the spearhead of the US Navy's airpower.

The US Naval Staff realised that there was a limit to the life of any
ship's hull, especially those that had been subjected to enemy attack and
continuous wartime service, and that sooner or later all would have to be
replaced. As early as 1948, when the Navy had an embarrassingly large
number of carriers, it was decided to instigate designs for a post-war
generation of entirely new ships under the designation '6A Carrier
Project'. Some seventy-eight proposals were examined before final
approval was given, on 24 June 1948, Congress then voting the funds;
and a contract to build the ship was awarded to Newport News
Shipbuilding and Dry Dock Company. The '6A Project' design was for a
65,000 ton carrier with a 1,030 feet flight deck and four catapults.
Curiously, the ship went right back to the early flush-deck concept; a
small island was to be provided but it was to be fitted to an elevator
which could be lowered when flying was in progress. That concept was
used first in HMS *Argus* in 1919. Profiting from wartime experience, all
openings in the flight deck were to be avoided, the aircraft lifts being
mounted on the side of the hull. For the first time in the design of a US
warship, the ability to pass through the Panama Canal — so long a
limiting factor — was abandoned. The hangar, as in the British fleet
carriers, was to be a strong armoured box, devoid as far as possible of
openings. There was disquiet on the part of BuShips on the question of
ducting the boiler gases, but construction was authorised and President
Truman approved the name: USS *United States* (CVB-58).

Events then moved with bewildering speed. On 15 April 1949, just
two days after the building was sanctioned by the House of Represent-
atives, the then Secretary of Defense, Louis Johnson, wrote to General
Eisenhower, at that time Presiding Officer of the Joint Chiefs of Staff,
questioning the need for a new carrier. While this letter was being
considered, on 18 April the keel of the ship was laid; on 23 April Johnson
ordered all work on the carrier to cease. John L. Sullivan, the Secretary
of the Navy, promptly resigned in protest.

As to what would have been the fate of the carrier and any subsequent
ones had the Korean War not supervened, who can say? On 12 July 1951
the work resumed, though the name was to be changed to USS *Forrestal*
and a new number, CVB-59, issued. The keel was relaid on 14 July 1952.
The flush-deck concept proposed for the *United States* was abandoned;

by the time *Forrestal* was laid down the advent of the angled deck had removed the objections to a conventional starboard island layout.

USS *Forrestal* was commissioned into the United States Navy on 1 October 1955, and she was by any standards impressive. With a displacement of 59,650 tons and 4 acres of flight deck, her rated horse-power was over 200,000, which gave the ship a speed in excess of 30 knots.

The US Navy's philosophy with regard to aircraft carriers was put by the Assistant Secretary of the Navy, James H. Smith, at the *Forrestal's* commissioning ceremony:

> If our way of life is to survive, we must maintain these two alternative military postures: the first is to maintain a powerful and relatively invulnerable reprisal force which will signal a potential enemy to stop, look and listen before he risks an all out atomic war. The second is to insure that we ourselves will not be forced to change the character of a limited war because of fear of ultimate defeat in a series of them.
>
> Fortunately, we need not maintain a completely separate set of forces for each posture. In this ship and the .... aircraft she can service we combine the two, [with] the ability to appear quickly at any .... far flung trouble spot ....

Forrestal was followed by a similar, though slightly faster, sister ship, USS *Saratoga* (CVA-60), commissioned 8 October 1955. The third of the class, USS *Ranger* (CVA-61), which had a longer and slightly different angled flight deck of 1,046 feet, compared with *Forrestal's* 1,039 feet, was commissioned on 10 August 1957. A fourth *Forrestal* class ship was built: *Independence* (CVA-62), commissioned 10 January 1959. The four *Forrestal* ships were followed by three essentially similar but sufficiently different carriers to merit a new designation: the *Kitty Hawk* class, which comprised *Kitty Hawk* (CVA-63) and *Constellation* (CVA-64). These two were commissioned in October and November 1961, the main difference that distinguished them from the earlier ships being their guided missile capability. Two additional ships of the class, the USS *America* (CVA-66) and the *John F. Kennedy* (CVA-67), were commissioned in January 1963 and September 1968 respectively.

The intermediate number, CVA-65, was allotted to the most revolutionary carrier then built — the world's first to be nuclear-powered, USS *Enterprise*, CVA(N)-65. With a displacement of 85,350 tons, she was at the time of commissioning the largest warship on earth; her eight nuclear reactors would, if required, enable the ship to cruise round the world twenty times before the need to refuel arose. *Enterprise* was equipped with four C-13 steam catapults which could provide an energy of 60,000,000 foot pounds, capable of accelerating a 78,000lb aircraft from a standstill to 160mph in 250 feet. Such catapults were, by the 1960s, essential, for the jets then in service could not take off unaided even from the *Enterprise*'s 1,040 feet flight deck. *Enterprise* was to take part with other US carriers in the highly unpopular Vietnam War.

It is difficult to imagine a larger, more powerful, carrier than *Enterprise*, but CVN-68, laid down on 22 June 1968, was such a ship. Named USS *Nimitz*, she was the first of three projected nuclear-powered, 95,000 ton carriers — the largest warships ever built or ever likely to be built. *Nimitz*, commissioned 3 May 1975, was followed by her sister, the USS *Dwight D. Eisenhower* (CVN-69), commissioned in October 1977. USS *Vinson*, the last of the class, was launched in March 1979 and is expected to enter service during 1981.

Constructed at a cost believed to be in excess of $1,000 million (£434 million) each, these three ships represent an enormous investment, and bring with them attendant problems. It is not only the staggering cost, which comfortably exceeds the gross national product of many Third World countries, it is also the difficulty of recruiting, training and retaining the nearly 6,000 highly skilled men required to crew each of these colossal ships. The commissioning of these carriers is, of course, a product of the continuing cold war, with a significant, new and potentially sinister development.

On 8 January 1972 the *Washington Post*, a paper with its ear particularly close to the capital's ground, announced that "the latest US reconnaissance satellite photos" revealed a very large ship under construction in Russia which, though possibly a supertanker, could be an aircraft carrier. The first official US announcement about the 'carrier' was made in March 1972, when Congress was informed that the Soviet Union was constructing the largest ship built in the Nikolayev Nosenko dockyard, a ship which was felt to be a carrier . . . .

The ship in question, the *Kiev*, first of a class, was launched in December 1972. Though at first it was thought that only anti-submarine helicopters would be embarked, photographs of the ship revealed an angled deck which indicated some sort of short take-off aircraft, which leads us back to Bill Bedford's landing on *Ark Royal* in 1963.

The background to the eventual development of that prototype vertical take-off aircraft into the now familiar Harrier is complex and full of stop-go political decisions. After a great deal of perseverance on the part of the manufacturers, the RAF and the Royal Navy, the Harrier finally entered RAF squadron service in 1969. They were to be closely followed by the US Marines.

The Marines' first involvement with the Harrier was informal: during the 1968 SBAC Air Show at Farnborough two Marine pilots turned up unannounced at the Hawker Siddeley chalet and calmly stated that they had come to England to fly the Harrier and could they please have a copy of the Pilot's Notes. Within two weeks of the end of the Farnborough Show, the two Marine officers, Colonel Tom Miller and Lt Colonel Bud Baker, had flown the Harrier at Dunsfold and returned with enthusiastic reports of its potential to America.

The results of the private enterprise was an official visit, three months later, by a US Navy test pilot team, who conducted a preliminary evaluation of the Harrier. Five months later the United States Marine Corps had obtained approval for the immediate purchase of twelve 'off the shelf' Harriers and declared an intent to acquire a further 110 by the mid-70s. In the event, Congress insisted that a fifteen-year licence agreement with the aircraft company McDonnell Douglas be signed for future US Harrier development.

The AV-8s, as the USMC Harriers are designated, were soon being flown to the limits — and at times beyond the limits — of their capability. The Marine pilots developed startling techniques by using the variable vectored thrust in flight. A Washington Review Board was told by one experienced Marine pilot: "This machine was just made for the Marine Corps. We took it . . . . expanded its envelope. It's a pilot's airplane . . . . it's way out and far beyond what Hawker Siddeleys ever figured the airplane was going to be used for . . . . this airplane will do anything . . . ."[4] In 1971 the Harriers were cleared for shipboard operation from the support carrier USS *Guam*.

Hawker Siddeley test pilots continued to demonstrate the Harrier aboard thirty ships of nine navies as far apart as India and Brazil. The Spanish Navy bought — via the USA — eight Harriers, which they renamed 'Matadors', for use aboard their wooden-deck carrier *Dedalo*. Other navies were interested, but the question frequently asked of the makers was simply 'if the Harrier is so good, why has the Royal Navy none?' A good question.

Pending the withdrawal of the last fleet carriers, *Eagle* and *Ark Royal*, the Navy were planning a new class of flat-decked ships, intending to operate helicopters. To avoid upsetting the politicians, these were euphemistically termed 'Through-deck cruisers' — the word 'carrier' was never mentioned. The first of these ships approved was HMS *Invincible*, 19,500 tons, which was laid down in 1972. With a new 'carrier' in prospect, the Navy began to press for Harriers and, possibly owing to an increased Russian naval presence in the Mediterranean and Indian

Oceans, coupled with the appearance of the 40,000 ton carrier *Kiev* with Yakovlev 36 (Forger) vertical take-off fighters embarked in May 1975, the UK Government sanctioned the development of a Sea Harrier variant.

HMS *Invincible* was launched by HM the Queen at Barrow-in-Furness in May 1977. If she was not the largest ship the Royal Navy had commissioned, she was, at £215 million, the most expensive. As this is written, *Invincible* is working up and her eight Sea Harriers of 800 Squadron will embark in late 1980. Altogether the Fleet Air Arm has forty-four Sea Harriers on order; some will probably be embarked on the existing support carrier HMS *Hermes*, in addition to *Invincible* and a sister ship *Ark Royal*.

Thus the British carrier, by whatever name, lives on, and it may observe, as Mark Twain once did: "Reports of my death are greatly exaggerated."

Notes and References

Chapter 1

1. PRO File AIR 1/674: *British Rigid Airships*, compiled by Captain D.C. Murray, 27 April 1921.
2. In point of fact, Ellyson was not the first naval officer to qualify; that distinction must go to Lt G.C. Colmore, RN, who gained British Aviator's Certificate No. 15 on 21 June 1910. However, Lt Colmore had trained as a private individual and at his own expense.
3. For comparison, a typical modern *ab initio* trainer, the Cessna 150, has a wing-loading of 10.2lb per square foot.
4. Later transferred to the Royal Air Force to become Air Vice Marshal Sir Arthur Longmore.
5. The record was established at Châlons by E. Nieuport of France, flying a machine of his own manufacture.
6. The first pilot to take off from water was the Frenchman Henri Fabré, in March 1910 in his floatplane at Martigues, near Marseilles.
7. The figure of 60 miles (in 1 hour 20 minutes) is given as the aircraft's longest recorded flight in the official US Navy's magazine *Naval Aviation News* for May 1976. In that issue there is a profile (p.20) entitled 'Navy No. A-1'. The Public Record Office, London, however, contains in 'Air 1/7' a contemporary review of foreign aircraft development and quotes a flight made on 25 October 1911 by the 'Curtiss Hydroaeroplane'. The machine was piloted by Lts Ellyson and J.H. Tower from Annapolis along Chesapeake Bay to a landing near Buckroe Beach — a distance of 145 miles. The time of the flight is quoted as 2 hours 27 minutes, which would indicate a nonstop journey.
8. The importance of the A-1 to the Navy was acknowledged when, in 1961, to celebrate fifty years of US naval aviation, a reproduction A-1 was built and flown at San Diego.
9. Quoted by Hugh Popham in *Into Wind* (Hamish Hamilton, London, 1969), p.3.
10. PRO File ADM 53/21953: Log of HMS *Africa*.
11. PRO File AIR 1/7.
12. PRO File ADM 53/16757: Log of HMS *Hibernia*.
13. PRO File ADM 53/31575: Log of Royal Yacht *Victoria and Albert*.
14. PRO File ADM 53/16759: Log of HMS *London*.
15. PRO File AIR 1/674: 'Early History of RNAS'.
16. PRO File AIR 1/625.
17. *The Times*, London 5 June 1914.

Chapter 2

1. The account of the action in the Rufigi Delta is based on an official report 'The Destruction of the enemy cruiser *Königsberg* in the Rufigi Delta' — PRO File AIR 1/674.

2. After the war, Cutler seems to have been asked by the Naval historical branch to write of his experience in the Rufigi. The Public Record Office in London contains a letter he wrote from a Hammersmith, London address on 22 June 1922, and which is quoted in part on p.35 (See PRO File AIR 1/2393/231/1: 'Operation in Rufigi by H.D. Cutler'.)

3. The *Dresden*, which was the only German ship to escape the Falkland Islands Battle, was scuttled when discovered and attacked by HMS *Kent* and *Glasgow* in March 1915. The German cruiser was at anchor off Mas a Fuera, out of coal.

4. Hugh Popham, *Into Wind* (Hamish Hamilton, London, 1969).

5. Scot McDonald, 'Evolution of Aircraft Carriers', *Naval Aviation News* special (Washington, DC, May 1962).

6. Quoted by Ernest Dudley in *Monster of the Purple Twilight* (Harrap, London, 1960), p.34.

7. Hugh Popham, *Into Wind*, p.21.

8. Figures quoted in *British Naval Aircraft* (Putnam, London, 1958).

9. Commander Samson, *Fights and Flights* (Putnam, London, 1922).

10. Sir Walter Raleigh, *War in the Air*, Vol. II, p.65.

11. PRO File AIR 1/17/122/36.

12. PRO File AIR 1/17/122/54.

13. PRO File AIR 1.

14. James McCudden, *Flying Fury* (John Hamilton, London 1930).

15. Sir Walter Raleigh, *War in the Air*, Vol I, p.486.

16. W.G. Moore, *Early Bird* (Putnam, London, 1963).

17. BBC Transcript of interview with Sir William Dickson, recorded 1975, in 'Pilots at Sea'.

18. W.G. Moore, *Early Bird*.

19. BBC Transcript, 'Pilots at Sea'.

20. W.G. Moore, *Early Bird*.

21. *Ibid*, p.101.

22. BBC Transcript, 'Pilots at Sea'.

23. PRO File AIR 1/344.

24. PRO File AIR 1/17/122/218.

25. BBC Transcript, 'Pilots at Sea'.

26. W.G. Moore, *Early Bird*.

27. PRO File AIR/344.

28. PRO File AIR/344/15/226/287.

29. Quoted by John W.R. Taylor, *Aerial Warfare* (Hamlyn, London, 1974).

30. H.H. Smith, *Yellow Admiral*, quoted by Hugh Popham, *Into Wind*.

31. PRO File AIR/1/344.

Chapter 3

1. A.J.P. Taylor, *The First World War* (Hamish Hamilton, London, 1963).
2. *United States Naval Aviation, 1910-1970*, 2nd Edition (Navair 00-80P-1, Washington DC, 1970), p.11.
3. Ibid., p.299, Appendix IV.
4. P.S. Dickey, *Smithsonian Annals of Flight*, Vol. I, No.3 (Washington DC, 1968).
5. *United States Naval Aviation, 1910-1970*.
6. R. Rausa, 'In 228 Days', *Naval Aviation News* (Washington, March 1976), p.18.
7. Quoted by Scot McDonald, 'Evolution of Aircraft Carriers', *Naval Aviation News* special (Washington DC, 1962), p.17.
8. Ibid.
9. Ibid.
10. Ibid, p.18.
11. R. Rausa, 'Turntables and Traps', *Naval Aviation News* (Washington DC, August 1976).
12. Ibid.
13. Ibid.
14. *United States Naval Aviation, 1910-1970*, p.48.
15. David Brown, *Aircraft Carriers of World War Two: Fact Files* (Macdonald and Janes, London, 1977), p.1.
16. FO File F1793 in PRO 371/8050.
17. See 'Note on Admiralty Programme for Fleet Air Arm' — PRO File AIR 19/105, 26 June 1930.
18. Ibid.
19. Ibid.
20. Ibid.
21. See Appendix IV, 'Aircraft on Hand', in *US Naval Aviation 1910-1970*.
22. This remark is contained in a 1930 paper which frankly reveals the attitude of the Air Ministry to the Royal Navy's claim to control of the Fleet Air Arm. See 'Resumé of the controversy about the existence of the Independent Air Force, with particular reference to the question of the Fleet Air Arm' — PRO File AIR 19/105.
23. '50 Years of Carriers', *Naval Aviation News* (Washington DC, March 1972).
24. Quoted from 'The Flagwavers', an article by Rear Admiral J.R. Tate, USN (Ret.), *Naval Aviation News* (Washington DC, April 1972).
25. Ibid.
26. See also 'Where Stands the LSO?', by Cdr Harborough, USNR, *Naval Aviation News* (Washington DC, January 1962), p.16.
27. *United States Naval Aviation, 1910-1970*, p.64.
28. Quoted by Owen Thetford in *The Fairey Flycatcher* (Profile Publications No. 56, London, 1965).
29. Quoted in J.V. Mizrahi, *Carrier Fighters* (Sentry Books, California, 1959).
30. Rear Admiral J.R. Tate, USN (Ret.), 'The Cinderella Ships', *Naval Aviation News* (Washington DC, March 1972).
31. Foreign Office Files in PRO 371/8050.
32. Ibid.

33. Ibid.
34. FO File F1065, 19 March 1922, in PRO 371/8050.
35. FO File F1793 in PRO 371/8050.
36. PRO File AIR 19/105.
37. Memorandum dated 15 May 1931. See Files of the Bureau of Aeronautics, National Archives, Washington.
38. David Brown, *Aircraft Carriers of World War Two: Fact Files*, p.18.
39. Quoted by Francis Mason in *The Fairey IIIF* (Profile Publications No.44, London, 1965).
40. Terence Horsley, *Find, Fix and Strike* (Eyre and Spottiswoode, London, 1943).

Chapter 4

1. Charles Lamb, *War in a Stringbag* (Cassell, London, 1977).
2. For a vivid eye-witness account of the fall of Neville Chamberlain, see Harold Nicolson, *Diaries and Letters 1939-45* (Collins, London, 1957).
3. A.J.P. Taylor, *The Second World War* (Hamish Hamilton, London, 1975).
4. Ibid.
5. Terence Horsley, *Find, Fix and Strike* (Eyre and Spottiswoode, London, 1943).
6. Charles Lamb, *War in a Stringbag*.
7. Admiral Cunningham, *A Sailor's Odyssey* (Hutchinson, London, 1951).
8. Ronald Lewin, *Ultra Goes to War* (Hutchinson, London, 1978).
9. These names were the ones adopted by the Americans after the commencement of the war. In general, fighters had male names, and bombers female.
10. For a comprehensive account of the solving of 'Purple', see David Kahn, *The Code Breakers* (Weidenfeld and Nicolson, London, 1973).
11. Winston Churchill, *The Second World War*, Vol III, p.589 (Cassell, London, 1949).
12. Appendix IV, 'Aircraft on Hand', in *US Naval Aviation, 1910-1970* (NAVAIR 00-80P-1, Washington DC, 1970).
13. Quoted in Scot Macdonald 'Early Attack Carriers', *Naval Aviation News* (Washington DC, November 1962).
14. The international dateline ran through the battle zone. The dates given in this narrative are appropriate to the US zone. The Japanese dates are one day in advance.
15. Interview with Captain J.P. Adams, US Navy (Ret), by the Author for BBC TV Programme 'Pilots at Sea', July 1975.
16. Ibid.
17. Figures from *US Naval Aviation, 1910-1970*, p.113.
18. The designation 'CVE' was adopted on 15 July 1943, replacing the earlier 'ACV', when escort carriers were put on a combat rather than an auxiliary commission.
19. Interview with author for BBC TV feature 'Pilots at Sea', July 1975.
20. Ibid.
21. Ibid.

22. Figures from the official *US Naval Aviation, 1910-1970*.
23. S.E. Morison, *Leyte, June 1944 – June 1945* (Atlantic Little, Boston, 1958).

Chapter 5

1. *United States Naval Aviation, 1910-1970*, 2nd Edition, (NAVAIR 00-80P-1, Washington DC, 1970).
2. Quoted in *Naval Aviation News* (October 1963).
3. Figures from *United States Naval Aviation, 1910-1970*.
4. Quoted from a lecture given before the Royal Aeronautical Society, January 1977, by J.W. Fozard, FRAes, Chief Designer, Harrier.

Appendices

Aircraft taking part in Spithead Review, July 1914

Seaplanes

A Flight — Isle of Grain

Aircraft Number	Engine and Type	Pilot
79		Flt Lt L.G. Brodrib
82	160 hp Short	Flt Lt B.D. Ash
119		Sqd Cmdr J.W. Seddon
120		Flt Cmdr J.T. Babington

B Flight — Dundee

Aircraft Number	Engine and Type	Pilot
74		Sqd Cmdr R. Gordon
75	100 hp Short	Flt Lt A.C. Baraby
76		Flt Cmdr D.A. Oliver
77		Flt Lt C.F. Kilner

C Flight — Yarmouth

Aircraft Number	Engine and Type	Pilot
139		Flt Cmdr F. Hewitt
141	120 hp	Flt Lt R. Bons
142	Horace Farman	Sqd Cmdr T. Courtney
143		Flt Lt H. Fawcett

D Flight — Felixstowe

Aircraft Number	Engine and Type	Pilot
113		Sqd Cmdr C. Risk
114	100 hp	Flt Cmdr C. Rathbone
115	Horace Farman	Flt Lt E. Nanson
95		Flt Lt C. Edmonds

E Flight — Calshot

Aircraft Number	Engine and Type	Pilot
93	200 hp Sopwith	Flt Cmdr Bigsworth
151	100 hp Sopwith	Flt Lt R. Ross
118	90 hp Sopwith	Flt Lt J.L. Travers
126	160 hp Short	Unknown

Aeroplanes — Based at Hilsea for the Review

Aircraft Number	Engine and Type	Pilot
50	70 hp BE	Cmdr C.R. Samson
153	80 hp Bristol	Sqd Cmdr I.T. Courtney
49	70 hp BE	Flt Lt E. Osmond
64	80 hp Short	Flt Lt H.A. Litterton
104	80 hp Sopwith	Flt Lt R.L. Marix
43	80 hp Bristol	Flt Lt H.W. Darlrymple-Clark
41	50 hp Short	Flt Lt E.T. Newton-Clare
152	80 hp Short	Flt Lt D.C. Young

In addition to the above list, which was printed in the London *Times* for Saturday, 18 July 1914, based presumably on an official release, the paper also included the following list of spare aircraft and four airships:

Spare Seaplanes

78	100 hp Short
19	160 hp Short
170	200 hp Short
52	100 hp Avro
53	100 hp Avro

Spare Aeroplanes

1	50 hp Short
2	50 hp Short
70	70 hp Farman
154	100 hp DFW

Airships

Astra-Torres (No 3), *Parseval* (No 4), *Gamma* and *Delta*.

Signal Code

The following code was used during the Rufigi Campaign for W/T signalling between ships and aeroplanes:

Call Signs	Ship firing A S	Aeroplane B S

Signals

BA	Am ready to observe.
BC	Cannot observe fall of shot.
BD	Observation doubtful.
BF	Last shot of salvo not observed.
BJ	Observe for batteries firing at me.
BV	Visibility too bad to see target.
BW	Am going to look for target.
F	Shall be ready in ten minutes.
FT	Fly, or will fly, vertically over target.
	(When over target aeroplane makes shorts.)
G	Guns now firing or carry on firing.
H	Aeroplane cannot start; will call up when ready.
HO	Aeroplane can start now.
HT	Hit.
J	Observe for range.
K	Observe for line.
LL	More than 400 yards left.
LC	Line correct.
LS	Line spread (i.e. salvoes much spread laterally).
M	Aeroplane started out.
MQ	Wait.
NG	(Ship) — Cannot fire on account of weather.
	(Aeroplane) — Cannot observe on account of weather.
OK	Target destroyed.
Q	Have ceased fire for today.
RR	More than 400 yards right.
RC	Range correct.
RS	Range spread (salvoes much spread along the line of fire).
SU	Carry out instructions already given.
T	I am on my way (from aeroplane).
U	Can you read my signals.
W	Your signals not being received.
WJ	Wireless is being jammed, please repeat.
WO	Going, or go, home.
Y	Yes.

The Cuxhaven Raid

The aircraft and pilots who took part in the Cuxhaven raid on 25 December 1914 were as follows:

HMS *Engardine*   119   Short 'Folder' 160hp Gnome Rotary
                                  Flt Cmdr R.P. Ross
                          120   Short 'Folder' 160hp Gnome
                                    Flt Lt A.J. Miley

HMS *Empress*   814   Short Type 74, also known as 'Improved 41', 100hp Gnome Rotary
                                    Flt Sub-Lt V. Blackaby
                          815   Type 74
                                    Flt Cmdr Douglas Oliver

HMS *Riviera*   135   Type '135', 135hp Salmson radial
                                    Flt Cmdr F. Hewlett
                          136   Type '135', 135hp Salmson radial
                                    Flt Cmdr C.F. Kilner
                        (135 and 136 were the only two Short '135s' built).
                          811   Type 74
                                    Flt Lt G.K. Edmonds

The two non-starters were:
122   A short 'Folder', embarked on *Engardine*.
812   A Type 74, embarked on *Empress*.

United States Navy Carriers in Commission during World War II

Original Classes: of the first seven ships, *Langley, Ranger* and *Wasp* were single types, *Lexington* and *Saratoga* were sisters as were *Yorktown* and *Enterprise*. *Hornet*, originally announced as a third sister of *Yorktown* and *Enterprise*, was modified enough to be considered a single.

Essex Class: seventeen ships, Hull Nos 9-21, 31-35, 37-39. Of these numbers, 14, 15, 19, 21, 32-34, 36-39 are sometimes referrred to as 'Long-Hull' *Essex* Class.

Independence Class: (CVL), nine ships, Hull Nos 22-30.

Hull No.	Name	Commissioned	Fate
1	Langley	20 March 1922	Bombed and sunk 27 Feb, 1942 whilst ferrying AAF P40s to Tjilatjar, Java.
2	Lexington	14 Dec. 1927	Lost, enemy action, 8 May 1942.
3	Saratoga	16 Nov. 1927	Sunk, Operation Cross-Roads, 26 July 1946 (Bikini Atoll).
4	Ranger	4 June 1934	
5	Yorktown	30 Sept. 1937	Lost, enemy action, 7 June 1942.
6	Enterprise	12 May 1938	
7	Wasp	25 April 1940	Lost, enemy action, 15 Sept. 1942.
8	Hornet	20 Oct. 1941	Lost, enemy action, 26 Oct. 1942.
9	Essex	31 Dec. 1942	
10	Yorktown	15 April 1943	
11	Intrepid	16 Aug. 1943	Rebuilt post-war with angled deck.
12	Hornet	29 Nov. 1943	
13	Franklin	31 Jan. 1944	
14	Ticonderoga	8 May 1944	Rebuilt post-war with angled deck.
15	Randolph	9 Oct. 1944	
16	Lexington	17 Feb. 1943	Rebuilt post-war with angled deck.
17	Bunker Hill	25 May 1943	
18	Wasp	24 Nov. 1943	
19	Hancock	15 April 1944	Rebuilt post-war with angled deck.

Hull No.	Name	Commissioned	Fate
20	Bennington	6 Aug. 1944	
21	Boxer	16 April 1945	

Hull numbers 22-30 CVL of *Independence Class*. (See separate table below)

Hull No.	Name	Commissioned	Fate
31	Bon Homme Richard	26 Nov. 1944	Rebuilt post-war with angled deck.
36	Antietam	28 Jan. 1945	
38	Shangri-La	15 Sept. 1944	Rebuilt post-war with angled deck.
39	Lake Champlain	3 June 1945	

CVL – Independence Class

Hull No.	Name	Commissioned	Fate
22	Independence	14 Jan. 1943	
23	Princeton	25 Feb. 1943	Lost, enemy action, 24 Oct. 1944.
24	Belleau Wood	31 Mar. 1943	
25	Cowpens	28 May 1943	
26	Monterey	17 June 1943	
27	Langley	31 Aug. 1943	
28	Cabot	24 July 1943	
29	Bataan	17 Nov. 1943	
30	San Jacinto	15 Dec. 1943	

Essex Class completed post-War

Hull No.	Name	Commissioned
32	Leyte	11 April 1946
33	Kearsage	2 March 1946
34	Oriskany	25 Sept. 1950
35	Not allocated	

Escort Carriers — CVE

Original designation AVG changed to ACV on 20 August 1942 and to CVE on 15 July 1943.

Classes:

Long Island: two ships, Hull Nos 1 and 30.

Bogue: eleven ships, Hull Nos 9, 11-13, 16, 18, 20-23, 25, 31. (Hull Nos 32-54 transferred to UK).

Sangamon: four ships, Hull Nos 26-29.

Casablanca: fifty ships, Hull Nos 55-104.

Commencement Bay: nineteen ships, Hull Nos 105-123. (Includes eight ships built during the war but not commissioned until after VJ Day. Hull Nos 121 and 123, *Rabaul* and *Tinian*, not commissioned).

Hull No.	Name	Commissioned		Fate
1	Long Island	2 June	1941	
9	Bogue	26 Sept.	1942	
11	Card	8 Nov.	1942	
12	Copahee	15 June	1942	
13	Core	10 Dec.	1942	
16	Nassau	20 Aug.	1942	
18	Altamaha	15 Sept.	1942	
20	Barnes	20 Feb.	1943	
21	Block Island	8 March	1943	Lost, enemy action, 29 May 1944.
23	Breton	12 April	1943	
25	Croatan	28 April	1943	
26	Sangamon	25 Aug.	1942	
27	Suwannee	24 Sept.	1942	
28	Chenango	19 Sept.	1942	
29	Santee	24 Aug.	1942	
30	Charger	3 March	1942	
31	Prince William	9 April	1943	
55	Casablanca	8 July	1943	
56	Liscome Bay	7 Aug.	1943	Lost, enemy action, 24 Nov. 1943.
57	Anzio (ex-Coral Sea)	27 Aug.	1943	
58	Corregidor	31 Aug.	1943	
59	Mission Bay	13 Sept.	1943	
60	Guadalcanal	25 Sept.	1943	
61	Manila Bay	5 Oct.	1943	
62	Natoma Bay	14 Oct.	1943	
63	St Lo (ex-Midway)	23 Oct.	1943	Lost, enemy action, 25 Oct. 1944.
64	Tripoli	31 Oct.	1943	
65	Wake Island	7 Nov.	1943	
66	White Plains	15 Nov.	1943	
67	Solomons	21 Nov.	1943	
68	Kalinin Bay	27 Nov.	1943	
69	Kasaan Bay	4 Dec.	1943	
70	Fanshaw Bay	9 Dec.	1943	
71	Kitkun Bay	15 Dec.	1943	
72	Tulagi	21 Dec.	1943	
73	Gambier Bay	28 Dec.	1943	Lost, enemy action, 25 Oct. 1944.
74	Nehenta Bay	3 Jan.	1944	
75	Hoggatt Bay	11 Jan.	1944	
76	Kadashan Bay	18 Jan.	1944	
77	Marcus Island	26 Jan.	1944	
78	Savo Island	3 Feb.	1944	
79	Ommaney Bay	11 Feb.	1944	Lost, enemy action, 4 Jan. 1945.

Hull No.	Name	Commissioned		Fate
80	Petrof Bay	18 Feb.	1944	
81	Rudyerd Bay	25 Feb.	1944	
82	Saginaw Bay	2 March	1944	
83	Sargent Bay	9 March	1944	
84	Shamrock Bay	15 March	1944	
85	Shipley Bay	21 March	1944	
86	Sitkoh Bay	28 March	1944	
87	Steamer Bay	4 April	1944	
88	Cape Esperance	9 April	1944	
89	Takanis Bay	15 April	1944	
90	Thetis Bay	21 April	1944	
91	Makassar Strait	27 April	1944	
92	Windham Bay	3 May	1944	
93	Makin Island	9 May	1944	
94	Lunga Point	14 May	1944	
95	Bismarck Sea	20 May	1944	Lost, enemy action, 21 Feb. 1945.
96	Salumaua	26 May	1944	
97	Hollandia	1 June	1944	
98	Kawajalein	7 June	1944	
99	Admiralty Islands	13 June	1944	
100	Bougainville	18 June	1944	
101	Matanikau	24 June	1944	
102	Attu	30 June	1944	
103	Roi	6 July	1944	
104	Munda	8 July	1944	
105	Commencement Bay	27 Nov.	1944	
106	Block Island	30 Dec.	1944	
107	Gilbert Islands	5 Feb.	1945	
108	Kula Gulf	12 May	1945	
109	Cape Gloucester	3 March	1945	
110	Salerno Bay	19 May	1945	
111	Vella Gulf	9 April	1945	
112	Siboney	14 May	1945	
113	Puget Sound	18 June	1945	
114	Rendova	22 Oct.	1945	
115	Bairoko	16 July	1945	
116	Badoeng Strait	14 Nov.	1945	
117	Saidor	4 Sept.	1945	
118	Sicily	27 Feb.	1946	
119	Point Cruz	16 Oct.	1945	
120	Mindoro	4 Dec.	1945	
121	Rabaul	Not Commissioned		
122	Palau	15 Jan.	1946	
123	Tinian	Not Commissioned		

Hull numbers omitted: 2-5 not used; 6-8, 10, 14, 15, 17, 19, 22, 24, 32-54 transferred to United Kingdom; 124-39 cancelled.

General Index

Index

Aero Engines, Aircraft, Airships, Ships and Submarines